PRACTICAL R FOR BIOLOGISTS
An Introduction

PRACTICAL R FOR BIOLOGISTS
An Introduction

Donald L.J. Quicke

Buntika A. Butcher

and

Rachel A. Kruft Welton

CABI is a trading name of CAB International

CABI
Nosworthy Way
Wallingford
Oxfordshire OX10 8DE
UK

Tel: +44 (0)1491 832111
Fax: +44 (0)1491 833508
E-mail: info@cabi.org
Website: www.cabi.org

CABI
WeWork
One Lincoln St
24th Floor
Boston, MA 02111
USA

Tel: +1 (617)682-9015
E-mail: cabi-nao@cabi.org

© Donald Quicke, Buntika A. Butcher and Rachel Kruft Welton 2021. All rights reserved. No part of this publication may be reproduced in any form or by any means, electronically, mechanically, by photocopying, recording or otherwise, without the prior permission of the copyright owners.

References to Internet websites (URLs) were accurate at the time of writing.

A catalogue record for this book is available from the British Library, London, UK.

Library of Congress Cataloging-in-Publication Data

Names: Quicke, Donald L.J., author. | Butcher, Buntika A., author. | Welton, Rachel A. Kruft, author.
Title: Practical R for biologists : an introduction / Donald L.J. Quicke, Buntika A. Butcher, and Rachel A. Kruft Welton.
Description: Wallingford, Oxfordshire ; Boston, MA : CAB International, [2021] | Includes bibliographical references and index. | Summary: "A new textbook showing beginners how to use the free programming language R for fundamental biostatistical analysis, graphical display, and experimental design. The book takes a simple step-by-step approach to give a good grounding in the use of R for undergraduate/beginning postgraduate biology students"-- Provided by publisher.
Identifiers: LCCN 2020025099 (print) | LCCN 2020025100 (ebook) | ISBN 9781789245349 (paperback) | ISBN 9781789245356 (ebook) | ISBN 9781789245363 (epub)
Subjects: LCSH: Biometry--Data processing. | R (Computer program language)
Classification: LCC QH323.5 .Q56 2021 (print) | LCC QH323.5 (ebook) | DDC 570.1/5195--dc23
LC record available at https://lccn.loc.gov/2020025099
LC ebook record available at https://lccn.loc.gov/2020025100

ISBN-13: 9781789245349 (paperback)
        9781789245356 (ePDF)
        9781789245363 (ePub)

Commissioning Editor: Ward Cooper
Editorial Assistant: Lauren Davies
Production Editor: Tim Kapp

Typeset by SPi, Pondicherry, India
Printed and bound in the UK by Severn, Gloucester

This book is dedicated to all the people who help look after street dogs throughout the world. This is the first author with Ma Pao, who is cared for by those working at noodles shop and a coconut stall near Saphan Taksin station, Bangkok.

Contents

About the Authors	xv
Preface	xix
Acknowledgements	xxxi
1. How to Use This Book	**1**
Setting Up Your Computer	1
Running Code as You Go Along	1
Chapter Structure	2
2. Installing and Running R	**3**
Downloading and Installing R onto Your Computer	3
Installing Packages	7
3. Very Basic R Syntax	**9**
4. First Simple Programs and Graphics	**13**
Basic R Features	13
Commas, Brackets and Concatenation	14
The Colon Character	15
Raise to the Power of Symbol	15
Exiting from R	16

Help Pages	16
Beginning with Simple R Code to Get Used to the Command Line System	16
Playing with Graphics	19
Working with Character Variables	23
Built-in R Datasets	27
The **table** Function	27
Ragged Data	28
5. The Dataframe Concept	**31**
Combining Sets of Tables for Data Collected on Different Dates	34
Converting Factors in a Dataframe to Numeric or Character	34
6. Plotting Biological Data in Various Ways	**37**
Example 1 – Bryophytes up a Mountain	37
Troubleshooting 1	41
Adding a Legend to a Plot	43
Troubleshooting 2 – Vector Lengths Differ	45
Troubleshooting 3 – Missing Data and NAs	46
Incorporating More Types of Data on the Same Graph	48
Example 2 – Tropical Forests, Rural Population, Logarithmic Axes and Installing Packages	49
Example 3 – Creating a Barplot: Bryophytes Side-by-side	52
Example 4 – Stacked Bar Chart, with Different Colours, Fills and Legends	53
Example 5 – Dietary Differences between Hornbill Species – Entering Data as a Table	57
Example 6 – Horizontal Bar Plot of Camera Trap Data and More Troubleshooting	60
Example 7 – Adding Error Bars to a Barplot or Plot: Fly Ommatidea	62
Example 8 – Creating Pie Charts Using **pie** and **circlize**	64
Example 9 – Fish Metacercarial Load and Box and Whisker Plots	69
Adding Notches to a Boxplot	73
Tukey's Honest Significant Difference Test	74
7. The Grammar of Graphics Family of Packages	**79**

8.	**Sets and Venn Diagrams**	**85**
9.	**Statistics: Choosing the Right Test**	**95**
	Explanatory and Response Variables, Experiments and Surveys	97
	Parametric versus Non-parametric Tests	98
	Difference between Linear Models and Generalized Linear Models	98
	Our Basic Aim Is to Achieve a Near-linear QQ Plot and Even Variance	102
10.	**Commonly Used Measures and Statistical Tests**	**103**
	Normality, Skew and Kurtosis	103
	Testing Whether Proportions Agree with Null Expectations	104
	The Special Case of Contingency Tables	106
	Hardy-Weinberg Equilibrium	107
	Alternatives to the Chi-squared Test under Some Circumstances	110
	Testing Whether Two Means Are Significantly Different	111
	Single-sample t-test	111
	Two-sample t-test	112
	Paired t-test	113
	Testing Whether Three or More Means Differ from One Another	113
	Comparing Two Variances	114
	Non-normally Distributed Data with Small Sample Sizes – Mann-Whitney U Test	114
	Non-parametric Two-sample Tests	116
	Binomial Test	117
11.	**Regression and Correlation Analyses**	**119**
	Linear versus Non-linear Regression	120
	Log-log Plot Example Correlation of Numbers of Species with Area	121
	Linearizing Data with No Known Underlying Model	123
	Errant Points and Leverage	125
	QQ Model Plot from the **car** Library	129
	Comparing Regression Slopes and Intercepts Using t-test	130
	Non-linear Regression	134
	Multiple Regression	137

Pairwise Plots of Explanatory Variables to Visually Inspect Interactions	138
Polynomial Regression and Model Simplification	140
Model Simplification	143

12. Count Data as Response Variable — 147

Example 1 – Fledgling Numbers in Relation to Clutch Initiation Date	148
Example 2 – Pollinator Flower Visits in *Passiflora* in Relation to Flower Size	151

13. Analysis of Variance (ANOVA) — 155

Example 1 – A One-way ANOVA, the InsectSprays Dataset	155
Example 2 – ANOVA with Proportion Data as Response Variable Using Arcsine Transformation	157
Example 3 – Analysis with Proportion Data as Response Variable Using Logit Transformation	163

14. Analysis of Covariance (ANCOVA) — 166

Example 1 – Growth of Tagged Gobies	166
Example 2 – Fitting through the Origin and Count Data as Response Variable	168

15. More Generalized Linear Modelling — 171

Model Inspection	171
Binary Response Variable with One Continuous Explanatory Variable	172
Example 1 – Logistic regression of gall former predation	172
LD50s	176
Example 2 – Pollinator counts – showing importance of deviance	177
Example 3 – Proportion data with N known	182

16. Monte Carlo Tests and Randomization — 187

Random Number Generator Code	187
Example 1 – Flower Visits by Thai Honey Bee Species	188
Randomizing Cells in a Matrix	191

17.	**Principal Components Analysis**	**194**
	Example 1 – Rock Oyster Allozymes	194
	Example 2 – The Iris Dataset	197
18.	**Species Abundance, Accumulation and Diversity Data**	**200**
	Species Accumulation Data	200
	Species Accumulation Curves and Randomization	202
	Species Richness Estimation	208
	Species Diversity Indices	208
	A Note to Be Cautious about Logarithms in Functions	210
	Broken-stick Models	211
	A Much Faster Approach Using Vectorization	214
19.	**Survivorship**	**218**
	Example 1 – Survival of Killdeer Nests	218
20.	**Dates and Julian Dates**	**227**
	Problem with Two-digit Dates and POSIX: A Date of Burial Example	232
	Phenology and the **density** Function	234
	Extracting Day and Month from Julian Days	236
	Seasonal Patterns and Other Smoothing Curves	238
21.	**Mapping and Parsing Text Input for Data**	**240**
	Creating Our Own Map from Digitized Coordinates	247
22.	**More on Manipulating Text**	**257**
	Example 1 – Standardizing Names in a Phylogenetic Tree Description	257
	Method 1 with Wildcards	259
	Method 2 Based on Fixed Character String Length	262
	Method 3 Using a Vector of Positions	262
	Example 2 – Substrings of Unknown Length	264
	Trimming White Spaces and/or Tabs	268
	Using Wildcards to Locate Internal Letter Strings	268
	Finding Suffixes, Prefixes and Specifying Letters, Numbers and Punctuation	269

Manipulating Character Case	271
Ignoring Character Case	272
Specifying Particular and Modifiable Character Classes	273

23. Phylogenies and Trees — 275

Branch Lengths	279
Random Trees	280
Different Types of Plots in **ape**	281

24. Working with DNA Sequences and Other Character Data — 284

Sequential Runs of Base Types	288
Downloading DNA Sequences from GenBank	290
Translating DNA to Amino Acids	292
Prettifying a Table	293
Easy Ways to Extract Taxon Names from a Phylogenetic Matrix	295
Replacing Specified Ambiguity Codes with a Question Mark	296

25. Spacing in Two Dimensions — 297

26. Population Modelling Including Spatially Explicit Models — 303

Example 1 – Ricker Population Growth Model, Plotting as You Go	303
Example 2 – Host–Parasitoid Population Modelling – Discrete Time Version	306
Example 3 – Spatial Host–Parasitoid Model	310
Example 4 – Genetic Drift, a Program Aimed at Teaching Students about Evolution	318

27. More on *apply* Family of Functions – Avoid Loops to Get More Speed — 322

Using **apply**	323
Using **tapply** to Calculate Values Based on Factors	324

28. Food Webs and Simple Graphics — 326

A Parasitoid **foodweb** Example	326
Foodweb and **Community** Packages	328

29. Adding Photographs — 332

30. Standard Distributions in R — 335

- The Normal Distribution — 335
- Student's t Distribution — 338
- Lognormal Distribution — 340
- Logistic Distribution — 341
- Poisson Distribution — 342
- Gamma Distribution — 343
- The Chi-squared Distribution — 344

31. Reading and Writing Data to and from Files — 348

- Appending Data to an Existing File — 349
- Using **read.delim** with Non-tab Separator — 350
- Choosing a File to Read Interactively — 350
- Using Excel for Data Entry — 351
- The **readxl** Function and Tibbles — 352
- Reading PDF Files for Data Mining — 354
- Writing Graphics Directly to Disc — 354

Appendix 1: Summary of Graphical Parameters — 357

- Arguments Passed Directly to **par** Function — 357
- Arguments Applied Directly to the **plot** Function as well as in Some Others — 357
- Arguments for the **lines** Function — 358
- Having Multiple Graphics Windows Open at the Same Time — 358
- Macintosh-specific Graphics — 359
- Using the **layout** Function — 359
- Using the **split.screen** Function — 359

Appendix 2: General Housekeeping R Functions and Others Not Covered in the Main Text — 360

- General Housekeeping Functions — 360
- Setting or Changing the Working Directory — 360
- Finding What Files Are in a Directory — 361
- Graphical Functions and Parameters — 361

Interaction with User	361
Mathematical Functions	361
Writing Concatenated Data Straight to File (in the Working Directory) Using **cat**	362
Troubleshooting Package Installation	362

Appendix 3: Some Useful Statistical and Mathematical Equations — 364

Logical Mathematical Operators	364
Descriptive Statistics	364
Distributions	365
Correlation Coefficients	365
Statistical Tests	365
Logarithms and Exponents	366
Logistic Functions	366
Weibull and Gompertz Equations	366
Trigonometric Functions	367
Convert Radians and Degrees Functions	367

Bibliography — 369

Web Resources — 375

Index — 377

Online Supplementary Appendices

1. Online Resources: Data Files

2. Online Resources: Complete R Codes Used for Graphs, Analyses and Simulations

3. Online Resource: Suggested Answers to Exercises

These Online Resources can be found at: www.cabi.org/openresources/45349

About the Authors

Prof. Dr Donald Quicke has had more than 40 years' experience teaching undergraduate and postgraduate biology students, initially at Sheffield University, UK and then at Imperial College London.

Buntika Butcher gained her PhD at Imperial College and is currently Associate Professor in the Biology Department at Chulalongkorn University, Bangkok, with 20 years of teaching experience.

About the Authors

Dr Rachel Kruft Welton did her master's degree at Imperial College, London and a PhD at University of Birmingham, UK before qualifying as a teacher. She has been a professional biology and science tutor for nearly 20 years, including mentoring undergraduates as part of Birmingham University's alumni scheme.

Collectively the authors have a vast amount of teaching experience which they apply here to make the passage into R programming as gentle and easy as possible, whilst guiding the reader to tackle quite complicated programming.

Preface

There are many easy ways to create simple graphs using common computer software. The popular Microsoft Excel for example, offers a variety of choices that can be called simply by highlighting blocks of cells. It even offers some built-in basic statistical tests that are often introduced in biology A-level courses in some western countries (they certainly were in the senior author's day in the UK). However, there are very many limitations to just doing that. Not only do you have virtually no control over the presentation of the graphics, the simple stats provided will often give completely misleading answers. There are other potential problems too. With programs such as Excel it is very easy to mess things up; it is easy to fail to sort all the columns or to select all the rows. There is an infamous case of just that, the so-called Reinhart and Rogoff effect, a wrongly sorted spreadsheet that led to totally wrong economic conclusions (Herndon et al., 2014) that arguably led to the whole western austerity drive from 2010 onwards. Because R is an explicit, scripted language it is far harder to make such devastating mistakes with datasets.

What Is R?

R is an open-source statistical environment modelled after the previously widely used commercial programs S and S-Plus, but in addition to powerful statistical analysis tools, it also provides powerful graphics outputs and, as we are all taught at an early age, 'a picture is worth a thousand words'. R users with expertise are constantly adding new associated packages, but the range already available is immense. R is now used by a very wide range of people, biologists (who are the main targets of this book) for sure, but also other scientists plus economists, market researchers, medical professionals, epidemiologists, historians, geographers and those working in numerous other fields.

In addition to its statistical and graphical capabilities, R is a programming language suitable for medium-sized projects. It is not suitable for things like

CGI (computer generated imagery) or gaming for which very powerful specific software applications exist, but for modelling things like population dynamics, it offers a very fast and free solution.

In this book we will work our way through a set of studies that collectively represent almost all the R operations that beginners, analysing their own data up to perhaps the early years of doing a PhD, need. Although the chapters are organized around topics such as graphing, classical statistical tests, statistical modelling, mapping and text parsing, we have chosen examples based largely on real scientific studies at the appropriate level and within each we nearly always cover the use of more R functions than are simply necessary just to get a *p*-value or a graph.

R comes with around a thousand base functions which are automatically installed when R is downloaded. This book covers the use of those of most relevance to biological data analysis, modelling and graphics. Throughout each chapter the functions introduced and used in that chapter are summarized in Tool Boxes, and throughout the book we also show the user how to adapt and write their own code and functions. A selection of base functions relevant to graphics that are not necessarily covered in the main text are described in Appendix 1, and additional housekeeping functions in Appendix 2. These have been chosen to appeal to users from a variety of disciplines.

The idea is to talk users through an approach to analysing a given type of data and how to deal with many issues, such as error messages, that confront the inexperienced when they first start using R, or any other stats/graphics package for that matter. We hope that this heuristic approach will be especially beneficial to those just embarking on their scientific careers, whilst the book should also provide a framework and methods appropriate to the analysis and presentation of a wide range of data.

In many of our examples we build up the code bit by bit, trying to illustrate the sorts of logical steps that will be needed. In most cases probably, experienced programmers and users of R will be able to create far shorter, far quicker and far more elegant code. Whilst this is of course nice, novices might find some of the shortcuts rather confusing. By progressing step by step, the logic of writing code will be far more apparent, and like everything, with ever more experience students will start to invent their own ways of doing things. Biologists don't often have such vast datasets that speed is of major concern, though large simulations might take some time.

The choice of material to present in this sort of book is always hard because all authors will have an idea of what is best for the students that they teach, but here we are aiming at the needs of a wider audience. Therefore, we also want to make sure that we cover the sorts of tests and procedures that are in common use across a wide range of biological areas, to make sure they are carried out appropriately and sometimes to present better, and even easier, alternatives.

There are plenty of detailed and excellent statistics books for biologists based around R, though many of these are not appropriate for beginners who have never had to program before nor carry out any proper statistical analyses. The latter is essential in most of biology, any student wanting to go on to have biological research as part of their career (or just get a good degree) is going to have to get to grips with some of the basics. Loaiza *et al.* (2011) examined the types of analyses reported in publications in two independent journals of tropical biology approximately between 2008 and 2009. The lists were amazingly similar, so there is a subset of the plethora of available tests that are favoured, probably largely due to what the authors were taught during the early stages of their college years. For general interest, the top five procedures (analysis of variance (ANOVA), chi-square test, Student's t test, linear regression, Pearson's correlation coefficient) in each journal were even in exactly the same rank order. Computing power is no longer a real limitation anywhere. So we will make sure these 'top tests' are covered.

To a large extent, our approach here is to present plausible and real analysis jobs that cover a wide range of scenarios. Most books teach the way to do it, but we include many examples of errors that commonly occur, especially when starting to get to grips with R. We also provide many examples of troubleshooting. Because a lot of our readers will come to the book to get advice on doing a particular operation, we repeat a lot of things in different places. Nevertheless, we really encourage readers to start at the beginning and to run all the code examples in order. There is logic to the order.

One of the most powerful things about R from a user's perspective is that it is a 'functional' language – you manipulate and analyse data applying functions to it, be the functions simple, built-in arithmetic operations such as calculating means, or more complex user-defined ones. Functions can be nested in one another so in a single line of code you might calculate, e.g. the range of the logarithms of the means of population sizes. The syntax of R is generally simple and logical, though the only way to really learn R is through trial and error.

Here we try to do two things. There are so many R packages and their included functions that a biologist can use them to do almost anything that a biologist or ecologist could ever want to do, certainly up to postgraduate level. However, simply plugging blocks of data into programs teaches the student almost nothing. It doesn't teach them the logic behind the order needed to perform functions, and they often go away without knowing what the function has actually done. They might even have entered the data inappropriately and received a wrong answer and walk away without seeing it as such. Also, although these packages/functions work fine, there are many steps to go through to get from your freshly typed-in raw data to the forms that these ready-made functions can handle. We want to cover both.

We hope that by the end of this book, the reader will be able to do things for themselves. There are many packages, which are often of a 'plug-in' type, which do not allow much or any modification of the output. We show the user how to program functions themselves, in order to create your own desired output. In doing this, we introduce not just recipes to get a statistical answer, but tuition on programming, together with biologically relevant examples.

Each chapter is meant to take the reader along a gentle progression from simply performing arithmetic and creating easy graphs, through increasingly sophisticated levels of data analysis (statistics), then through a range of techniques to cope with the nitty-gritty of writing code, followed by refinements to polish the presentation of the analyses. It tries to stress important things that the serious user, perhaps aiming to perform statistical analyses for publication or to get good grades, must take into account, and read around more, using the many excellent online resources to make sure you know what you are doing. Some of our examples are not easy to analyse statistically and there is not always a definitive best way. This reflects the real world and despite the very clean examples presented in some textbooks, some degree of user judgement is often needed.

There are many useful online resources for R. The CRAN website provides a very handy short reference card PDF file that we suggest everyone should download: https://cran.r-project.org/doc/contrib/Short-refcard.pdf.

Another very useful little aid is 'R Reference Card v2.0' by M. Baggott (based on V1 by T. Short, 2004). This is a public domain document available at https://cran.r-project.org/doc/contrib/Baggott-refcard-v2.pdf.

R is such a rapidly developing oeuvre that it might be that libraries change over time. Some certainly have disappeared since we started writing this, but a quick Google search will usually solve your problem. A quick Google search of 'R program' plus a few key words will normally take you to some very helpful pages on stackoverflow.com (accessed 22 April 2020), r-bloggers.com (accessed 22 April 2020) or similar sites where programmers and experienced users exchange information and provide solutions to problems. You almost certainly will not be the first person to have experienced your current issue. Quite a lot of the solutions presented here come from such sources but it is not currently possible to cite responses to give credit to them individually, so we express our enormous thanks to the many experts who helped us along the way.

Packages used or mentioned in the book are summarized in the table below.

Chapter 2	base	Chapter 12	vcd	Chapter 23	ape
					maps
					phytools
Chapter 4	akima	Chapter 15	aod	Chapter 24	ape
			MASS		BiocManager
					Biostrings
					sequinr
Chapter 6	circlize	Chapter 16	picante	Chapter 26	beepr
	sfsmisc				simecol
Chapter 7	ggplot2	Chapter 17	ade4	Chapter 28	cheddar
	ggplotly		amp		foodweb
	ggpubr		biplot		
	gridExtra		ggplot2		
	MASS		FactoMineR		
	Stats2Data		stats		
Chapter 8	BiocManager	Chapter 18	coexist	Chapter 29	jpegtiff
	car		devtools		
	ellipse		MBI		
	limma		SPECIES		
	plotrix		Stat2Data		
			vegan		
Chapter 9	DescTools	Chapter 19	dplyr	Chapter 31	openxlsx
	ggpubr		MASS		pdftools
	MASS		survival		readxl
	survival		survminer		tibble
Chapter 10	Deducer	Chapter 20	erer	Appendix I	svglite
	DescTools		ggplot2		
	MASS		ggpubr		
	moments				
	nortest				
Chapter 11	car	Chapter 21	dismo	Appendix II	DescTools rJava
			maptools		
			raster		
			rworldmap		

Finally, in most chapters we set some exercises so the reader can practise analysing data, writing their own code, and using their own inventiveness. For each one there is an online resource providing model answers. There are always many ways of carrying out the exercises, so do not expect that your solution will necessarily be the same as ours, and that actually points to the versatility of the R language.

Three sets of files are available from the CABI website by accessing the links below. Firstly we suggest that readers download all of the data files to a dedicated folder on their computer and use this as the R working directory. Secondly, there is a list of files containing suggested answers to the questions in the Exercise Boxes as well as some of the graphical 'Results' that the answers give. Thirdly, because a lot of the code examples in the text are dispersed among the text so that we can explain what each part is doing, we provide files containing the complete working R code for these.

Data Files

A zip of all the files: https://site.cabi.org/wp-content/uploads/datafiles.zip

https://site.cabi.org/wp-content/uploads/ActiniaXYdata.csv
https://site.cabi.org/wp-content/uploads/Apis_data.txt
https://site.cabi.org/wp-content/uploads/Append_NG_butterfly_transect.txt
https://site.cabi.org/wp-content/uploads/A_placidus.txt
https://site.cabi.org/wp-content/uploads/British_ladybirds.txt
https://site.cabi.org/wp-content/uploads/Burial_Register_HOPE.xlsx
https://site.cabi.org/wp-content/uploads/Caribb_widespread.csv
https://site.cabi.org/wp-content/uploads/CarribTot.csv
https://site.cabi.org/wp-content/uploads/CHAP14_Sirindhornia_count.csv
https://site.cabi.org/wp-content/uploads/Chapter_14_Malone_gobies.csv
https://site.cabi.org/wp-content/uploads/CornBorer_1992.txt
https://site.cabi.org/wp-content/uploads/Corsican_blue_tit.csv
https://site.cabi.org/wp-content/uploads/culicid_DT.txt
https://site.cabi.org/wp-content/uploads/Doi_woodpeckers.txt
https://site.cabi.org/wp-content/uploads/European_corn_borer_2003.txt
https://site.cabi.org/wp-content/uploads/fledge_number.csv
https://site.cabi.org/wp-content/uploads/friends.csv
https://site.cabi.org/wp-content/uploads/HOPE_Date_Order_Table_1.csv
https://site.cabi.org/wp-content/uploads/Insect.tre
https://site.cabi.org/wp-content/uploads/KaengKrachan_woodpeckers.txt
https://site.cabi.org/wp-content/uploads/KhaoYai_woodpeckers.txt
https://site.cabi.org/wp-content/uploads/krill.txt
https://site.cabi.org/wp-content/uploads/mantid.txt
https://site.cabi.org/wp-content/uploads/Monilobracon.fas
https://site.cabi.org/wp-content/uploads/nodes.csv
https://site.cabi.org/wp-content/uploads/Oriental_Pied_Hornbill.jpg
https://site.cabi.org/wp-content/uploads/Panamanian_cycads.txt
https://site.cabi.org/wp-content/uploads/Panamanian_cycads_colon.txt
https://site.cabi.org/wp-content/uploads/Passiflora_visits.csv
https://site.cabi.org/wp-content/uploads/Pet-Cats-Australia.csv
https://site.cabi.org/wp-content/uploads/PNG_transects.txt
https://site.cabi.org/wp-content/uploads/PrimersTable.csv
https://site.cabi.org/wp-content/uploads/PrimersTable.txt
https://site.cabi.org/wp-content/uploads/PrimersTable.xlsx
https://site.cabi.org/wp-content/uploads/properties.csv

https://site.cabi.org/wp-content/uploads/Rapp.history
https://site.cabi.org/wp-content/uploads/RData
https://site.cabi.org/wp-content/uploads/Rhistory
https://site.cabi.org/wp-content/uploads/Thailand_Border.csv
https://site.cabi.org/wp-content/uploads/trophic_links.csv
https://site.cabi.org/wp-content/uploads/Tuatara.txt
https://site.cabi.org/wp-content/uploads/Tuatara_extra.txt
https://site.cabi.org/wp-content/uploads/Tunjai_regeneration.txt
https://site.cabi.org/wp-content/uploads/Tunjai_regeneration.xlsx
https://site.cabi.org/wp-content/uploads/Wildlife-at-Doi-Inthanon-National-Park.pdf
https://site.cabi.org/wp-content/uploads/Wildlife-at-Khao-Yai.txt
https://site.cabi.org/wp-content/uploads/Wildlike-at-Kaeng-Krachan.txt
https://site.cabi.org/wp-content/uploads/Wright_y_Muller_Landau_2006.csv

Exercise Answers

A zip of all the files: https://site.cabi.org/wp-content/uploads/exerciseanswers.zip

https://site.cabi.org/wp-content/uploads/ExerciseBox_4.1_adjusting_friends_names_arrows.txt
https://site.cabi.org/wp-content/uploads/ExerciseBox_4.2_fill_matrix_with_ragged_data.txt
https://site.cabi.org/wp-content/uploads/ExerciseBox_6.1_beaver-temperature.txt
https://site.cabi.org/wp-content/uploads/ExerciseBox_6.2_bryophyte_barplot_exercise.txt
https://site.cabi.org/wp-content/uploads/ExerciseBox_6.3_piechart_exercise_solutions.txt
https://site.cabi.org/wp-content/uploads/ExerciseBox_7.1_Fitch_mammal_skull_data.txt
https://site.cabi.org/wp-content/uploads/ExerciseBox_8.1_Venn_exercise.txt
https://site.cabi.org/wp-content/uploads/ExerciseBox_10.1.1_faithful.txt
https://site.cabi.org/wp-content/uploads/ExerciseBox_10.1.2_Odds_ratio_exercise.txt
https://site.cabi.org/wp-content/uploads/ExerciseBox_10.2_SexRatio_deer.txt
https://site.cabi.org/wp-content/uploads/ExerciseBox_11.1_mammal_skull_linear_model.txt
https://site.cabi.org/wp-content/uploads/ExerciseBox_11.2.1_Krill-and-dip.txt
https://site.cabi.org/wp-content/uploads/ExerciseBox_12.1_cardinal_lay_date.txt
https://site.cabi.org/wp-content/uploads/ExerciseBox_12.2_Passiflora_quasibinomial.txt
https://site.cabi.org/wp-content/uploads/ExerciseBox_13.1_seed_establishment.txt
https://site.cabi.org/wp-content/uploads/ExerciseBox_14.1_garlic_Aspergillus_plot.txt
https://site.cabi.org/wp-content/uploads/ExerciseBox_14.2_Sirindhornia_pollinaria.txt

https://site.cabi.org/wp-content/uploads/ExerciseBox_15.1_menarche_age.txt
https://site.cabi.org/wp-content/uploads/ExerciseBox_15.2_snails.txt
https://site.cabi.org/wp-content/uploads/ExerciseBox_16.1_stratified.txt
https://site.cabi.org/wp-content/uploads/ExerciseBox_17.1_Iris_species_data.txt
https://site.cabi.org/wp-content/uploads/ExerciseBox_18.1_Species_Area_islands.txt
https://site.cabi.org/wp-content/uploads/ExerciseBox_18.2_Styracaceae.txt
https://site.cabi.org/wp-content/uploads/ExerciseBox_19.1_leukaemia.txt
https://site.cabi.org/wp-content/uploads/ExerciseBox_20.1_Corsican_Blue_tit.txt
https://site.cabi.org/wp-content/uploads/ExerciseBox_20.2_Hope_age_at_death_by_sex.txt
https://site.cabi.org/wp-content/uploads/ExerciseBox_20.3_European_corn_borer.txt
https://site.cabi.org/wp-content/uploads/ExerciseBox_21.1_cobra_map.txt
https://site.cabi.org/wp-content/uploads/ExerciseBox_21.2_Australian_cat_roaming.txt
https://site.cabi.org/wp-content/uploads/ExerciseBox_22.1_tidying_DNA-dataset.txt
https://site.cabi.org/wp-content/uploads/ExerciseBox_22.2_capital_genus_names.txt
https://site.cabi.org/wp-content/uploads/ExerciseBox_23.1_Tetrapod_dated_phylogeny.txt
https://site.cabi.org/wp-content/uploads/ExerciseBox_24.1_revcomp.txt
https://site.cabi.org/wp-content/uploads/ExerciseBox_24.2_pairing.txt
https://site.cabi.org/wp-content/uploads/ExerciseBox_24.3_amino_acid_translation.txt
https://site.cabi.org/wp-content/uploads/ExerciseBox_25.1_random_WeissPlot.txt
https://site.cabi.org/wp-content/uploads/ExerciseBox_26.1.txt
https://site.cabi.org/wp-content/uploads/ExerciseBox_26.2.1.txt
https://site.cabi.org/wp-content/uploads/ExerciseBox_26.2.2.txt
https://site.cabi.org/wp-content/uploads/ExerciseBox_26.3.1_allele-decline1.txt
https://site.cabi.org/wp-content/uploads/ExerciseBox_26.3.2_allele-decline_diploid.txt
https://site.cabi.org/wp-content/uploads/ExerciseBox_27.1_smokers.txt
https://site.cabi.org/wp-content/uploads/ExerciseBox_30.1_compare_normal_binomial.txt
https://site.cabi.org/wp-content/uploads/Result_4.1_friends_heights.pdf
https://site.cabi.org/wp-content/uploads/Result_4.2_ragged.png
https://site.cabi.org/wp-content/uploads/Result_6.1_beaver_temperature.pdf
https://site.cabi.org/wp-content/uploads/Result_6.2_green_bryophyte.pdf
https://site.cabi.org/wp-content/uploads/Result_7.1_Fitch_plot.pdf
https://site.cabi.org/wp-content/uploads/Result_7.1_ggplot_mammal_skull_palate.pdf

https://site.cabi.org/wp-content/uploads/Result_8.1-3_coloured-ellipse_Venn.pdf
https://site.cabi.org/wp-content/uploads/Result_8.1.1_4_venn_circles.pdf
https://site.cabi.org/wp-content/uploads/Result_10.1.1_Faithful.pdf
https://site.cabi.org/wp-content/uploads/Result_12.1_cardinal_lay_date_model2.pdf
https://site.cabi.org/wp-content/uploads/Result_12.2_passiflora_text_angles-precise.pdf
https://site.cabi.org/wp-content/uploads/Result_14.1.pdf
https://site.cabi.org/wp-content/uploads/Result_13.1_seedestablishment_model-plots.pdf
https://site.cabi.org/wp-content/uploads/Result_15.1_menarche.pdf
https://site.cabi.org/wp-content/uploads/Result_16.1_stratified_sample.pdf
https://site.cabi.org/wp-content/uploads/Result_17.1.1_iris_PCAs.pdf
https://site.cabi.org/wp-content/uploads/Result_17.1.2_scatterplot.pdf
https://site.cabi.org/wp-content/uploads/Result_17.1.3_scatterplotPCA1-vs-2.pdf
https://site.cabi.org/wp-content/uploads/Result_18.1_Sunda_Islands.pdf
https://site.cabi.org/wp-content/uploads/Result_19.1.1_leukaemia.pdf
https://site.cabi.org/wp-content/uploads/Result_20.2_Hope_Burials.pdf
https://site.cabi.org/wp-content/uploads/Result_20.3_European_corn_borer.pdf
https://site.cabi.org/wp-content/uploads/Result_21.1_Naja_map.pdf
https://site.cabi.org/wp-content/uploads/Result_21.2_ozzy_cats_1.pdf
https://site.cabi.org/wp-content/uploads/Result_21.2_ozzy_cats_2.pdf
https://site.cabi.org/wp-content/uploads/Result_23.1_Tetrapod_dated_phylogeny.pdf
https://site.cabi.org/wp-content/uploads/Result_25.1.1_random_points.pdf
https://site.cabi.org/wp-content/uploads/Result_25.1.2_100_random_Weiss.pdf
https://site.cabi.org/wp-content/uploads/Result_25.1.3_accounting-for-pdd.pdf
https://site.cabi.org/wp-content/uploads/Result_26.1_host_parasitoid_4_plots.pdf
https://site.cabi.org/wp-content/uploads/Result_26.1.1_host_parasite_4plots.pdf
https://site.cabi.org/wp-content/uploads/Result_26.2.1_hostParasitoid_ratio.pdf
https://site.cabi.org/wp-content/uploads/Result_26.2_population_variance-with-generations.pdf
https://site.cabi.org/wp-content/uploads/Result_26.3.1_allele_decline.pdf
https://site.cabi.org/wp-content/uploads/Result_30.1_compare_distributions.pdf

Complete working R code

A zip of all the files: https://site.cabi.org/wp-content/uploads/workingrcode.zip

https://site.cabi.org/wp-content/uploads/CHAP4_friends_R_code.txt
https://site.cabi.org/wp-content/uploads/CHAP4_ragged.dataRcode.txt
https://site.cabi.org/wp-content/uploads/CHAP6_bryophytes_R_code.txt
https://site.cabi.org/wp-content/uploads/CHAP6_bryophyte_barplot_R_code.txt
https://site.cabi.org/wp-content/uploads/CHAP6_cameratrap_histograms_R_code.txt
https://site.cabi.org/wp-content/uploads/CHAP6_Circlize_mammals_R_code.txt
https://site.cabi.org/wp-content/uploads/CHAP6_Dipterocarp_ants_R_code.txt

https://site.cabi.org/wp-content/uploads/CHAP6_Hornbill_foraging_barplot_besideTrue_R_code.txt
https://site.cabi.org/wp-content/uploads/CHAP6_Marsh_pie_chart.txt
https://site.cabi.org/wp-content/uploads/CHAP6_Metacercaria_R_code.txt
https://site.cabi.org/wp-content/uploads/CHAP6_Ommatidia_R_code.txt
https://site.cabi.org/wp-content/uploads/CHAP10_Box_whisker_ant_damage_normtest_R.code_.txt
https://site.cabi.org/wp-content/uploads/CHAP10_chisqu_Clonorchis_infection_R_code.txt
https://site.cabi.org/wp-content/uploads/CHAP10_frog_boxplot_sexes_R_code.txt
https://site.cabi.org/wp-content/uploads/CHAP10_HardyWeinberg_moth_R_code.txt
https://site.cabi.org/wp-content/uploads/CHAP10_HardyWeinberg_plot_R_code.txt
https://site.cabi.org/wp-content/uploads/CHAP10_PAOD_chisq_R_code.txt
https://site.cabi.org/wp-content/uploads/CHAP11_AntilButterflies_Lomolino.txt
https://site.cabi.org/wp-content/uploads/CHAP11_Ants_burnt_unburnt_R_code.txt
https://site.cabi.org/wp-content/uploads/CHAP11_Channa_snakehead_growth_R_code.txt
https://site.cabi.org/wp-content/uploads/CHAP11_island_rodents_R_code.txt
https://site.cabi.org/wp-content/uploads/CHAP11_Knor_krill_O2consump_R_code.txt
https://site.cabi.org/wp-content/uploads/CHAP11_leaf_beetle_damage_R_code.txt
https://site.cabi.org/wp-content/uploads/CHAP11_Lucanidae_boxplot_R_code.txt
https://site.cabi.org/wp-content/uploads/CHAP11_Ochlero_R_code.txt
https://site.cabi.org/wp-content/uploads/CHAP11_weaver_ant_red_green_R_code.txt
https://site.cabi.org/wp-content/uploads/CHAP12_Cardinal_Poisson_R_code.txt
https://site.cabi.org/wp-content/uploads/CHAP12_Passiflora_R_code.txt
https://site.cabi.org/wp-content/uploads/CHAP13_Tunjai_regeneration_R_code.txt
https://site.cabi.org/wp-content/uploads/CHAP14_ANCOVA_Asperillus_garlic_R_code.txt
https://site.cabi.org/wp-content/uploads/CHAP14_Sirindhornia_R_code.txt
https://site.cabi.org/wp-content/uploads/CHAP15_psyllid_R_code.txt
https://site.cabi.org/wp-content/uploads/CHAP15_snail_binomial.txt
https://site.cabi.org/wp-content/uploads/CHAP16_Thai_bees_Montecarlo_R_code.txt
https://site.cabi.org/wp-content/uploads/CHAP17_Iris_PCA_R_code.txt
https://site.cabi.org/wp-content/uploads/CHAP17_oyster_allozyme_PCA_R_code.txt

https://site.cabi.org/wp-content/uploads/CHAP18_brokenstick_Cocc_R_code.txt
https://site.cabi.org/wp-content/uploads/CHAP18_Fisherfit_PNG_butterflies_R_code.txt
https://site.cabi.org/wp-content/uploads/CHAP18_PNG_butterfly_species_accum_R_code.txt
https://site.cabi.org/wp-content/uploads/CHAP18_simpson_shannon_R_code.txt
https://site.cabi.org/wp-content/uploads/CHAP18_Styracaceae_brokenstick_R_code.txt
https://site.cabi.org/wp-content/uploads/CHAP19_Atuo_Killdeer_survival_R_Code.txt
https://site.cabi.org/wp-content/uploads/CHAP20_cornborer_R_code.txt
https://site.cabi.org/wp-content/uploads/CHAP20_Hope_cemeteryR_code.txt
https://site.cabi.org/wp-content/uploads/CHAP20_JonkersKucera.txt
https://site.cabi.org/wp-content/uploads/CHAP21_DismoBhutanitis_GBIFF_R_code.txt
https://site.cabi.org/wp-content/uploads/CHAP21_Maptools_Butterflies_per_country_R_code.txt
https://site.cabi.org/wp-content/uploads/CHAP21_placidus_R_code.txt
https://site.cabi.org/wp-content/uploads/CHAP21_rworldmap_R_code.txt
https://site.cabi.org/wp-content/uploads/CHAP22_fish_get_umbra_R_code.txt
https://site.cabi.org/wp-content/uploads/CHAP22_process_subfamily_names_in_text_R_code.txt
https://site.cabi.org/wp-content/uploads/CHAP22_put_MRS_first_R_code.txt
https://site.cabi.org/wp-content/uploads/CHAP22_suffixes_etc_R_code.txt
https://site.cabi.org/wp-content/uploads/CHAP22_text_manip_R_code.txt
https://site.cabi.org/wp-content/uploads/CHAP23_Ape_and_Phytools_R_code.txt
https://site.cabi.org/wp-content/uploads/CHAP23_Tuatara_R_code.txt
https://site.cabi.org/wp-content/uploads/CHAP24_DNA_data_R_code.txt
https://site.cabi.org/wp-content/uploads/CHAP24_prettiying_Xenarcha_matrix_R_code.txt
https://site.cabi.org/wp-content/uploads/CHAP25_Actinia_spacing_R_code.txt
https://site.cabi.org/wp-content/uploads/CHAP26_Beddington_R_code.txt
https://site.cabi.org/wp-content/uploads/CHAP26_Geneticdrift_R_code.txt
https://site.cabi.org/wp-content/uploads/CHAP26_Ricker_R_code.txt
https://site.cabi.org/wp-content/uploads/CHAP26_spatial_parasitoid_model_R_code.txt
https://site.cabi.org/wp-content/uploads/CHAP27_tapply_R_code.txt
https://site.cabi.org/wp-content/uploads/CHAP28_Cheddar_R_code.txt
https://site.cabi.org/wp-content/uploads/CHAP28_food_web_R_code.txt
https://site.cabi.org/wp-content/uploads/CHAP29_jpeg_image_rasterImage_R_code.txt
https://site.cabi.org/wp-content/uploads/CHAP30_Poisson_R_code.txt

Acknowledgements

We thank Yves Basset (Smithsonian Tropical Research Institute, Panama) for kindly providing unpublished raw butterfly transect data. Luca Börger (Swansea University) helped with analysis of proportion data (Chapter 13); Eoin O'Gorman (University of Essex, Colchester, UK) helped with getting nice output from the cheddar package (Chapter 25). William Pearse (Imperial College) provided helpful comments, particularly on the statistical chapters; Rob Kirkwood (formerly University of Loughborough) and Melbryn Kruft Welton kindly tested parts of R code for us; Idonea Kruft Welton and Andrew Davies (University of Goettingen) helped proof-read, and collectively offered many useful suggestions and corrections on an earlier draft.

Copyright permissions

We thank Professor Chatchawan Chaisuekul (Chulalongkorn University) for permission to use unpublished data from Nuraemram (2011) on ant diversity in dipterocarp forests (Chapter 6); Jennifer S. Powers, Editor-in-Chief, for permission to use data from Wright and Muller-Landau (2006) (Chapter 6); Dr Wichase Khonsue (Chulalongkorn University, Bangkok, for permission to use unpublished bullfrog data from his research group (Chapter 10); Dr Tuantong Jutagate for permission to use snakehead fish growth data published in Jutagate et al. (2013) (Chapter 11); Jennifer S. Powers, Editor-in-Chief, for permission to use data from Offenberg et al. (2004) published in *Biotropica* (Chapter 10); Vincente Gomez for permission to use data from Van Ngan et al. (1997) (Chapter 11); Professor Daniel P. Shustack (Massachusetts College of Liberal Arts, MCLA) for permission to use fledging data from Shustack and Rodewald (2011) (Chapter 12); Dr Krzysztof Raciborski, Managing Editor of Finnish Zoological and Botanical Publishing Board, for permission to use data from Srimuang et al. (2010) published in *Annales Botanici Fennici* (Chapter 14).

1 How to Use This Book

Setting Up Your Computer

The online resources include all the datasets that we show how to analyse. We recommend that readers create a folder on their computer desktop into which all these data files can be downloaded. In addition, we recommend that readers create a subfolder within that folder into which all the complete working R codes for all worked examples are placed; these are in online resource url www.cabi.org/openresources/45349. The outer folder can then be set to be R's working directory, so that all file reading and writing is done from and to this folder. Choose the working directory using the pull-down menus or use **setwd** as explained in Appendix 2 (p. 360).

Running Code as You Go Along

We strongly recommend that readers cut and paste the lines (or small groups of lines) of R code into R as they go along, and, at every stage, make sure that you understand what you are telling R to do with the data. There is a little repetition in different chapters to facilitate understanding, and much of the code is annotated with comments, but it will still be best for beginners to start at the beginning and progress chapter by chapter. When blocks of code run over a page boundary, simply first copy and paste the first part into R, without pressing carriage return, then immediately after that, copy and paste the second part of the code. If you are copying the code out of the ebook and pasting into R, you may find the lines do not break at the correct points. To ensure the code runs correctly, make sure it is formatted exactly as shown in the text. As an alternative, all the code can be found in the online resources in the text file https://site.cabi.org/wp-content/uploads/rcode.txt

Chapter 1. How to Use This Book

Each chapter takes you through how to download, code and use different functions. We troubleshoot many of the common problems and illustrate each function and package with real data examples. Finally, we show you how to produce publication quality graphics.

Function names are always shown in bold.

R input is identified by blue, sans serif font.

`R output` is indicated by purple Courier font.

R error messages are shown in bright red.

Information Box 📖

These provide further explanation of how some functions work, details of syntax, etc. or information on related functions or packages.

Exercise Box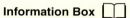

We give you the opportunity to practise your coding skills, with some carefully planned exercises. Sample answers are given in the online Appendix at www.cabi.org/openresources/45349.

Tool Box ✘

At the end of various large sections of chapters, the uses of main functions and arguments included in that section/chapter are summarized, giving you a handy checklist for easy reference.

At the start of each chapter, you can find a list of PACKAGES and FUNCTIONS introduced.

2 Installing and Running R

Summary of R Functions Introduced	
available.packages	R.version
chooseCRANmirror	length
demo	library
	unique, but see Chapter 8

Downloading and Installing R onto Your Computer

With your computer online go to r-project.org (accessed 22 April 2020) and click download. You will then see a page of CRAN (= Comprehensive R Archive Network) mirror sites cran.r-project.org/mirrors.html (accessed 22 April 2020) from which you should select the one nearest to your location (e.g. Fig. 2.1), though all of them should work. CRAN has a network of servers hosting R code, many packages and their documentation. Here is part of the list of CRAN mirror URLs (accessed 18 May 2019).

Taiwan	
https://ftp.yzu.edu.tw/CRAN/	Department of Computer Science and Engineering, Yuan Ze University
http://ftp.yzu.edu.tw/CRAN/	Department of Computer Science and Engineering, Yuan Ze University
http://cran.csie.ntu.edu.tw/	National Taiwan University, Taipei
Thailand	
http://mirrors.psu.ac.th/pub/cran/	Prince of Songkla University, Hatyai
Turkey	
https://cran.pau.edu.tr/	Pamukkale University, Denizli
http://cran.pau.edu.tr/	Pamukkale University, Denizli
https://cran.ncc.metu.edu.tr/	Middle East Technical University Northern Cyprus Campus, Mersin
http://cran.ncc.metu.edu.tr/	Middle East Technical University Northern Cyprus Campus, Mersin
UK	
https://www.stats.bris.ac.uk/R/	University of Bristol
http://www.stats.bris.ac.uk/R/	University of Bristol
https://cran.ma.imperial.ac.uk/	Imperial College London
http://cran.ma.imperial.ac.uk/	Imperial College London

© Donald Quicke, Buntika A. Butcher and Rachel Kruft Welton 2021. Practical R for Biologists: An Introduction (D. Quicke, B.A. Butcher and R. Kruft Welton)
DOI: 10.1079/9781789245349.0002

After selecting a mirror, you will be given the option to download versions of R for Linux, Mac OSX and Windows. Select your platform – this will almost certainly be either PC or Mac, then click the 'Download…' link and follow the instructions.

Having installed R turn it on (double click the R icon) and a window should appear. The window you see will differ depending whether you are using a PC or a Mac. The PC version of R automatically starts with RGui (R graphical user interface), either as a 32 bit version or a 64 bit version.

Installing and Running R

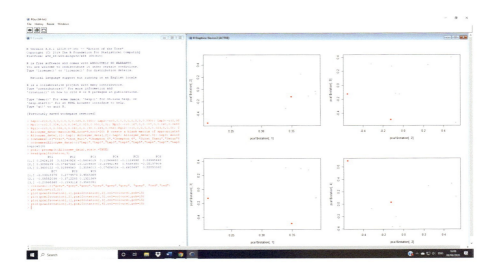

Whereas the Mac version opens a plain R console window.

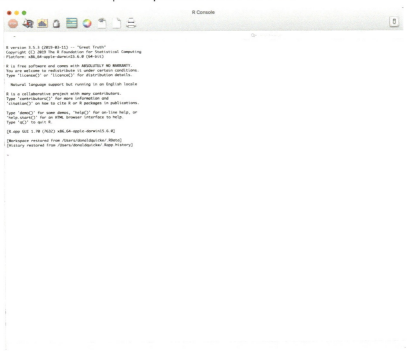

At the bottom of the window is a prompt sign >. This is where you start telling R what you want it to do. This may look daunting but it is the great thing about command line programming: it gives the user full control of all that the program is capable of doing. Further this is essential in the case of complicated statistical analyses.

Almost always you will type the name of a variable (a term that we use here for most of the types of object that R handles), followed by the prompt (or assign operator) '<-', which is usually spoken as 'gets'. Then you type what you want it to get. The syntax is like *a<-b+c*, but obviously, at the beginning R doesn't know what *a*, *b* or *c* are.

One of the first things newbies might want to type in is
demo(image)
or
demo(colors)
and click through the example pages until you have seen enough. To exit from a demo or any other process press the 'escape' key (sometimes you may need to do this several times because each press only escapes from a single process that is being carried out). Type 'demo()' to see list of other available demonstrations.

In addition to R, we recommend that you turn on a simple text editor (preferably not Word, but something like 'TextWrangler' or 'TextEdit' on a Mac or 'TextPad' or 'EditPad Lite' on a PC. If you use Word you may have problems, particularly with some characters such as inverted commas (used for handling text), because Word always tries to put intelligent ones in (different for opening and closing quotes) whereas R needs straight ones.

The PC RGui window shows all the code lines can be seen at the same time as output. Once you have produced a graphic, the graphics window can be tiled vertically or horizontally next to the coding window, using the pull-down menus. Many users will find this layout too spartan. A more comprehensive set of windows can be set up using the special R interface program, RStudio, which automatically opens R and displays the R code in one pane, the graphics window in another and a list of your variables and their current contents in another, and various other options. It is available free under AGPL licence from rstudio.com/products/RStudio/#Desktop (accessed 22 April 2020). You can find out more about RStudio at support.rstudio.com/hc/en-us/articles/200549016-Customizing-RStudio (accessed 23 April 2020).

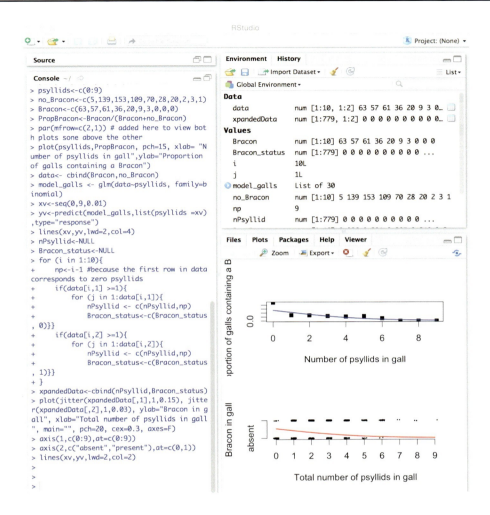

Installing Packages

Most of the functions that we describe in this book are in the basic R package, but for some things such as mapping the world or working with phylogenies, you really need to install various free R packages. There are very many R packages available, some for very specific purposes, some more generally useful to biologists. The base R installation already includes a number of packages. These will not always be the same on every platform. To see what has been installed as part of the package **base**, enter

library(help="base")

To see all your currently installed packages use **installed.packages**(). This may be a very long list if you have been using R for a while. Alternatively, you can use the function **available.packages**, which downloads a complete list with extra information. Package names are in the first column. Most are duplicated because of the file structure.

a<-available.packages()
length(unique(a[,1])) # we will explain this syntax later
[1] 14814

but this shows that at the time of writing there are quite a lot of packages.

We introduce installing a new package first in Chapter 6 (p. 49). The key thing is that to use functions in a new package is a two-step process. First you must install the package, which means getting R to download it off the www and putting it in the folder where R stores these things, and then telling R that in your current session you want to have access to the functions in that package. Installing a package from the CRAN depository is done using the function **install.packages** and the name of the package must be in inverted commas. Then you must use the function **library** to tell R to make the functions available for use (i.e. attach it to the function search list). A very few important packages frequently used by biologists have to be downloaded from other websites (e.g. github.com, see Chapter 18, also see Chapter 24).

It is inevitable that not all packages (perhaps some mentioned here) will stay identical or even be available for ever. Sometimes when trying to install a package or use a function in a package you will get an error message that they are 'deprecated'. That means that they are in the process of being removed, usually to be replaced by a newer version. In general, the deprecated version should be avoided, and usually, a quick search on Google or Stack Overflow will lead you to a newer (better) version.

Finally, you may sometimes have to report or know what version of R you have installed and for that there is the simple function **R.version**.

3 Very Basic R Syntax

> *Summary of R Functions Introduced*
> log
> sum

Computers are an everyday part of life, especially for science students. For many they are put to use for surfing the web, writing essays on a word processor and sometimes handling data from lab or field research experiments. For the latter task many students rely on what appear to be simple options such as Excel or maybe software especially designed for carrying out various statistical analyses. This might appear to be the easiest solution but, in all honesty, usually fails to facilitate the student's understanding of what they are doing. This book is about R, a programming language that has a huge range of inbuilt statistical and graphical functions. In easy steps it can allow the user to analyse data appropriately (something that simple tests in Excel fail to do) and to produce easily understood, informative and beautiful graphical outputs. R is immensely powerful and can, for example, take in the whole text of the Bible (the King James Authorized Bible has 783,137 words) and with a one-line command tell you in a few seconds how many occurrences there are of the word 'and' – yes, it is very good for handling text too, though it isn't a word processor.

R is not a compiled language and the computer executes your input line-by-line, which makes it easy to find any mistakes you may have made. These will often be just typos but sometimes the mistake may be in trying to handle the wrong type of data (maybe letters where numbers are required) or sometimes the brackets are in the wrong place. R is probably fast enough for nearly anything an undergraduate or MSc student would care to do, but because it is not compiled like C++, for example, it is thousands of times slower. Put simply it is not suitable for handling the outputs of large hadron collider experiments or doing CGI, but will handle and plot tens of thousands of data points quickly.

Firstly, we show how R works by talking you through a number of exercises, often producing graphical output, so you will get to know how to write simple code and become familiar with some of the most commonly used R functions for manipulating data and doing simple calculations. For ease we'll firstly use a non-biological type of example. Thereafter we will enter, display and analyse a number of real biological or medical datasets as might be obtained in student class experiments or fieldwork projects. As we expect students to dip in and out of this book, some things will be necessarily repeated or cross-referenced.

Further on, we present an outline of statistical tests appropriate to various types of data that you will come across. This book is not intended as a course in statistics, and certainly before publishing any results, readers should ensure that they have a full understanding of the data and have chosen appropriate analysis methods. Some statistical analyses are very intricate and wrongly formulating the data or analysis can lead to very misleading conclusions. Nevertheless, you have to start somewhere and therefore we have tried to provide fairly reasonable descriptions of some of the ways data commonly collected by biologists should be analysed.

There is quite a lot of jargon about R and its different sorts of variables, but we have tried to minimize this and mostly use terms that you will be familiar with. R handles quite complex types of data: simple numbers, characters, lists of numbers, lists of character strings, tables and matrices, and more complex 'structures' containing various combinations of these. Single values are called variables, but we are often dealing with sets of variables which are called 'vectors', but please note that the use of this word in R is not the same as a mathematical vector (a direction and magnitude). Technically vectors are a sequence of variables with two or more components all of the same type (i.e. all numbers or all character strings, or all TRUEs and FALSEs) though we will use the term more loosely to include items of length 1 for simplicity. The types (modes) of things stored in vectors relevant to biologists are usually numeric, character, factor and logical. Factors are stored as numeric values, and numeric values may be integers (whole numbers) or real numbers (which are always double precision in R). Unlike in C++, they are simply created as and when needed.

Any text on a line following the hash character, #, is ignored and that makes it easy to annotate the code. This also means that you can cut and paste the blue code and #'d comment into R without getting an error.

Each variable, vector, etc. is identified by a name that you give it using the *assign* <- (pronounced 'gets') command. The structure is: X<-{some value} or X<-{the results of some operation on values using R functions}, e.g.

X<-3 # X now has the value 3 (a number); it doesn't matter if there are spaces before or after the <-

X<-"3" # X now has the character 3, which cannot be used in mathematical operations

X

[1] "3"

Single apostrophes can also be used in this situation. Note that the apostrophes are simple, plain, straight ones, not the intelligent curly type that many word processors will give you by default; R does not like those, and you will usually get an error message: Error: unexpected input in ...

X<-log(3) # X now has the value the logarithm of 3 to the base e (the default base)

X<-3+4 # X now has the value 7

X<-sum(1:53) # X is now the sum of the numbers (integers) from 1 to 53, i.e. 1431

X<-X*3 # X is now 1431 multiplied by 3, i.e. 4293

X<-sum(1:53)/3 # X is now 1431 divided by 3, i.e. 477
At any point you can find the value of X by typing X at the prompt
X
[1] 477
The '[1]' that is displayed means it is the first (and in this case only) value stored in that variable.

If you enter a calculation that gives infinity (or −infinity), R will usually recognize it as such and return 'Inf'.
X<-1/0
X
[1] Inf
and these can be compared too (here we use ==, which reads as 'exactly equal')
2/0==1/0
[1] TRUE
X<-"My aunty's name is Jennifer" # X is now a character string, a single letter is also a string
X
[1] "My aunty's name is Jennifer"
Within a character string, if you enter intelligent (curly) quotes "'" or "'" as below, you may get that character interpreted rather differently from expected– how they are treated is version dependent.
X<-"My aunty's name is Jennifer"
X
[1] "My aunty\342\200\231s name is Jennifer"
Each time you assign something to a variable such as 'X' it replaces what was stored under that name previously.

You can also use R in 'calculator mode' by typing what you want to calculate at the prompt, e.g.
(117+117^2+log(54))/(1+log(54)) # logarithms are given in base e as default
[1] 2768.096
but in this case the value is now forgotten whereas if you wanted to use the exact result (the value R calculated is rounded to 7 significant digits by default), you should assign it to a variable name,
u<-(117+117^2+log(54))/(1+log(54))
or in reverse if you like.
(117+117^2+log(54))/(1+log(54))->u
u
[1] 2768.096
Both assign a value to 'u', though the reverse method is not recommended.

Similarly, the way we have written this book shows many of the possible stages that will lead to the final R code which will give the desired output. We deliberately show examples of code that will not give the beautiful and informative output wanted because this is how learning programming goes. The more mistakes you make, the better you become at writing the code and achieving your ends. Thus, we also give the final polished copy examples in an electronic appendix (www.cabi.org/openresources/45349) and these can be copied to provide the final output.

The selection of examples used here is rather eclectic and biased towards our own personal interests and expertise, but nevertheless it is quite broad. Rather than going into immense detail in areas such as phylogenies, graphics, statistical analyses, etc. we have tried to give the basics that will allow any reader to move on into the field in greater depth – the sort that will ultimately be required if you wish to publish results in learned journals or even just to do a good project write-up for your diploma or degree. This book is definitely not meant to be a 'be all and end all' for any serious and publishable research, but should enable most students working in the correct direction.

The type(s) of statistical analysis approach that we use is currently what most biological researchers use. This could quite well change in coming years where analytical methods and computing software are leading to an increasing role for Bayesian approaches. We do not attempt to cover these and general use of those methods at undergraduate level is probably some years away.

4 First Simple Programs and Graphics

Summary of R Functions Introduced	
arrows	median
as.factor	min
as.integer	par
axis	pdf
c	plot
colours	q
dev.new	quantile
for	range
graphics.off	rep
gsub	repeat
help	sd
ifelse	summary
lengths	table
list	text
matrix	var
max	which
mean	while

This chapter presents you with the basics for handling text, numbers and simple data files. Type the text in the examples below and see what happens, then read the explanation. To prove that it is not too difficult we will very soon show how to do some basic statistics and graphs.

Basic R Features

Variable names (and names of other types of R object) are:

1. Case sensitive.
2. Must start with a letter (not a number or symbol).
3. Must not include a blank space, people often use an underscore '_' or full stop '.' to make variable names more intelligible.

© Donald Quicke, Buntika A. Butcher and Rachel Kruft Welton 2021. Practical R for Biologists: An Introduction (D. Quicke, B.A. Butcher and R. Kruft Welton)
DOI: 10.1079/9781789245349.0004

4. Should not include a minus sign '–' either because R will interpret this as meaning take the second part from the first part! Ditto '+', '/' or '*'.
   5. Avoid 'T' and 'F' as you may well accidentally overwrite TRUE/FALSE class abbreviations (this can be recoverable, but it is easier not to have to).

Variable names can be long or just a single letter. In practise you should invent names that help you to understand what you have written, but not so long that it takes ages to type them. This is especially true if you write code and then need to come back to it some days or months later – by then you will certainly have forgotten what x, y, z, etc. mean, but if you use variable names such as 'number_species', 'sizes', 'month', it will make your life far easier.

Commas, Brackets and Concatenation

Commas and brackets are important in R

() – values passed to functions are enclosed in round brackets after the function name, e.g. the trigonometric sine or cosine functions, which in R are **sin** and **cos**. When multiple values or arguments are assigned to a variable or argument, they are separated by commas, e.g. **max**(6,9,1,21,4). It is optional whether you add a space after the commas, we have not done so here. A very convenient thing in R is that the order in which named arguments are passed to functions does not matter.

[ ] – the positions (indices) of values in a list, matrix, array, table or data-frame are specified in square brackets, e.g. if X is a list with four elements 'p', 'q', 'r', 's',
X<-c("p","q","r","s")
then X[3] returns 'r' (third value in list); X<-X[–3] deletes the third character from X.

The function **c** (= concatenate) is incredibly important. It concatenates or combines a series of values into a list (i.e. a dimensional array). In matrices, row numbers are specified first, followed by a comma and then the column number, for example, 'X[72,6]' means the contents of the cell in the 72nd row and 6th column if matrix 'X'.

{ } – curly brackets are used to enclose blocks of instructions that are always performed together as part of a loop (e.g. **for** {x in some range}, **while** {some test is TRUE} or **repeat** {until some condition is satisfied}), and also in **if** statements and user-defined (home-made) functions.

Some examples are:
If x is an empty vector to start with (x<-NULL) then
for(i in 20:30) {x<-c(x,i)} # creates a list of the integers from 20 to 30.
Curly brackets are necessary if there is more than one line of code operations that need to be done as a block but are optional if it is only a single operation on the same row.
for(i in 20:30) x<-c(x,i)
x
[1] 20 21 22 23 24 25 26 27 28 29 30

First Simple Programs and Graphics

The next example has two separate operations. Every pass around this loop calculates the next value of x^2 and adds it to the growing list of numbers in the list called y. After this, x is incremented by one. At the start of each pass around the loop, the value of x is checked to see if it is less than 10. If it isn't the loop stops and the square values of x are now in the vector y.

```
x<-1 # the initial (starting) value of x
y<-NULL
while(x < 10){ # calculates x² for values of x from 1 to 9
        y<-c(y,x^2 )
        x<-x+1 # incrementing x by one each cycle
        }
y
[1]  1  4  9 16 25 36 49 64 81
```

Note that there is a danger of using simple variable names. If we had previously used a variable called x and it had a value > 9 before entering this code, then the loop would not be run, so it would be best to set x<-1 (or some other starting value) before the **while** statement. Here we see the output at the end by inspecting the contents of y. If we want to see the results at each pass through the loop we can include an extra command 'print(x^2)' within the while loop.

In the following x is incremented by 1 each cycle of the repeat loop until it reaches 10001. The loop is terminated when the criterion for executing break is met.

```
x<-0
repeat{
        x<-x+1
        if(x>10000) break # the break command exits from the loop
}
```

Tool Box 4.1 ✹

- Use () to apply a function to something.
- Use [] to ask about position in a list.
- Use {} to enclose blocks of functions.

The Colon Character

The ':' character is used to indicate that all discrete values between two extremes are also included in the sequence, e.g. x<-1:6 is the same as x<-c(1,2,3,4,5,6). The values are incremented by one from the starting value, so c(0.2:4) gives 0.2 1.2 2.2 3.2.

Raise to the Power of Symbol

^ means raise 'to the power of', e.g. x^3 is x cubed, x^0.5 is the square root of x. Note that if it doesn't work you may have used a circumflex accent instead, which looks similar but is smaller.

Exiting from R

To quit the R program you can use either pull-down menus or use the function **q**.
q()

Help Pages

For the many in-built functions in R you can call up a help page by typing 'help({*name of function*})' or just '?{*name of function*}'. The function **help** by default only searches for functions within the R base package, however, adding 'try.all.packages=TRUE' searches all the installed libraries.
help("diversity",try.all.packages=TRUE)

Help for topic 'diversity' is not in any loaded package but can be found in the following packages:

Package	Library
igraph	C:/Users/Rachel/Documents/R/win-library/3.6
phangorn	C:/Users/Rachel/Documents/R/win-library/3.6
vegan	C:/Users/Rachel/Documents/R/win-library/3.6

Here we find out that there are three different R functions among our installed packages called **diversity**, and by clicking on each package name we will be taken to the relevant help page for the function **diversity** in that package. The functions **help.search** and **??** look for a word or phrase in all packages, even if they are not loaded.

Beginning with Simple R Code to Get Used to the Command Line System

Let us assume we have eight friends called Adam, Alice, Brian, Barbara, Charles, Carla, Christine and David (to make typing the example easier, you can use the names of your own friends or colleagues or anything). We can assign this list of names to a variable like this following the R prompt '>' using the function **c**.
names<-c("Adam","Alice","Brian","Barbara","Charles","Carla","Christina","David")
Note that each name is enclosed between simple (straight line) inverted commas, and each is separated by a comma. We use double inverted commas but single ones will probably also work. You will soon find that if you try writing R code in some word processors, they will default to using 'intelligent', i.e. curly, inverted commas, "x". However, R does not understand these and will error. To edit these out by hand is a pain, and so it is best to use a simple text editor to create your code.

First Simple Programs and Graphics

To check how many friends you have in your list, use the function **length**
length(names)
[1] 8

Then we might create three other lists, for their sexes, ages and heights. In each case the order of the values (sex, age, height) must be the same as the order that each friend appears in your list of names.
sex<-c("M","F","M","F","M","F","F","M")

R doesn't understand gender so we could use any values to represent males and females, e.g. 0 and 1, 'a' and 'b', 'male' and 'female', 'George' and 'Freda', etc. However, ages are numeric so it is easiest to use numerals, in this case whole years
ages<-c(18,19,22,21,20,29,20,19)
and similarly heights
heights<-c(167.5,160,170.5,154,179,166,158,160.5)

Now we have four vectors. We can now practise a few built-in R functions on these data. For example, to find the arithmetic average of their ages use the function **mean**
mean(ages)
[1] 21

or mean of their heights
mean(heights) # for the arithmetic mean of their heights
[1] 164.4375

We can also obtain all other common descriptors such as
max(ages) # the oldest (maximum) age
[1] 29
min(ages) # the youngest (minimum) age
[1] 18
range(ages) # the range of ages
[1] 18 29
var(ages) # the variance of the ages
[1] 12
quantile(heights) # quantiles, but normally with many more data points
       0%   25%   50%   75%   100%
   154.00 159.50 163.25 168.25 179.00

Two other useful summary statistic functions in R with self-explanatory names are **median** and **sd** (standard deviation = square root variance).

> **Information Box 4.1**
>
> *A note on means.* The arithmetic mean is a standard measure of central tendency. It is not the only such measure, the median, mode and geometric mean are others. With large sample sizes drawn from an approximately bell-shaped distribution, the mean is a good measure, but with bimodal distributions or small samples that might, by chance, be skewed, it can be misleading. In the example of the ages of our eight friends, we obtained a mean of 21, which is pretty central. Consider, however, what would happen if we included dear Aunt Amy (age 94) among our friends. The mean of the nine ages now goes up to slightly more than 29, older than all but one of our original eight. Clearly Aunt Amy's age is an outlier, which makes our age distribution very skewed. In other words, a few extreme points may make the mean less meaningful. Statisticians often overcome this by excluding a small number of the most extreme values on either side of the mean, so the value we calculate is based more on those values closer to the centre – this is called a 'trimmed mean'. In R the function **mean** can accept an argument called 'trim=' which removes a small proportion of values from both the upper and lower ends of the distribution. The value of trim typically used is 0.2, technically the nearest integer less than 0.2 × the total number of observations. To apply trim in our example use mean(ages, trim=0.2). With the eight original ages the trimmed mean is 20.17, and with Aunt Amy included it is 21.42, both values being close to the original calculation. Some people say that by trimming you are discarding data (true) but it all depends on your purpose. The tails of the data may be very interesting but they are not necessarily helping you get an estimate of central tendency.
>
> In our example of 9 values (including Amy's), the trim function calculates 9 × (1 − 0.2) = 7.2. The nearest integer to 7.2 is 7, so the trimmed mean only uses the 7 most central values to calculate the mean, so in our small dataset case it omits one value at each end of the distribution.

We can find out how old each friend is as follows. If we want to know Barbara's age we need to know what position Barbara is in the list of friends' names and then retrieve her age from the vector we called 'ages'. A long-hand way to do this would be to create a new variable, 'pos' to get Barbara's position.

pos<-which(names=="Barbara")
pos
[1] 4

Barbara is the fourth name in the list. Now for her age we enter the result into 'ages'

ages[pos]
[1] 21

Note that positions in a list (indices), rows or columns in a matrix or table, are always entered in square brackets '[]', whereas the things we pass to functions (such as **min**, **max**, **mean**, **log** ...) are always in simple rounded brackets '()'.

We can write the code to get Barbara's age in one line using **which** and exactly equals, ==

ages[which(names=="Barbara")]

We can also use negative logic, which is represented in R by an exclamation mark, **!**. To ask which friends' ages are not 20,

not20<-which(ages !=20)
names[not20]

```
[1] "Adam"   "Alice"   "Brian"   "Barbara"   "Carla"   "David"
```
To calculate the total ages of our friends we use **sum**
sum(ages)
```
[1] 168
```
so, our friends' ages add to 168 years.

We used **length** to find out how many friends are in our list (i.e. how many elements there are in the vector), and we can combine that with **which** to find out how many of our friends are male or female,
Male.friends<-length(which(sex=="M"))
Male.friends
```
[1] 4
```
Female.friends<-length(which(sex=="F"))
Female.friends
```
[1] 4
```
Alternatively, we can tell R to treat 'M' and 'F' as factors and ask for a summary of 'sex'. Factors are the different levels of a categorical variable.
summary(as.factor(sex))
```
F M
4 4
```
This again tells you that you have four Ms and four Fs.

Tool Box 4.2 ✖

- Use c() to link lists of values into an ordered vector.
- Use length() to see how many items there are in your vector.
- Use mean() to calculate the arithmetic mean of a set of numbers.
- Use max() to find the maximum value.
- Use min() to find the minimum value.
- Use : to define the range of values to use.
- Use ^ to raise something to a power.
- Use var() to calculate the variance.
- Use range() to calculate the range.
- Use quantile() to calculate the quartile values.
- Use median() to calculate the median.
- Use == to test exact quality of two characters, values or vectors.
- Use which() to get the position(s) of a character or value in a list.
- Use != to mean values that are not equal to something.
- Use summary() to create a table of variables.
- Use as.factor() to coerce a numeric vector into factors.
- Use print() to see the result.

Playing with Graphics

If we want to see our friends' names and heights in technicolour we can make a little graph and print them on it. The command line in the function **plot** below tells R to open a graphics window, with no labels (the x axis title argument 'xlab=' and the *y* axis title argument 'ylab=' are both set to be empty represented by a pair of "" with nothing between them, i.e. representing a text variable that contains no text), and no axes are drawn (the 'axes=' argument is set to the logical operator FALSE) ('axes=F' also works but is not recommended). However, we do have to define the *x* axis and *y* axis limits ('xlim=' and 'ylim='); here we have set the *x* axis range to be from 0 to 2 and the *y* axis range from the minimum to the maximum heights of our friends. Note that all ranges are given in the form **c**({*lowest value you want*},{*highest value you want*}). The argument called 'main' puts an optional title above the plot.

```
plot(NULL,axes=FALSE,xlim=c(0,2),ylim=c(min(heights),max(heights)),
     xlab="",ylab="",main="My friends' heights")
```

My friends' heights

Yes, it is meant to be a blank plot window with just the title.

In R the word NULL means nothing or empty, and NA means not applicable, and although approximately the same they are used in different contexts. These two words are very important and will be used later quite a lot, though in different contexts. You will get to understand their sometimes subtle difference best by seeing how they are used in examples.

If you do not specify that the axis labels are empty using inverted commas, R will put its default axis names according to the information it has been given – try it without putting the 'xlab' and 'ylab' arguments and see what happens.

Graphic functions in R such as **plot** have many optional arguments that can be passed to the function that specify axis names, axis extents, font type, etc. (see Appendix 1). In this chapter we use only a few of them and we will introduce more as we progress through the book.

In our first example we are controlling the xlim and ylim values by passing a two-element numeric vector, **c(**{*lower limit*},{*upper limit*}**)**. However, if we are using **plot** to show data that are specified when it is first called, R will automatically determine the ranges of the *x* and *y* axes. Nevertheless, you can still override this by specifying your own preferred **xlim** and **ylim** values.

The function for adding text to a plot is called **text**: it requires arguments specifying the *x* and *y* positions where to write the text, and the text itself. For all the graphic functions we must enter the *x* values first, then the *y* values, separated by a comma. Here we will put all names centred on (the default) *x* = 1, and plot them along the *y* axis in accordance with our friends' heights. Each variable passed to **text** (or **plot**) must have the same number of elements, i.e. have the same length.

There are eight names, and eight heights so we need eight *x* positions. We could write '**text(c**(1,1,1,1,1,1,1,1), heights, names)' but that is tedious, so instead we will use the function **rep** (=repeat) to create a list of eight 1s for the *x* axis

text(c(rep(1,8)),heights,names)

rep has produced a vector with the number 1 repeated eight times

rep(1,8)
[1] 1 1 1 1 1 1 1 1

Obviously, we probably want to see their actual heights and we can do this in two ways – instead of plotting completely blank axes we could add the name of the *y* axis

plot(NULL,xlim=c(0,2),ylim=c(min(heights),max(heights)),xlab="",ylab=
    "Height (cm)",main="My friends' heights")

but this will also put numerical values along the *x* axis (we usually do want to do that, but not on this occasion). Instead we used the argument 'axes=FALSE' to specify that we do not want axes drawn. Then separately we use the function **axis** to draw just one axis for our height values. Which axis to draw is specified by a number from 1 to 4: 1 = bottom, 2 = left, 3 = top, 4 = right. We want a vertical axis on the left so we will pass the number 2 to **axis** function. Then there are two more arguments required: the *y*-values (heights) we want printing at the side of the *y* axis, and where along the axis we want to place them; in our case these will be the same. Our shortest friend is 154 cm tall and the tallest 179 cm so we might want to show heights 150, 155, 160, 165, 170, 175 and 180 cm.

plot(NULL,xlim=c(0,2),axes=F,ylim=c(min(heights),max(heights)),xlab="",
    ylab="Height (cm)",main="My friends' heights")
axis(2,c(150,155,160,165,170,175,180),at=c(150,155,160,165,
    170,175,180))

Notice that the lowest value has been chopped off, and this is because we used the minimum actual height (154 cm, Barbara again), and 150 cm is too far below this for the **plot** function to show. We can of course still show it by

incorporating the very useful argument 'xpd=NA' in our command line, which allows plotting/drawing outside of the active plot area. So now we enter
axis(2,c(150,155,160,165,170,175,180),at=c(150,155,160,165,170, 175,180),xpd=NA)
text(rep(1,8),heights,names)

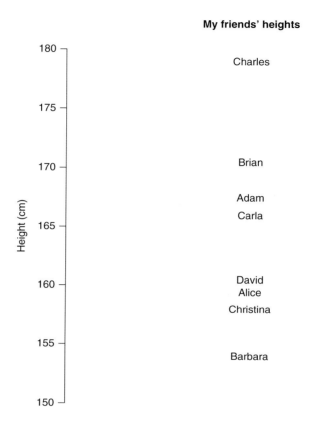

We are getting there, but our graph still is a bit ugly because the names are a long way from the *y* axis. So we can choose to make the plot taller and narrower. We can specify the margins of the plot area (see Chapter 6 and Appendix 1) but an easy way to do this is to specify that we want there to be two graphs side-by-side in the plot window – we will leave the second one blank. We do this bypassing the argument 'mfrow=' (which stands for 'MultiFrame ROWwise layout') to the graphical parameters function **par**. The structure is par(mfrow=c({*number of rows of plots per page*}, {*number of columns of plots per page*})). **par** can be passed many parameters (see Appendix 1) or use ?**par** to see the whole range of options.

par(mfrow=c(1,2)) # telling the plotting device that we want one row of plots with two columns

You will need to re-run the whole plot to make the change.

First Simple Programs and Graphics

> **Tool Box 4.3** 🛠
> - Use plot() to plot points on a graph.
> - Use xlim=c(min,max) to set the length of the *x* axis.
> - Use ylim=c(min,max) to set the length of the *y* axis.
> - Use xlab=" " to name the *x* axis.
> - Use ylab=" " to name the *y* axis.
> - Use main=" " to name the graph.
> - Use rep() to repeat.
> - Use text() to add text to a graph.
> - Use axis() to define which axis is affected 0 = S, 1 = W, 2 = N, 3 = E.
> - Use par() to set graphical parameters.
> - Use par(mfrow=c(number rows, number columns)) to set the rows and columns per page.

Working with Character Variables

Because we used 'M' and 'F' to indicate gender we have first to convert these to numerical values that can be plotted. It seems appropriate here to introduce you to your first bit of R programming though there are other simpler ways of doing this. We need to create a new variable for 'sex' that can be used to position things on the graph – specifically here we want to make a new variable with 1 replacing 'M' and 2 replacing 'F'. There are several ways to do this but here we are going to use the loop function **for**. Initially we want our new variable to be empty so we will set it to NULL

sex.as.number<-NULL

We determined above that we have eight friends and we want to look at each one's gender one by one as we go through the list of values in the variable 'sex'. To do this we use a counting variable '*i*' (it can be called anything, but for loops it is traditional to use *i*) and the **for** command to make add 1 to *i* each time it goes around the loop. Doing nothing each time this would look like 'for(i in 1:8){*put the operations you want to perform multiple times here*}'.

The contents of all our variables can be accessed by their position in the list of values, from one to eight in our example. To find the sex of the fifth person in the list for example, we type

sex[5]

Then for each value of *i* as it progresses from one to eight we test whether 'sex' is 'M' or 'F' using the function **if**, and if it is an 'M' we concatenate a 1 onto 'sex.as.number', and then check whether it is an 'F' in which case we concatenate a 2.

```
for(i in 1:8){
    if(sex[i]=="M") sex.as.number[i]<-1
    if(sex[i]=="F") sex.as.number[i]<-2
    } # end of i loop
sex.as.number
 [1] 1 2 1 2 1 2 2 1
```

There is an easier function to use in the loop: instead of the two separate **if** statements we can replace them with the function **ifelse**, which does exactly the same thing; the syntax is **ifelse**(test, TRUE, FALSE). Assigning the first value if sex[i]=="M" is TRUE, and the second value if it is FALSE
ifelse(sex[i]=="M",sex.as.number[i]<-1,sex.as.number[i]<-2)
When processing large sets of values when there are only two alternatives, **ifelse** is quicker as it only makes one evaluation rather than two, and is more elegant.

Information Box 4.2

Ifelse. The **if** and **else** parts of **ifelse** can be separated and that can sometimes be convenient. For example,
```
q<-1
if(q==0) r<-"bad" else r<-"good"
r
[1] "good"
```

Even simpler is to use the substitution function **gsub** (=get and substitute), which doesn't require you to write a loop, but will require coercing (forcing) the numbers we wanted to represent gender to be numbers rather than character strings. Create a new vector called 'sex.as.number' as a copy of sex
```
sex.as.number<-sex
sex.as.number<-gsub("M",1,sex.as.number)
sex.as.number<-gsub("F",2,sex.as.number)
sex.as.number
[1] "1" "2" "1" "2" "1" "2" "2" "1"
```
Here you see that substitutions we made in 'sex.as.number' using **gsub** have not been treated as numbers but as characters – i.e. they are shown in "". That is because a vector (a list in this case) must only contain one particular type of value. When we substituted the 'M' for 1 there were still character elements in the vector, the 'F's, so therefore the 1s also had to be treated as characters, hence '1'. Having done both (all) the substitutions and now all the values are character representations of numbers, we can coerce them to be numbers using **as.integer**
```
sex.as.number<-as.integer(sex.as.number)
sex.as.number
[1] 1 2 1 2 1 2 2 1
```
Now you can do arithmetic with these values or use them in a plot to specify axis positions or colours. For example, if the sex.as.number character is 1 (male) we might want the names to appear in blue and if 2 (female) then in pink. So we define a colour vector and use the 1 or 2 in sex.as.number to choose which colour for each name:
```
colours1<-c("blue","deeppink")
text(c(rep(1,8)),heights,names,col=colours1[sex.as.number])
```
Finally, we might want to add some arrows to show our friends' heights more accurately against the *y* axis. We will use the function **arrows**, which takes the form arrows({*starting x value*},{*starting y value*},{*end x value*}, {*end y value*}, code={*arrow at end*}, length={*size of arrow head*})

```
for(i in 1:8){
    arrows(0.65,heights[i],0,heights[i],code=2,col=colours1[sex.as.
        number[i]],lwd=2, length=0.07)} # end i loop
```

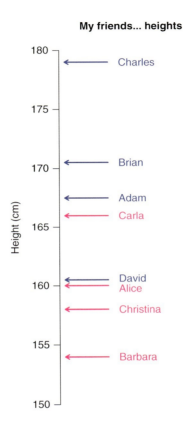

Now that we have a nice plot in the plot window we can save it by going to the file menu (or right click on it), and then choose what to call it. There is also a built-in R function **pdf** that can save a plot.
pdf("rplot.pdf") # you can call it what you like
{put the code here that generates your graphic and run it}
graphics.off() # **turns off the current open graphics device**
So you need to also know how to turn it back on again!
dev.new() # **new [graphics] device**
This command also allows you to have more than one graphics window at a time, re-enter it and a second window will open. Each window will have a different name and plotting can be directed to specified ones.

> **Exercise Box 4.1**
>
> 1) The argument 'adj=' tells **plot** and **text** how to justify the text, the default is centred achieved by 'adj=0.5', then 'adj=0' means left justify (or bottom if the text is printed vertically), 'adj=1' means right justify or top. Use this to right justify the names of the friends. Redraw the graph moving all names to the right (change the x coordinates).
> 2) A bit more complicated, change the arrows statement by using different values for the right-hand end of the arrows (currently 0.65) for each friend. Hint: you can change the font to one like courier with all characters of equal width by adding the argument 'family="mono"' and you can count the characters in each name using the **nchar** function.

> **Tool Box 4.4**
>
> - Use NULL to define an empty set (you cannot use 0).
> - Use xpd=NA to allow plotting outside of the current plot area.
> - Use family= to change the font; options are sans, serif, mono, which correspond to Arial, Times New Roman and Courier in Windows.
> - Use for(i in {one value}:{another value}) to step from one value of *i* to another in steps of 1.
> - Use ifelse() to define alternative actions given a yes/no question.
> - Use gsub() to get and substitute values or characters.
> - Use as.integer() to coerce characters into integer values.
> - Use col= to set the colour of text, lines, symbols or other plot items.
> - Use arrows() to add arrows to a graph, e.g. arrows(start x value, start y value, end x value, end y value, code=shape of arrow at end, length=size of arrow head).
> - Use lwd= to set the width of plotted lines.
> - Use pdf("filename.pdf") to save the plot as a pdf with that filename.
> - Use graphics.off() to close all open graphics devices at once.
> - Use dev.new() to open a new graphics window.

R can colour all elements of a plot any colour you like – there are 256 × 256 × 256 (i.e. 16,777,216 or approximately 16 million) possible colours, but for simplicity, R has 657 colour names, which you can see by typing
colours() # or colors()
There are slightly fewer named ones than might first appear because R accepts both UK and US spellings of grey. We show how to use **rgb** in Chapter 21.

While we are typing in R, the program remembers all the values of the variables that they were last set to. Also, after R has executed a command or series of commands and the cursor is shown, pressing the up arrow will automatically enter the last entered line or lines of commands, and these can be edited before pressing return. Suppose one of us or you wanted their name on the plot to be much bigger and bolder, we can alter that without having to type the entire command line in again.

In addition to text, we can plot lines, points, various shapes, maps, 3D surfaces etc. Some of these require downloading and installing additional free library units – all will be explained. As well as the function **plot** there are special in-built ones to create barplots, boxplots, curves, pie charts, balloon charts, etc. For more sophisticated plotting such as 3D topologies, the reader should first explore the functions **interp** and **wireframe** within the package **akima**.

Built-in R Datasets

R has several built-in datasets, mostly small or fairly small, that allow you to experiment with functions, e.g. BOD (Biochemical Oxygen Demand) and HairEyeColour (correlations of hair and eye colour for a group of students, by sex). These datasets provide a range of excellent examples for practising writing/running R code without having to laboriously enter your own artificial data. Other available datasets can be seen at https://stat.ethz.ch/R-manual/R-devel/library/datasets/html/00Index.html (accessed 22 April 2020) or by entering 'data()'. Typing the names of these datasets after the R prompt shows you their contents.

```
BOD
   Time  demand
1    1     8.3
2    2    10.3
3    3    19.0
4    4    16.0
5    5    15.6
6    7    19.8
HairEyeColour
, , Sex = Male
        Eye
Hair     Brown Blue Hazel Green
  Black     32   11    10     3
  Brown     53   50    25    15
  Red       10   10     7     7
  Blond      3   30     5     8
, , Sex = Female
        Eye
Hair     Brown  Blue  Hazel  Green
  Black     36     9      5      2
  Brown     66    34     29     14
  Red       16     7      7      7
  Blond      4    64      5      8
```

The *table* Function

We have introduced factors earlier (factors are the different levels of a categorical variable, such as 'M' or 'F'), and here we show how **table** can be used to automatically cross reference values and a factor according to their order in two vectors. This is the same as counting the occurrences of different factors in a list.

```
data<-list(c(1,2,2,4,1),c("A","A","B","C","A"))
data
[[1]]
[1] 1 2 2 4 1
[[2]]
[1] "A" "A" "B" "C" "A"
```

```
table(data)
            data.2
data.1 A B C
     1 2 0 0
     2 1 1 0
     4 0 0 1
```
The three values of data[[2]] ('A', 'B' and 'C') are the column headers of the table, and the matching values in data[[1]] are the row names. The 2 in the top left shows that the value 'A' in data[[2]] corresponds with the number 1 data[[1]] twice.

We can use this, for example, to look at the correlation of hair colour and eye colour on a per individual basis.

```
friends<-c("Jack","Piya","Lek","Ali","Peter","Robert","Sue","Emma","Mike")
eyes<-c("grey","brown","brown","brown","blue","brown","brown","blue",
    "blue")
hair<-c("brown","black","black","black","blonde","brown","brown","blonde",
    "blonde")
table(list(c(eyes),c(hair)))
               .2
.1     black  blonde  brown
blue       0       3      0
brown      3       0      2
grey       0       0      1
```

Ragged Data

A particular problem that students and researchers encounter is that they do not have equal sample sizes. For example, Møller (1988) surveyed data on various male reproductive features such as sperm concentration, ejaculate size and testes size among primates. Not surprisingly, some species had been investigated more times than others. Here we will look at just the sperm concentration data (millions ml^{-1}) for the apes. Rather than running separate code to obtain such values as means, variances, maximums and minimums for each of our samples a, b, c, etc., we could be interested in combining them into a single R object. The way to do this normally is to combine them into a dataframe (see Chapter 5). For instructive purposes though we will look at some other issues with these types of ragged data. The important thing is to note that R does not have an object category for ragged data and some default things R does could corrupt your data if you are not careful.

```
Hylobates_lar<-152.4
Gorilla_gorilla<-c(171,191,48.4,107.9,195.2)
Pan_troglodytes<-c(548,608.8,391.1,258.3,142.9,274.5)
Pan_paniscus<-456.1
Pongo_pygmaeus<-61
Homo_sapiens<-63.5
```

names<-c(Hylobates_lar,Gorilla_gorilla,Pan_troglodytes,Pongo_pygmaeus, Homo_sapiens)

If we try to combine these into a table we cannot do it

x<-as.table(Hylobates_lar,Gorilla_gorilla,Pan_troglodytes,Pongo_pygmaeus, Homo_sapiens)

x

```
A
152.4
```

we only get the value for the first vector, Hylobates_lar. If we try the **table** function we see the issue clearly from the error message.

table(Hylobates_lar,Gorilla_gorilla,Pan_troglodytes,Pongo_pygmaeus, Homo_sapiens)

```
Error in table(Hylobates_lar, Gorilla_gorilla, Pan_troglodytes,
    Pongo_pygmaeus, : all arguments must have the same length
```

If we try to coerce them into a matrix, it does not work either

x<-as.matrix(Hylobates_lar,Gorilla_gorilla,Pan_troglodytes,Pongo_pygmaeus, Homo_sapiens)

x

```
     [,1]
[1,] 152.4
```

Again, only the first vector is given, and the same happens if we try to specify the number of rows by including the argument 'nrows=5'.

So we have to use simple fudges to create an array of these data. One way is first to create an empty matrix (a matrix filled with NAs) with one column (or row) for each species and the number of rows (or columns) equal to the largest sample. We cannot use **length** to determine the largest sample because length takes just one input, however, R provides another function, **lengths**, which returns the lengths of multiple vectors if they are in a list.

samples<-lengths(list(Hylobates_lar,Gorilla_gorilla,Pan_troglodytes, Pongo_pygmaeus, Homo_sapiens))

samples

```
[1] 1 5 6 1 1
```

So to create the necessary matrix, get the length of the longest vector

L<-max(samples)

sperm_conc<-matrix(nrow=L,ncol=5) # by default, cells of the matrix are filled with NA

sperm_conc

```
     [,1] [,2] [,3] [,4] [,5]
[1,]  NA   NA   NA   NA   NA
[2,]  NA   NA   NA   NA   NA
[3,]  NA   NA   NA   NA   NA
[4,]  NA   NA   NA   NA   NA
[5,]  NA   NA   NA   NA   NA
[6,]  NA   NA   NA   NA   NA
```

Now we can fill each column with just our data. However, we must specify how many rows are to be filled (which must be the same as the length of the vector)

```
sperm_conc[1:length(Hylobates_lar),1]<-Hylobates_lar
sperm_conc[1:length(Gorilla_gorilla),2]<-Gorilla_gorilla
sperm_conc[1:length(Pan_troglodytes),3]<-Pan_troglodytes
sperm_conc[1:length(Pongo_pygmaeus),4]<-Pongo_pygmaeus
sperm_conc[1:length(Homo_sapiens),5]<-Homo_sapiens
sperm_conc
       [,1]   [,2]   [,3]   [,4]   [,5]
[1,]  152.4  171.0  548.0    61   63.5
[2,]    NA   191.0  608.8    NA    NA
[3,]    NA    48.4  391.1    NA    NA
[4,]    NA   107.9  258.3    NA    NA
[5,]    NA   195.2  142.9    NA    NA
[6,]    NA     NA   274.5    NA    NA
```

Now we can perform actions on each column such as getting their means *if* we tell R to remove the NAs. If we don't do that (i.e. we try **mean**(sperm_conc[,2]), we get an answer NA). So we use na.rm (= remove NAs) so that our numerical function works.

```
mean(sperm_conc[,2],na.rm=TRUE)
[1] 142.7
```

See **tapply** and Chapter 27 to find out how to generate the means of 'sperm_conc' values of each species in one go.

Exercise Box 4.2

1) A simpler alternative to creating an empty matrix as we did above is to add extra NAs to each ape vector, by using the function **length** to assign a length value, for example: **length**(Hylobates_lar)<-max(samples). Use this method to generate the matrix of these data with species as rows. NOTE: this is not as difficult as some help pages would suggest. We show two methods in the answer.

Tool Box 4.5

- Use colours() to bring up a full list of named colours.
- Use lengths() to get lengths of multiple elements.
- Use list() to coerce into a list.
- Use table(filename) to present a data correlation summary as a table.
- Use as.table() to convert variables of equal length into a table.
- Use as.matrix() to convert variables of equal length into a matrix.
- Use matrix() to create a matrix.
- Use ncol= to specify number of columns.
- Use nrow= to specify number of rows.
- Use na.rm=TRUE to omit NA values from numerical functions.

5   The Dataframe Concept

Summary of R Functions Introduced	
as.character	detach
as.data.frame	read.csv
as.numeric	read.table
attach	textConnection
cbind	typeof
class	with
	write

R objects come in a variety of types – dataframes, matrices, vectors, and arrays for example. Dataframes are an important concept in R, allowing vectors of different types to be combined column-wise into a single object. This is different from matrices all elements of which have to be of a single type, e.g. all numbers, all characters, all logical, etc. You can use the function **class** to check the class of an R object.
class(BOD)
[1] "data.frame"
However, the in-built dataset HairEyeColour is not, it is a table
class(HairEyeColour)
[1] "table"
Dataframes are almost central to using R, they are a special case of a structure called a list that comprise a number of vectors, which we can see using the function **typeof**:
typeof(BOD)
[1] "list"

In dataframes, each row may contain a mix of data types, such as character data like names (which are character strings) or states that will be treated as factors (male or female, juvenile or adult, wet or dry), numbers and logical variables (TRUE or FALSE). However, each column only contains a single type. Both rows and columns can have names. If you create a matrix from columns of numbers and one or more of text then all the numbers will be converted to text but not if you create a dataframe.

If you are going to keep appending new data to a dataframe, as you collect more samples it is better to start with a dataframe containing at least one row of data rather than a blank one. Again using our friends' names, we have placed each friend and their characteristics in a separate row. The names and their characteristics are separated by commas – this is called a csv (=comma separated values) format, and we will come across this a lot later on. All rows must have the same number of elements. If there are missing data points these should normally be replaced with an NA, which R recognizes as missing or 'not applicable'.

We will now write the data to a file and then read that file we will name 'friends.csv'. The file will be saved in R's working directory (see Appendix 2 about choosing what that is; ideally for this book it should be a folder that you download the online resources data and code files to).

```
f<-"Adam,male,18,167.5
Alice,female,19,160
Brian,male,22,170.5
Barbara,female,21,154
Charles,male,20,179
Carla,female,29,166
Christina,female,20,158
David,male,19,160.5"
write(f,file="friends.csv") # for more on writing and reading files see Chapter 31.
Fr<-read.csv("friends.csv",header=FALSE)
Fr
        V1      V2 V3    V4
1      Adam    male 18 167.5
2     Alice  female 19 160.0
3     Brian    male 22 170.5
4   Barbara  female 21 154.0
5   Charles    male 20 179.0
6     Carla  female 29 166.0
7  Christina female 20 158.0
8     David    male 19 160.5
```

Note that **read.csv** is a special case of the function **read.table**, and is the same as including the argument 'sep=","' inside the latter (see Chapter 31). The file we saved is just a flat text file, rows of characters. However, when we read our saved files from disc specifying that it is a csv file (**read.csv**), R recognizes that the input contains columns of character strings and columns of numbers so the object created which we called 'Fr' is a dataframe which we can check using the function **class**.

```
class(Fr)
[1] "data.frame"
```

Note that column names (the R default is V followed by column number) have been added automatically to the saved file. Because this is a small file we can actually read it as a csv table directly from the entered value of 'f' using the function **textConnection**. With either read operation we can choose to assign column names in the function **read.csv**, e.g.

The Dataframe Concept

Ff<-read.csv(textConnection(f),header=FALSE,col.names=c("name","sex", "age","height"))

For the dataframe 'Fr' it is also easy to overwrite (change) the default column names given by R using the function **colnames** (there is no full stop in the name of this function), thus

colnames(Fr)<-c("name","sex","age","height")

Examine the classes of each of the columns using the $ sign to specify them by name.

class(Fr$name) # **class**(Fr$V1) if we have not changed the names
```
[1] "factor"
```
class(Fr$sex)
```
[1] "factor"
```
class(Fr$age)
```
[1] "integer"
```
class(Fr$height)
```
[1] "numeric"
```

Our dataframe comprises two different factors, an integer (whole number) and a numeric (real number) column. If any cell in either column 3 (age) or column 4 (height) contained any letters, then the whole column would have been interpreted as factors. The special case exception being if the character string was precisely 'NA'. The data in each column is a vector, i.e. a set of values of a single class. So a dataframe combines as columns a set of vectors that are not all of the same class, at least one is usually a factor.

We have already seen that you can access columns or rows using indexing within square brackets. Here we are using the dollar sign notation to access the individual columns within our dataframe. There are other ways of doing this. Some people use **attach**("{dataframe name}") (and after they are finished, **detach**("{dataframe name}")). When a dataframe is attached you can directly use the column names alone, e.g.

attach(Fr)

mean(age) # age is the age of the third column in our dataframe F
```
[1] 21
```
detach(Fr)

After detaching the dataframe the vector 'age' is lost

mean(age)
```
Error in mean(age) : object 'age' not found
```

We personally do not recommend attaching dataframes, though many people do. If you forget to detach them you may easily run into problems because you have vectors still in memory, often with common names such as 'age', 'height', 'name', 'x', 'y'… This can also happen with packages that maybe use the same name for a function. Sometimes you will get the R error message

```
The following objects are masked from…
```

which tells you that R has encountered one or more duplicated vector or function names.

An alternative to attaching a dataframe is to use the function **with**. The format is **with**({*data name*},{*function to apply to it*}). For example,

```
with(Fr,mean(age))
[1] 21
```
　　Using **with** allows us to access columns using their name, but that column name is only available within the context of **with**(dataframe...) rather than globally. Therefore without specifying the dataframe using $ does not find anything
```
age
Error: object 'age' not found
```

Combining Sets of Tables for Data Collected on Different Dates

　　Often at the start of a study we will not know how many data points or sets of data we will end up with, or even perhaps what all the types of data we are going to collect are. Some people leave analyses to the end of their investigation, something that many students foolishly do. This is extremely bad practise for several reasons. Firstly, you may be doing something wrong that would be easily revealed by early analyses. Secondly, there may be some unexpected factor which might invalidate your final result, but which could again have been revealed by progressive analyses soon after the start. Thirdly, analysis as you go along might reveal some interesting new feature worthy of deeper/special investigation, an opportunity that might be otherwise missed or at least have to wait for a new experiment of the next field season. When you already have a dataframe and you want to add more rows to it, use the function **rbind**.

Converting Factors in a Dataframe to Numeric or Character

　　Sometimes we may find that during conversion of an object to a dataframe either numbers or characters are being treated as factors rather than as what we intended. This is usually because we have done something inappropriate in how we created our dataframe as we will show below, but it might also be because of the format in which we were originally given the data, perhaps in an email. Here we use data on the relationship between whole animal resistance to tetrodotoxin (TTX) for newts at two locations in Oregon, USA (from fig. 4 in Geffeney *et al.*, 2002). We start with two vectors, the locations (sites), which are factors, and level of resistance (ttx_resist), which is numeric, i.e. **class**(ttx_resist) = numeric.
```
sites<-c(rep("Benton",7),rep("Warrenton",5))
ttx_resist<-c(0.29,0.77,0.96,0.64,0.7,0.99,0.34,0.17,0.28,0.2,0.2,0.37)
```
If we use the function **cbind** (=column bind) to form them into a 2 × 12 array we end up with a matrix object
```
TTXmatrix<-cbind(sites,ttx_resist)
TTXmatrix
     sites     ttx_resist
[1,] "Benton"  "0.29"
[2,] "Benton"  "0.77"
```

```
[3,]   "Benton"     "0.96"
[4,]   "Benton"     "0.64"
[5,]   "Benton"     "0.7"
[6,]   "Benton"     "0.99"
[7,]   "Benton"     "0.34"
[8,]   "Warrenton"  "0.17"
[9,]   "Warrenton"  "0.28"
[10,]  "Warrenton"  "0.2"
[11,]  "Warrenton"  "0.2"
[12,]  "Warrenton"  "0.37"
```
class(TTXmatrix)
```
[1] "matrix"
```
We cannot use **class** to determine how the column ttx_resist is being treated because this is not a dataframe, but we can see from the double inverted commas that these are not numeric. The first obvious thing to do is to coerce this into a dataframe using the function **as.data.frame**.

TTXdf<-as.data.frame(TTXmatrix)

However, now

class(TTXdf$ttx_resist)
```
[1] "factor"
```
So we now need to convert this column's class to numeric, but the obvious solution of using the function **as.numeric** on the column, does not work.

TTXdf$ttx_resist<-as.numeric(TTXdf$ttx_resist)
TTXdf$ttx_resist
```
[1] 4 9 10 7 8 11 5 1 3 2 2 6
```
These are definitely not the numbers we want! This apparently strange behaviour is because applying **as.numeric** to a factor is meaningless. The names of the factors are character representations of numbers but the factors themselves are not. The solution therefore, is to apply **as.numeric** to the factor names (or the factor names as characters) using the function **as.character**, which here converts the levels of the factor to character strings representing the numbers, and then these are converted to number values with **as.numeric**

TTXdf<-as.data.frame(TTXmatrix) # starting again with our original dataframe
TTXdf$ttx_resist<-as.numeric(as.character(TTXdf$ttx_resist))
class(TTXdf$ttx_resist)
```
[1] "numeric"
```
This is a bit fiddly and, of course, if we had initially used the function **data.frame** to combine our two original vectors, 'sites' and 'ttx_resist', the columns would be the correct classes from the beginning.

Information Box 5.1

A note on different R data types.
- Vector: a list of elements all of the same type.
- Matrix: a 2D array of numbers, characters or other variables all of the same type; it is a special form of vector with the attribute of having two or more columns and rows.
- Array: similar to a matrix but can have more than two dimensions.
- List: a set of elements which do not all have to be the same type.
- Dataframe: a 2D array which can have columns of different data types.

Table: a generic form of matrix and array; if you create an array using **as.array** the object will have class 'table' and simultaneously be a 'matrix' and an 'array'. If, however, you create a matrix using the functions **matrix** or **array**, it will pass the **is.array** and **is.matrix** tests but not **is.table**, which will give a FALSE. You can usually use the 'as' family functions (**as.matrix**, **as.array**, **as.table**, **as.data.frame**) to coerce objects from one class to another, e.g.

```
a<-matrix(1:6,nrow=3,ncol=2,dimnames=list(c("x","y","z"),c("p","q")))
a
  p q
x 1 4
y 2 5
z 3 6
class(a)
[1] "matrix"
is.array(a)
[1] TRUE
is.table(a)
[1] FALSE
b<-as.table(a)
b
  p q
x 1 4
y 2 5
z 3 6
is.matrix(b)
[1] TRUE
is.array(b)
[1] TRUE
is.table(b)
[1] TRUE
```

Tool Box 5.1

- Use class() to check the class of an R object.
- Use read.csv() to turn a comma separated text file into a dataframe.
- Use col.names= in read.csv to specify column names.
- Use colnames() to specify or view column names.
- Use attach(column name) to only work with that column.
- Use detach(column name) to stop working with that column.
- Use $ to specify a component such as a column of an object, e.g. function(filename$column name) will apply that function only to that column.
- Use as.numeric() to convert numbers as characters to numeric objects.

6 Plotting Biological Data in Various Ways

Summary of R packages introduced
   circlize
   sfsmisc

Summary of R functions introduced
   aov, but see Chapter 12
   barplot
   colnames
   cumsum
   draw.sector in "circlize"
   eaxis in "sfsmisc"
   factor
   head
   install.packages
   is.na
   levels
   library
   lines
   matrix
   mtext
   na.omit
   p.adjust
   pie
   points
   rbind
   rownames
   title
   TukeyHSD

A picture is worth a thousand words is an old English adage, and it is very appropriate for scientists too. Plotting your data is really the first thing you should do because it helps you identify possible trends and to spot obvious errors, and when presenting data to others, good graphics quickly convey the important trends (or lack thereof). R is an extremely powerful system for generating informative graphic output.

In this chapter we introduce plotting line graphs, bar charts, pie charts, box and whisker plots. We troubleshoot the main areas where you are likely to encounter problems. We show you how to create log plots, add legends, error bars, notches and confidence limits, and introduce confidence limits and statistical testing.

Example 1 – Bryophytes up a Mountain

Let us move swiftly on to a rather more biological example. For the moment we will stick to entering the data that we want to examine within the text file that we write and paste into R, which is expedient if the number of values is small and do not need to be updated. For larger datasets, one will almost invariably have entered data initially in a spreadsheet such as Excel, and we will describe how to save data from Excel and import into R data in Chapter 31.

In this first example we examine some data on the numbers of records of mosses and liverworts (collectively often referred to as bryophytes) up an altitudinal gradient in Khao Nan National Park (Chantanaorrapint, 2010). Surveys were carried out in order of increasing altitude with six altitudes being sampled. Thus, we can enter the numbers of records for each group as ordered lists called vectors. We will use the concatenate function, **c**, for this again here.

mosses<-c(53,85,69,98,130,90)
liverworts<-c(47,67,85,113,190,239)
If we now simply type
plot(mosses)

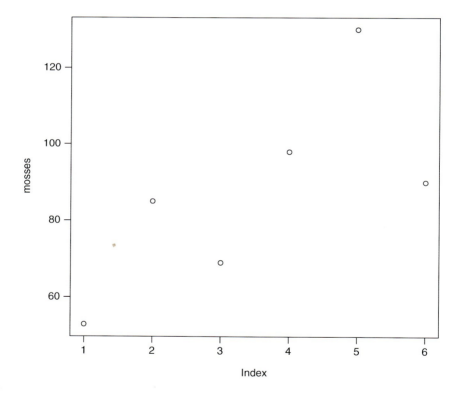

you will get a graph of the numbers in the vector (this time a list) against their position in the list (index) using the default open circle character. It's not very pretty but it demonstrates how very easy it is to obtain a graph using R. Note that the y axis has automatically been labelled 'mosses', the name of the variable, but we can change this using the 'ylab =' (y-label) argument. We will do another plot inserting 'ylab="Number of records"', as well as a label for the x axis, and a title for the graph, which is done with the parameter main. We might also want to change the shape and colour of the symbol that we use for plotting the moss records, and to do that we will use the arguments 'col=({*colour*})' and 'pch=({*plot character*})'.

> **Information Box 6.1**
>
> *Symbols in R plots.* To see symbols associated with the numbers passed to pch, run the following short code, which plots them using the function **points**.
> ```
> plot(NULL,axes=F,ylab="",xlab="",main="",xlim=c(0,6),ylim=c(0,10))
> text(rep(0,10),c(0:9),c(0:9),cex=2)
> points(rep(1,10),c(0:9),pch=c(0:9),cex=2)
> text(rep(2,10),c(0:9),c(10:19),cex=2)
> points(rep(3,10),c(0:9),pch=c(10:19),cex=2)
> text(rep(4,6),c(0:5),c(20:25),cex=2)
> points(rep(5,6),c(0:5),pch=c(20:25),cex=2)
> ```
>
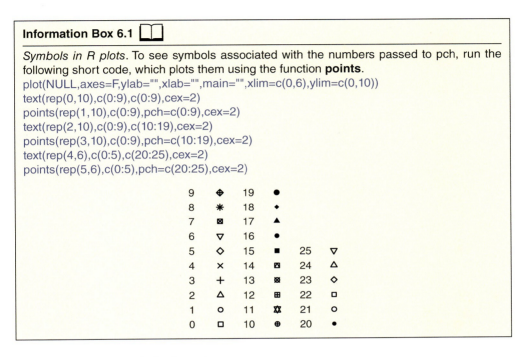

So let's try with the bryophytes:
plot(mosses,ylab="Number of records",xlab="Site number",pch=15,col="blue",
    main ="Altitudinal distribution of bryophytes")

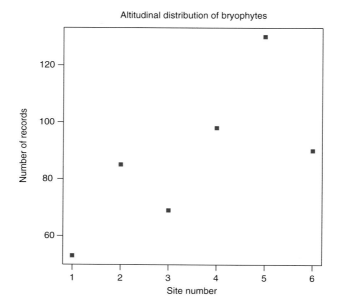

To compare the distribution of mosses with those of liverworts visually we will want to add the liverwort data points onto the same graph, and distinguish them from those of the mosses. To do this we will use the function **points**, which adds points onto an already open plot window. If the above plot window has not been closed, simply type

points(liverworts,pch=16,col="brown")

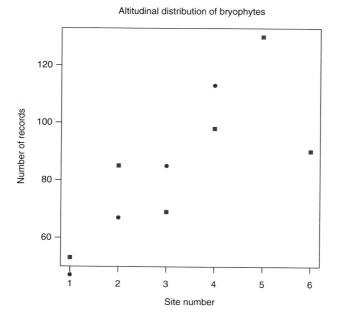

otherwise, re-enter the mosses data plot command and then add the liverwort points.

Troubleshooting 1

Look at the plot above, and in particular, the liverworts points for sites 5 and 6 – they are missing. Why is this? Well, there were more than 130 individuals recorded at those sites but the upper limit of the *y* axis is 130, because it was automatically generated by R when plotting only the mosses data. R uses 'pretty' limits for *x* and *y* axes by default. We can overcome this in several ways. We could manually define the graph axis ranges using the ylim arguments after having eye-balled our data to see what the largest y value was, e.g.
plot(mosses,ylab="Number of records",xlab="Site number",pch=15,col="blue",
    main ="Altitudinal distribution of bryophytes",ylim=c(0,250))
points(liverworts,pch=16,col="brown")

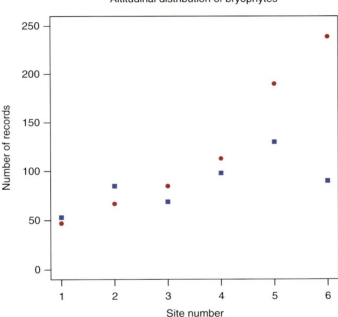

If you have a large dataset this might not be easy to do by eye. Automating this is easy using the functions **min** (=minimum) and **max** (=maximum) and then setting 'ylim=c(min(mosses,liverworts), max(mosses,liverworts))' as follows: you will get the same graph. In this case **min**(mosses,liverworts) gives the smallest value out of the combined moss and liverwort vectors; similarly **max** gives the maximum value
plot(mosses,ylab="Number of records",xlab="Site number",pch=15,col="blue",main=
    "Altitudinal distribution of bryophytes",ylim=c(min(mosses,liverworts),
    max(mosses,liverworts)))
points(liverworts,pch=16,col="brown") # figure will be same as the previous

We can also present the data for our two species to **plot** in a single line by concatenating the moss and liverwort data using **c**(mosses,liverworts) for the *y* values, and repeating the site list twice for the *x* axis

plot(c(1:6,1:6),c(mosses,liverworts),pch=c(rep(15,6),rep(16,6)),col= c(rep("blue",6), rep("brown",6)),ylab="Number of records",xlab="Site number",main ="Altitudinal distribution of bryophytes") # figure will be same as previous

The first two elements of the plot statement are the lists of x and y values respectively. We run 1 to 6 (in increments of 1) twice, the first for mosses the second for liverworts. The point types (pch=) and colours of the points (col=) are presented in the same order as the two classes of bryophytes.

The plot is still a bit of a mess, so we might want to add lines connecting the points for each species to make any pattern more apparent. We do this by using the function **lines** and simply type

lines(c(1:6),mosses,col="blue",lty=1) # lty= specifies the line type: 1 for normal, 2–6 for different types of dashes
lines(c(1:6),liverworts,col="brown",lty=2)

Our next job is to label the *x* axis in metres above sea level (a.s.l.), the recording sites being at 400, 600, 800, 1000, 1200 and 1300 m a.s.l. we create a vector of these sample site altitudes

altitude<-c(400,600,800,1000,1200,1300)

Our moss and liverwort data are now in a single concatenated list of *y* axis values (c(mosses, liverworts)), therefore we need to do the same with the sample site altitudes by repeating the altitude vector twice (c(altitude, altitude))

plot(c(altitude,altitude),c(mosses,liverworts),pch=c(rep(15,6),rep(16,6)), col=c(rep("blue",6),rep("brown",6)),ylab="Number of records",xlab= "Altitude a.s.l. (m)",main ="Altitudinal distribution of bryophytes")
lines(altitude,mosses,col="blue",lty=1)
lines(altitude,liverworts,col="brown",lty=2)

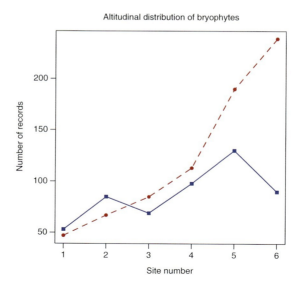

Adding a Legend to a Plot

We will want a legend explaining what the two sorts of points represent and for this there is the function appropriately called **legend**
A<-"Liverworts"
B<-"Mosses"
legend("topright",pch=c(15,16),col=c("blue","brown"),c(A,B))
This one (not shown) does not look good because the legend box overlaps some of the data points. Therefore, in the first instance, look for a gap where the legend box might fit. R has eight preset locations for a legend box (topleft, top, topright, left, center, right, bottomleft, bottom and bottomright). To refine the position of the legend box we can use the argument 'inset={*some proportion of the plot size*}', or we can specify the precise position relative to the values of the axes (see Appendix 1).

Information Box 6.2

Use of the type argument in plot. We have plotted points and lines separately, but if you want, you can use the in-built plot and points argument type=. For each set of points you can specify the default type="p" for points only, type="l" for lines only, type="b" for points connected by lines with a small gap between line and point, type="c" for empty points joined by lines} and type="o" for over-plotted points and lines (as we have created here).

---

legend("topleft",pch=c(15,16),col=c("blue","brown"),c(A,B))

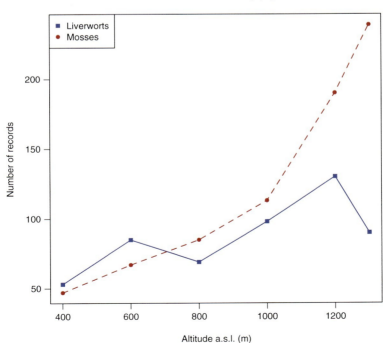

which looks better. If we think the legend is too obtrusive or not large enough, we can change its size using the character expansion argument 'cex='; values < 1 make the text smaller and values > 1 make it larger. If we don't like the box around it, use the argument 'bty=("n")' (bty is short for box type, and the 'n' means none), and/or if we want to move it a bit away from the far top corner we can use the argument 'inset=', for example,
legend("topleft",pch=c(15,16),col=c("blue","brown"),c(A,B),cex=0.8, inset=0.05) # inset is the proportion of the plot area to inset the legend by
You will need to re-run the plot to make any changes.

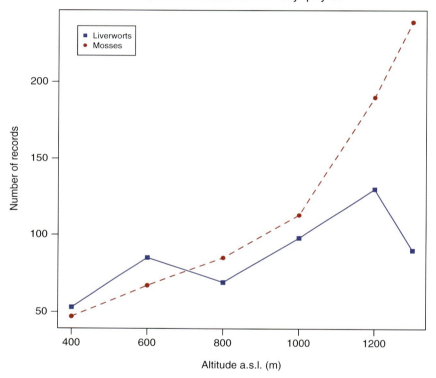

Altitudinal distribution of bryophytes

If the x axis label is clipped at the bottom of the screen resize it by dragging the lower-right corner. With the plot window at the front of the screen you can save it on a Mac by doing control+S or go to the file tab and select print whereby you can print out directly, or, normally more convenient, save as a .pdf (portable document format) or .ps (postscript) file, either of which can be imported into a Word document using insert or just dragging and dropping. Make sure your Word document is being viewed in page-layout mode or you might think that you haven't succeeded in importing the figure.

We have started by creating a graph of the data. This is important because you need to see that the data make sense. It is very easy when entering data into a spreadsheet to misplace a decimal point or miss one out altogether – increasing or decreasing one point by a factor of 10 or more. Quickly looking at a plot will show you whether you have any outliers that might have been the result of a human slip-up, and they can be corrected before you go any further with analysis.

Troubleshooting 2 – Vector Lengths Differ

Simply re-entering one of the groups' data with one value deleted, e.g.
mosses<-c(53,85,69,98,130)
instead of
mosses<-c(53,85,69,98,130,90)
Closing any open plot window and re-running the plot line will result in a blank plot window and an error message
```
Error in xy.coords(x, y, xlabel, ylabel, log) :
   'x' and 'y' lengths differ
```
This is very easy to interpret – the vector speciesA does not have the same number of elements as speciesB. You can see how many values are in each using the function **length**. In this example, we get an error message, but in a large body of code this could quickly scroll off the top of the screen and could be missed
length(mosses)
[1] 5
length(liverworts)
[1] 6
This is generally a sensible thing to do in any case, because with some functions, although an error message will appear, you might miss it, and R will roll over the values in the shorter vector, ultimately giving you the wrong answer. Therefore, to plot graphs or to do statistics, it is essential that the variables being plotted against one another have the same number of elements.

Information Box 6.3

What R does if you try to use an array with variables of different lengths
   Let us start with three vectors of different lengths and then try to fill an array with them using the row-binding function **rbind**.
a<-c(1:5); b<-c(1:5); c<-c(1:3)
rbind(a,b,c)

```
  [,1] [,2] [,3] [,4] [,5]
a  1    2    3    4    5
b  1    2    3    4    5
c  1    2    3    1    2  # the values of c have been rolled on
                            into the array
```

Looking at the bottom right you will see that R has filled our array by rolling over the shorter vector c. Fortunately we get a warning message:
```
Warning message:
In rbind(a, b, c) :
   number of columns of result is not a multiple of vector length
(arg 3)
```

Troubleshooting 3 – Missing Data and NAs

It may be the case that some data are missing, perhaps a sample was lost or a trap broken by a buffalo. In these cases it is important not to enter a value (such as zero) for the missing data because it is simply not known. Instead in R we can insert the 'value' NA, which stands for 'not applicable'. Suppose we have lost the count of the number of mosses at site 4, but still have the information for the liverworts at that site, we would enter the data as

mosses<-c(53,85,69,NA,130,90)
liverworts<-c(47,67,85,113,190,239)

Here we use NA to indicate missing data. We cannot enter a zero because that would be entering false data.

plot(altitude,mosses,pch=15,col="blue",ylab="Number of bryophytes",xlab="Altitude a.s.l. (m)",main ="Altitudinal distribution of bryophytes",ylim=c (min(na.omit (mosses,liverworts)),max(mosses,liverworts, na.rm=TRUE)))
   # see Information Box 6.4
points(altitude,liverworts,pch=16,col="brown")
lines(altitude,mosses,col="blue",lty=1)
lines(altitude,liverworts,col="brown",lty=2)

Information Box 6.4

Performing calculations when there are missing values
If you ask R to perform a calculation on a vector containing missing observations such as NA, it will not do so unless it is instructed to ignore or omit the NAs, e.g.
a<-c(1,2,3,4,5,6,NA,7,8)
mean(a)
[1] NA
To overcome this we include 'na.rm=TRUE' an argument in the function
mean(a,na.rm=TRUE)
[1] 4.5
Alternatively we can perform the function on the vector after omitting the NAs
mean(na.omit(a))
[1] 4.5
Here the function **na.omit** has removed the NAs from our vector *a*. However, if you assign the result of **na.omit** to a new vector it is a special object that contains more information than simply *a* with the NAs removed.

You can use **is.na** to find which elements of a vector are NAs
is.na(a)
[1] FALSE FALSE FALSE FALSE FALSE FALSE TRUE FALSE FALSE

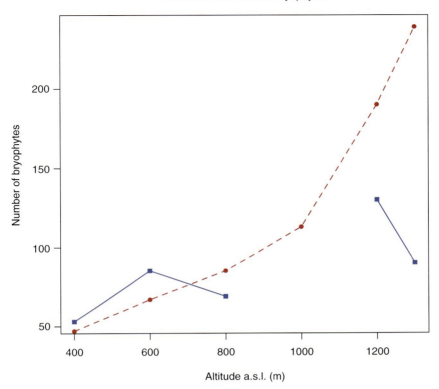

Altitudinal distribution of bryophytes

This is of course ugly, but we notice the missing point because we only have a few points and the absence is emphasized by the gap in the line.

We can plot both sets of points by first having combined our species data into a single vector, which we name 'bryos' and creating single vectors for the altitudes, symbol types and symbol colours. The reason to show you this now is because it puts the data into a format that can be converted easily into a dataframe.

To proceed we will regenerate the original data which we have edited above

mosses<-c(53,85,69,98,130,90); liverworts<-c(47,67,85,113,190,239)
bryos<-c(mosses,liverworts) # concatenating the two into a single vector
alt<-c(altitude,altitude) # concatenating the matched altitudes

and now we need a list of six 'blue' followed by six 'brown'. We could type this by hand but that would be tedious, and ridiculous if we had 200 sampling sites. So we will use the function **rep**

use_colours<-c(rep("blue",6),rep("brown",6))
use_symbols<-c(rep(15,6),rep(16,6))

These four vectors can now be used to create a plot as above.

Incorporating More Types of Data on the Same Graph

In addition to total mosses and liverworts, we might have other data that we would like to show on the same graph but which has a different range of *y* values. Here we will add information about the total numbers of species at each sampling site, but the additional data might be quite different things such as annual rainfall, or mean temperature, the principle is the same. In this case, the number of species is markedly lower than the number of records
moss_species<-c(12,19,20,24,27,19)
liverwort_species<-c(17,19,23,31,42,46)

Firstly, we need to increase the width of the right-hand margin to make space for the new *y* axis label – we pass the argument 'margins=' to **par**. The format is **par**(mar=c({*bottom margin in lines*},{*left margin, top margin, right margin*})). The default values are 5.1, 4.1, 4.1 and 2.1, respectively. We will increase the right margin to 4.1, which is a by-eye guess.
par(mar=c(5.1,4.1,4.1,4.1))
plot(alt,bryos,pch=use_symbols,col=use_colours,ylab="Number of records",
    xlab="Altitude a.s.l. (m)",main ="Altitudinal distribution of bryophytes",
    ylim=c(min(bryos),max(bryos)))
lines(altitude,mosses,col="blue",lty=1)
lines(altitude,liverworts,col="brown",lty=2)
legend("topleft",pch=c(15,16),col=c("blue","brown"),c(A,B),cex=0.8,
    inset=0.05)

However, to overlay the new data on species numbers, we cannot simply run **plot** again because that will wipe out the existing one, so we must first issue a command to tell R not to do that
par(new=TRUE) # tells R to plot the new graph on top of the existing one
plot(NULL,axes=F,ylab="",xlab="",main="",xlim=c(400,1300),ylim=
    c(min(c(moss_species,liverwort_species)),max(c(moss_species,
    liverwort_species))))
points(altitude,moss_species,pch=3,col="blue")
lines(altitude,moss_species,lty=3,col="blue") # using a different dotted line style
points(altitude,liverwort_species,pch=4,col="red")
lines(altitude,liverwort_species,lty=4,col="red") # using a different dotted line style
axis(4,c(10,20,30,40,50),at=c(10,20,30,40,50))
axis(4,las=0) # las=0 means print text vertically, las=1 means horizontally
mtext(side=4,line=2.5,"Numbers of species") # using mtext (=margin text) to print axis name on right ('side=4') and 2.5 lines from the axis

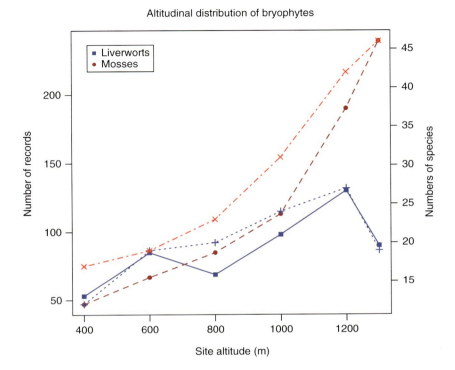

We now have a very presentable plot and we can see some things that are quite experimentally and biologically interesting: (i) for both mosses and liverworts, the number of species found correlates well with numbers of patches found; and (ii) liverworts seem to be particularly abundant and diverse at the highest elevation where mosses appear to be in decline.

Example 2 – Tropical Forests, Rural Population, Logarithmic Axes and Installing Packages

Wright and Muller-Landau (2006) provide data on the future of tropical forests and summarize the relationship between rural population size and the potential remaining intact forest. The data are provided in a .csv file. We will present the four plots together at the end of this example.

d<-read.csv("Wright_y_Muller_Landau_2006.csv",header=TRUE)
head(d)

```
  Rural_popn_density Potential_forest_remaining.   region
1          1.3116206                    99.72566 Americas
2          0.4431751                    95.61043 Americas
3          3.8332014                    98.49108 Americas
4          3.5091786                    85.45953 Americas
5          2.6586277                    84.49931 Americas
6          2.6586277                    81.06996 Americas
```

The functions **head** and **tail** allow us just to see the first or last few contents of a list or lines of an array or dataframe – very helpful if your files are massive. We can see what the column names are using **colnames**
colnames(d)
[1] "Rural_popn_density" "Potential_forest_remaining."
    "region"
Rural population density (people per km^2) varies greatly
range(d[,1])
[1] 0.4431751 767.2405000
and is highly skewed
mean(d[,1]); median(d[,1])
[1] 64.56797
[1] 22.99555
which are indications that we may want to plot it on a logarithmic scale. We will go through constructing four plots of various levels of sophistication to illustrate a number of plotting features
par(mfrow=c(2,2))
First we will plot the data without logging the *x* axis, and we will distinguish the three regions that the data come from, which are given in column three, to distinguish the shapes and colours of the plotting symbols. In the dataframe these are factors
levels(d[,3])
[1] "Africa" "Americas" "Indo_Malaya"
but if we convert them to numbers (integers) we can pass the latter directly to the symbol plotting arguments 'pch' and 'col'. Using the function **as.numeric**(d[,3]) produces a vector of integers and these have no levels, they are just numbers, and therefore can be used to assign plotting symbol types, colours, etc.
as.numeric(d[,3])
[1] 2 2 2 2 2 2 2 2 2 2 2 2 2 2 2 2 2 2 2 2 1 1 1 1 1
    1 1 1 1 1 3 3 3 3 3 3 3 3 3 3 3
levels(as.numeric(d[,3]))
NULL
plot(d[,1],d[,2],pch=as.numeric(d[,3]),col=as.numeric(d[,3]),xlab=colnames(d)
    [1],ylab= colnames(d)[2])
The points are now differentiated but the symbols and colours specified by 'pch=' 1..3 and 'col=' 1..3 are not quite what we want so we will add 14 and 3 respectively to these in the next plots to get colours that we prefer. We will also take the logarithm base 10 of the population density (*x* axis) values, and change the *x* axis label.
plot(log10(d[,1]),d[,2],pch=as.numeric(d[,3])+14,col=as.numeric(d[,3])+3,
    type="p",xlab="Log(rural popn density)",ylab=colnames(d)[2])
The default 'clever' *x* axis labels are not that intuitive to people unfamiliar with logs and also don't extend to –1 so next we will create a plot without axes and then specify the axes exactly how we want them using the function **axis** as we did in Chapter 4. This plot also illustrates a plot option of automatically specifying which axis you want logging, using the argument 'log="x"' in our case.

```
plot(d[,1],d[,2],log="x",pch=as.numeric(d[,3])+14,col=as.numeric(d[,3])+3,
    axes=FALSE,xlim=c(0.1,1000),ylim=c(0,100),xlab="Rural popn density",
    ylab="% potential forest remaining")
axis(1,c("0.1","1.0","10","100","1000"),at=c(10^-1,10^0,10^1,10^2,10^3),cex.
    axis=0.8)
axis(2,at=c(0:10)*10,cex.axis=0.8)
```
Using the same conversion of countries to numeric values that are used to select symbols and their colours in a legend though this is a little less straightforward than one might hope! Here is the code:
```
legend("bottomleft",pch=as.numeric(levels(as.factor(as.numeric(d[,3]))))+14,col=
    as.numeric(levels(as.factor(as.numeric(d[,3]))))+3,levels((d[,3])),bty="n")
```
and here is the explanation of the very long 'pch' and 'col' arguments. Working out from the inside of the brackets thus.

To obtain the levels of the numeric representation vector of regions we must convert them to factors
```
as.factor(as.numeric(d[,3]))
[1] 2 2 2 2 2 2 2 2 2 2 2 2 2 2 2 2 2 2 2 1 1 1 1 1 1
    1 1 1 1 1 1 3 3 3 3 3 3 3 3 3 3 3 3
Levels: 1 2 3
```
The function **levels** returns the three levels; however, these are now characters not numbers, which we can tell by each being inside inverted commas.
```
levels(as.factor(as.numeric(d[,3])))
[1] "1" "2" "3"
```
So again we use the function **as.numeric** to give us the three integers, which can then be used to pick plotting symbols (and colours).
```
as.numeric(levels(as.factor(as.numeric(d[,3]))))+14
[1] 15 16 17
```
We have specified the labels for the logged x axis as 0.1 to 1000, but if we had much smaller or bigger numbers these would be ugly, and it would be nicer to show values with exponents. The default in R for representing very big or very small numbers is the e notation, thus 1 million is 1e+06, and 1 billionth is 1e-09. The package **sfsmisc** has a special function **eaxis** (=exponential axis), which instead allows us to show exponents in the form 10^3, and we will use this for the fourth plot. First, re-run the basic plot command above without axes.

Obviously, to install packages your computer must be online. You can install packages either from the pull-down menus 'Packages & Data > Package Installer' and 'Package Manager', or from within R, or using the function **install.packages**. After having installed a package you need, each new session, to tell R to access the functions in the package using the function **library**, i.e. adding the functions to the current R library of functions. In the **install.packages** call, the name of the package must be in inverted commas, but not in the **library** call.
```
install.packages("sfsmisc")
library(sfsmisc)
eaxis(1,at=c(10^-1,10^0,10^1,10^2,10^3),cex.axis=0.8)
axis(2,at=c(0:10)*10,cex.axis=0.8)
```

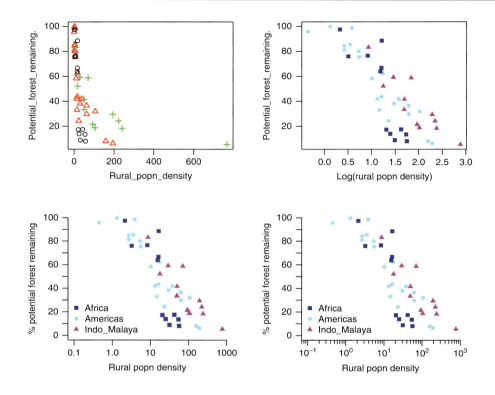

Exercise Box 6.1

1) Plot the data on body temperature and activity for two beavers in the datasets 'beav1' and 'beav2' in the **MASS** library on the same graph, differentiating the points according to beaver and active/inactive. Note: you need to set the x and y limits to encompass all the data.

Example 3 – Creating a Barplot: Bryophytes Side-by-side

Another way to present data such as the numbers of bryophytes is as a bar chart using the function **barplot**. Bar charts are similar to line graphs but the x variable is discrete as we have here – the sampling of both mosses and liverworts was carried out at particular altitudes along the transect. If the x axis is continuous and the precise values for each colony were measured (altitudes of each individual moss or liverwort patch in our example), the plot would be a histogram and we would use the R function **hist** (see Chapter 10).

There are two types of barplot that we can create from our data: side-by-side and stacked. The default **barplot** is the stacked type, for a side-by-side one you add the argument 'beside=TRUE' to the **barplot** command line. In stacked bar charts the total height of each bar is the total number of bryophytes at each site, and these bars will be separated into zones from top to bottom for each species – we can as before, specify the colours and other shading to be used.

Here we demonstrate **barplot** using the moss and liverwort data from an earlier section

data<-rbind(mosses,liverworts) # combining the original vectors into a matrix
barplot(data,beside=TRUE) # using the 'beside' argument to put bars side by side

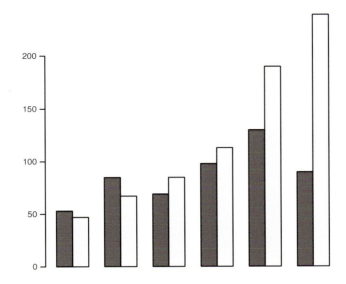

Here we do not need to worry about y axis scaling because R checks the whole of the data passed to it to determine the range automatically though we can override that if we want. R also determines the x axis based on the number of columns or groups of columns so you must not set an 'xlim' value with barplots unless you want to make adjustments to the total width of the plot (see below). To prettify the barplot is very similar to prettifying graphs made with the function **plot**, but with additional options for colouring and shading.

Exercise Box 6.2

1) Select colours for the bars, and add x axis label and legend. Hint: use the argument 'fill=' rather than 'pch=' to get boxes.

Example 4 – Stacked Bar Chart, with Different Colours, Fills and Legends

Nuraemram (2011) provided data on the abundance of ant species in unburnt and recently burnt dipterocarp forest in northern Thailand. Here we will plot the relative abundances of the seven most abundant species.
ants<-c("Oecophylla smaragdina","Pheidole spp.","Monomorium destructor",
    "Paratechina longicornis","Odontoponera denticulata","Tapinoma
    melanocephalum","Monomorium pharoensis")

```
unburnt<-c(2957,2020,556,445,418,401,308)
burnt<-c(696,1425,445,1450,1256,269,476)
data<-cbind(unburnt,burnt)
barplot(data,ylab="Numbers of individuals",sub="Dipterocarp forest condition")
```

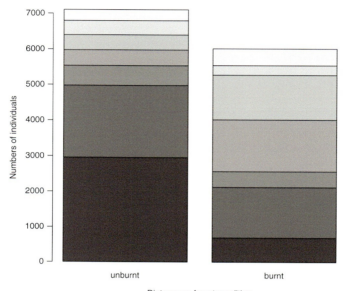

R has applied its default shading but we can specify to colour or shade the blocks according to our fancy. You can customize shading in R by varying the angle and density of shading lines or you can use colours

```
par(mfrow=c(2,2))
bardensity<-c(0,10,15,20,40,30,100)
barangle<-c(45,135,0,90,45,130,90)
barplot(data,ylab="Numbers of individuals",xlab="Dipterocarp forest condition",
    density=bardensity,angle=barangle,lwd=2,col="black")
barplot(data,ylab="Numbers of individuals",xlab="Dipterocarp forest
    condition",sub="effects on ant abundance",col=c("black","red","blue",
    "green","yellow","grey50","white")) # adding a subtitle to x axis using the
    'sub=' argument
```

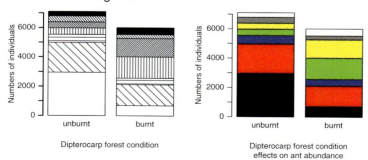

Use of horizontal shading is not recommended as it can be confusing.

We will obviously want to add a legend but the default barplot width, the similar heights of both bars and the long species names make this awkward. To make space an easy way is just to increase the extent of the y axis.

par(mfrow=c(1,1))
barplot(data,ylim=c(0,9500),ylab="Numbers of individuals",
xlab="Dipterocarp forest condition",col=c("black","red","blue","green",
"yellow","grey50","white"))
legend ("top",ants,fill=c("black","red","blue","green","yellow","grey50","white"),
cex=0.7)

As biologists we know that scientific names are always to be put in italics, but unfortunately we have to do this in different ways for text within a plot and text within a legend. For a **plot** command we add 'font=3' ('font=3' gives italics, 'font=2' gives bold) but in the function **legend** we have to use 'text.font=3'.

legend("top",ants,fill=c("black","red","blue","green","yellow","grey50","white"),
cex=0.7, text.font=3)

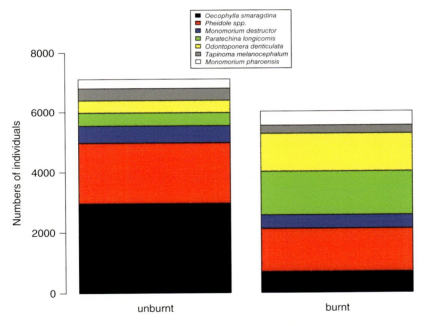

Alternatively, you can make space on the right-hand side by reducing the length of the x axis using 'xlim'

barplot(data,ylab="Numbers of individuals",xlab="Dipterocarp forest condition",
xlim=c(0,4.2),col=c("black","red","blue","green","yellow","grey50", "white"))
legend("right",ants,fill=c("black","red","blue","green","yellow","grey50","white"),
cex=0.7,text.font=3)

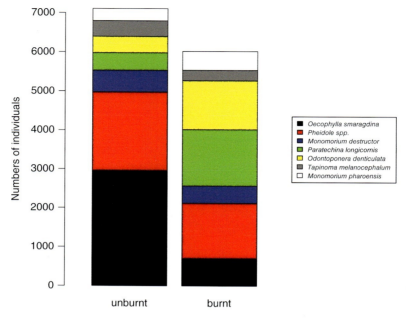

For more complicated legends, the argument 'legend=expression(...)' allows you to specify the font for each individual legend item separately as in:
legend(0,8100,pch=c(15,16),col=c("blue","brown"),legend=expression
   (italic("A. scientific name"),("other ants")),xpd=NA) # legend position
   specified here by x, y coordinates

Tool Box 6.1 ✖

- Use angle= in **barplot** to set the angle of the shading lines.
- Use barplot() to plot a bar chart.
- Use beside=TRUE in **barplot** to plot the bars next to each other.
- Use bty= to specify box type around a legend.
- Use cbind() to combine vectors, or bind them to matrices or dataframes by columns.
- Use density= in **barplot** to set the density of shading bars.
- Use fill= to put coloured boxes in a legend.
- Use font= in plot() to set the font type; see text.font below.
- Use legend=expression() to specify the font for each legend item separately.
- Use lwd= to set the line width.
- Use rbind() to combine vectors into a table or add them to a table.
- Use sub= to add a subtitles to the *x* axis label.
- Use text.font= in legend to set the font face. Font faces are 1 = plain, 2 = bold, 3 = italics, 4 = bold italics.

Example 5 – Dietary Differences Between Hornbill Species – Entering Data as a Table

In this section we will enter and plot some of the data from Poonswad et al.'s (1988) study of the feeding ecology of hornbill species in Khao Yai National Park. Four species of hornbill were included in their study: great hornbill (GH) *Buceros bicornis*, wreathed hornbill (WH) *Rhyticeros undulatus*, oriental pied (PH) *Anthracoceros albirostris* and brown hornbill (BH) *Anorrhinus (Ptilolaemus) tickelli*. The data collected for each species included the frequencies with which each bird species was observed using one of five foraging behaviours: (i) cracking tree bark; (ii) probing; (iii) hawking; (iv) plucking; and (v) snatching, and the types of fruit and animals consumed – clearly quite a lot of information. They then calculated behavioural and dietary similarities between the species which showed the way they have evolved to utilize different subsets of the total available hornbill food. We will enter some of their data in a way that facilitates analysis and replication of their graphics. At any point we can substitute scientific names and we can easily include these in any figure legends. The data are shown in Table 6.1; there is no need to enter percentages or totals since we can rapidly calculate all of these in R.

Just looking at the numbers we can see that all species employ plucking a lot, but that there are possibly significant differences between species in the use of other foraging methods. We can enter these data into R by hand as follows
r1<-c(23,5,9,0)
r2<-c(7,0,3,0)
r3<-c(0,0,12,21)
r4<-c(165,276,170,122)
r5<-c(0,0,7,12)
or we can enter data in an Excel spreadsheet, save it as either a tab-delimited text (.txt) or comma separated values (.csv) file and then use appropriate functions to read in the data (see Chapter 31). That is very convenient especially if we are likely to want to keep accessing and analysing the contents: or, depending on the source formatting, we may be able to grab the table data and paste it into R directly (see Chapter 5).

Table 6.1. Frequencies of foraging behaviours in four hornbill species (from Table 1 in Poonswad *et al.*, 1988).

Behaviour	GW	WH	PH	BH
cracking tree bark	23	5	9	0
probing	7	0	3	0
hawking	0	0	12	21
plucking	165	276	170	122
snatching	0	0	7	12

Having entered the separate row data manually in this case, we can combine these into a table using the function **rbind**.
bird.names<-c("GH","WH","PH","BH")
hornbill.forage<-rbind(r1,r2,r3,r4,r5)
hornbill.forage

```
     [,1] [,2] [,3] [,4]
r1    23    5    9    0
r2     7    0    3    0
r3     0    0   12   21
r4   165  276  170  122
r5     0    0    7   12
```

Note that the variable names have been applied to the rows, but the columns are only identified by their indices. To give the columns informative names we use the function **colnames**
colnames(hornbill.forage)<-bird.names
hornbill.forage

```
      OH   WH   PH   BH
r1    23    5    9    0
r2     7    0    3    0
r3     0    0   12   21
r4   165  276  170  122
r5     0    0    7   12
```

and for the row names we use the equivalent function **rownames**.
rownames(hornbill.forage)<-c("cracking","probing","hawking","plucking",
    "snatching")
hornbill.forage

```
           OH   WH   PH   BH
cracking   23    5    9    0
probing     7    0    3    0
hawking     0    0   12   21
plucking  165  276  170  122
snatching   0    0    7   12
```

However, this is not the way this sort of data would normally be entered because it is almost always going to be easier to work with the measured variables (the response variables) as separate columns. It is easy to transpose rows and columns using the function **t** (= transpose)
hornbill.forage<-t(hornbill.forage)
hornbill.forage

```
     cracking probing hawking plucking snatching
OH         23       7       0      165         0
WH          5       0       0      276         0
PH          9       3      12      170         7
BH          0       0      21      122        12
```

To present these data graphically we could simply use 'barplot' again
barplot(hornbill.forage, beside=TRUE)

Plotting Biological Data in Various Ways

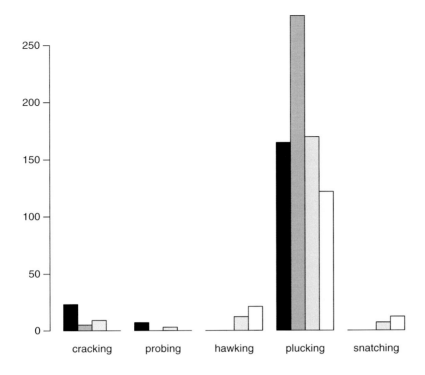

and obviously we can use **barplot** parameters to colour the columns, and **legend** to identify the species, 'ylab' (and 'xlab' if wanted) to label the axes. Here we used a 'barplot' parameter called 'beside'. The default barplot stacks the values. Setting 'beside=TRUE' (or 'beside=T' for short) places the bars side-by-side. The basic R package does not do 3D bar charts as you see in some papers, but they can be created using the R package 'ggplot2' (see Chapter 7). We dislike 3D bar charts because it is difficult to actually read the precise data values from them.

As an alternative to using **legend** separately, you can include relevant information within the **barplot** command line, for example
barplot(hornbill.forage,beside=TRUE,legend.text=bird.names,args.legend= list(x="topleft",bty="n")) # the legend position argument is called 'x'

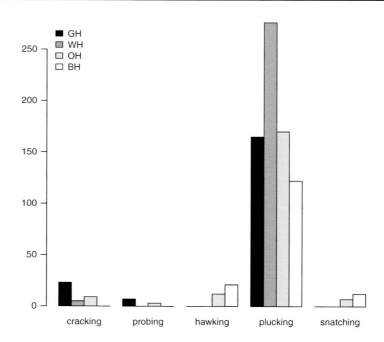

Example 6 – Horizontal Bar Plot of Camera Trap Data and More Troubleshooting

Instead of having bars rising vertically, we might prefer to show some data with horizontal bars. Here we use some of the data from Jenks *et al.*'s (2011) study of camera-trap images taken in Khao Yai National Park from their Appendix 2 (abridged). In this section we show how to do progressive troubleshooting to end up with the final image we want. We will show all four versions in a single plotting window at the end.

image<-c("poacher","domestic_dog","ranger","villager","tourist",
    "barking_deer"," wild_pig","macaque","porcupine","sambar_deer",
    "gaur","elephant","serow","pangolin","Indian_civet","black_bear","sun_
    bear","mongoose","dhole","clouded_leopard", "golden_cat","binturong")
number_of_images<-c(25,17,16,7,5,60,60,37,30,28,13,9,3,2,37,21,16,11,8,8,5,2)
Plotting the bars horizontally is easily achieved by incorporating the 'horiz=TRUE' parameter, but the result is not quite like the Jenks paper – the order has been reversed and the bar names run vertically with many omitted. We will put the four stages we are illustrating in a block of four to make the comparison easy.
par(mfrow=c(2,2))
barplot(number_of_images,col="black",names=image,horiz=TRUE)
We overcome the names orientation by including the 'las=2' argument, which specifies the orientation of axis numbers or other labels,
barplot(number_of_images,col="black",names=image,horiz=TRUE,las=2)

but we note that the names now extend way off the graph to the left, so we need to use the graphics function **par**(mar=c({*bottom,left,top,right*})) to move the plotting area to the right leaving a wider left-hand margin to accommodate the names. The default margin values (in lines) are 5.1, 4.1, 4.1 and 2.1, so we will try increasing the second of these (left) and plot again (just take a guess at a larger value).
par(mar=c(5.1,9.1,4.1,2.1))
barplot(number_of_images,col="black",names=image,horiz=TRUE,las=2)
This looks good, so now we just need to reverse our vectors using the reversing function, **rev**.
number_of_images<-rev(number_of_images)
image<-rev(image)
barplot(number_of_images,col="black",names=image,horiz=TRUE,las=2, xlab="Number of camera trap photos",ylab="'Animal' photographed")
Note that the horizontal axis is still the x axis, however, the y-label is printed over the 'animal' names and therefore we need to move it. One way is to not print a y axis label but to write some vertical text, e.g.
text(xpd=NA,-32,12,"'Animal' photographed",srt=90) # the argument **srt** is used to give a y axis label with 90 degree text rotation
We may have to fiddle with the offsets (you often have to in any case). A potentially better way is to write the ylab as part of a title call
barplot(number_of_images,col="black",names=image,horiz=TRUE,las=2, xlab="Number of camera trap photos",ylab="")
title(ylab="'Animal' photographed",line=7,xpd=NA) # within the title call we use 'line' to move the positions of the x labels

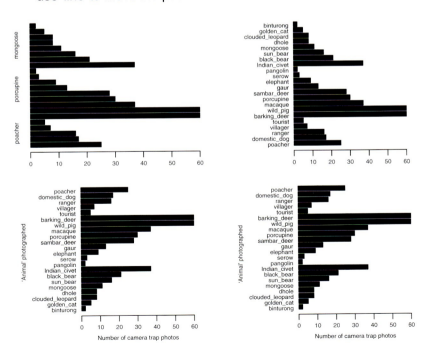

> **Tool Box 6.2**
>
> - Use colnames() to specify the columns in a table.
> - Use horiz=TRUE in **barplot** to plot bars horizontally.
> - Use line= in **title** to set the position of text.
> - Use rev() to reverse the order of a components in a vector.
> - Use rownames() to view or specify the rows in a matrix, dataframe, etc.
> - Use srt= to set the angle of text.
> - Use t() to transpose rows and columns in a matrix.
> - Use text() to add text to a plot.

Example 7 – Adding Error Bars to a Barplot or Plot: Fly Ommatidea

There is no built-in R function for error bars so we have to write our own code to add lines – well, in fact, many people have written such functions and made them available; they can be found by searching the web. Below we present the one given by Crawley (2008, p. 56). For experimental data, the error bars shown are normally standard errors, but you might choose to show 95% confidence limits or ranges. In most cases they will show the errors up the y axis. You need to calculate the lengths of each error bar above and below the heights of the bars in the barplot. If error bars based on standard errors of the means of two samples overlap, we can be confident that there is no significant difference between the means of the two samples. However, the converse is not necessarily true – non-overlapping error bars do not necessarily mean that the means are significantly different, the critical test is the t-test, and the error bars would need to have, as their length, the standard error of the differences between the means, which will be larger than the standard errors of the means. Obviously the bigger the gap is between non-overlapping error bars, compared to the lengths of the individual bars, the more likely the difference between means will be significant.

> **Information Box 6.5**
>
> *Standard deviation (SD) and standard errors (SE).* Standard deviation is a measure of the variation in the data, it is the square root of the variance. There are two versions of each depending on whether we are considering only a sample from a much larger population of data, or values actually of the whole population – when you are basing calculations only on a sample there will be some inaccuracy. The equations are given in Appendix 3. The sample standard deviation is denoted by s, and the population value by σ. So their respective variances are s^2 and σ^2.
>
> Standard errors are measures of the uncertainty in an estimated value, typically the mean. It increases with sample variance and decreases with sample size.

Here is Crawley's (2008) version of code for a function for drawing error bars which you simply type into R – we will talk more about home-made functions in Chapter 24.

```
error.bars<-function(xv,z,nn){
    xv<-barplot(yv,ylim=c(0,(max(yv)+max(z))),
    names=nn,ylab=deparse(substitute(yv)))
    g=(max(xv)-min(xv))/50 # the 50 determines the horizontal lengths of the
        bar at top and bottom of each error bar
    for(i in 1:length(xv)){
        lines(c(xv[i],xv[i]),c(yv[i]+z[i],yv[i]-z[i])) # vertical part of the line
        lines(c(xv[i]-g,xv[i]+g),c(yv[i]+z[i],yv[i]+z[i]))
        lines(c(xv[i]-g,xv[i]+g),c(yv[i]-z[i],yv[i]-z[i]))
        } # end i loop
    } # end function
```

However, a rather simpler way is to assign the 'barplot' to a vector, here called 'bardata' and then use this to pass the x axis values of the centres of the bars to the function **arrows** with the argument 'angle=90' to give 'T'-shaped arrows. We will take our data from Sukontason et al.'s (2008) study on the number of ommatidia (eye facets) in males and females of six species of nuisance flies.

```
flies<-c("L. dux","C. megacephala","C. rufifacies","C. nigripes","L. cuprina",
    "M. domestica")
males<-c(6032,4371,5367,4801,3678,3488)
females<-c(6073,5635,5207,4778,3620,3427)
standarddev_males<-c(385,344,385,323,155,110)*1.96 # data in the original
    paper are 1 S.D.
standarddev_females<-c(207,514,244,277,209,231)*1.96
```

Previously we combined lists using cbind or rbind but we can also use the function **matrix**, which requires us to specify the dimensions, and how the data should be entered (byrow=TRUE/FALSE).

```
data<-matrix(c(males,females),2,6,byrow=TRUE) # the 2 and 6 are the
    number of rows and columns in the matrix respectively
errors<-matrix(c(standarddev_males,standarddev_females),2,6,byrow=TRUE)
bardata<-barplot(data,beside=TRUE,ylim=c(0,max(data)+max(errors)),
    ylab="Mean number ommatidea (+/- 95% confidence)",names=flies,
    cex.names=0.7,xlab="Fly species",col=c("lightblue","pink"))
arrows(bardata,data+errors,bardata,data-errors,angle=90,code=3,
    length=0.1)
legend("topright",fill=c("lightblue","pink"),c("males","females"))
```

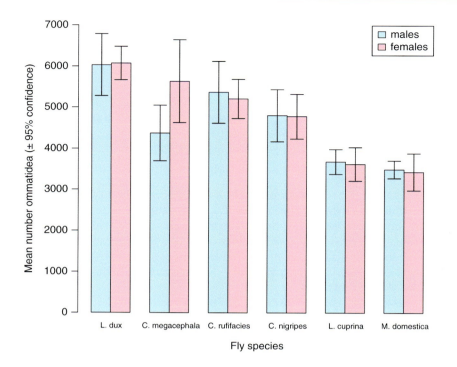

The character '±' is not in basic ASCII but may be available in R on some platforms as a unicode character that can be called \u00B1. In the code above we used the less elegant '+/−'. Replacing the barplot line with the following should work in most cases

bardata<-barplot(data,beside=TRUE,ylim=c(0,max(data)+max(errors)),
    ylab="Mean number ommatidea (\u00B1 95% confidence)",names=flies,
    cex.names=0.7, xlab="Fly species",col=c("lightblue","pink"))

Example 8 – Creating Pie Charts Using *pie* and *circlize*

Pie charts are a popular way of representing how a set of data is distributed among a number of categories. They are not approved of by everyone because they don't make it quite so easy to appreciate the relative quantities as in, for example, bar charts. Nevertheless, there is a function **pie** which plots pie charts. As an example we will use data from Marsh *et al.* (2009) on the numbers of genera of plants in 21 Taxonomic Databases Working Group (TDWG) Level 2 regions (https://www.tdwg.org) (Table 6.2).

We could save this as a table but we would have to remove spaces from column names; for ease we have already edited place names.

TWIG<-c("Andaman_Islands","Cambodia","Laos","Myanmar","Nicobar_Islands", "South_China_Sea","Thailand","Vietnam","Borneo","Christmas_Island", "Cocos_Keeling_Island","Jawa","Lesser_Sunda_Islands","Malaya","Maluku", "Philippines","Sulawesi","Sumatra","Bismarck_Archipelago","New_Guinea", "Solomon_Islands")

number_of_genera<-c(453,686,767,1304,277,39,1605,1401,1396,172,62, 1403,812,1457, 898,1406,1000,1300,509,1424,665)

number_of_genera_per_area<-c(943,598,622,812,779,291,1097,1061,841, 980,438, 1307,802,1354,1026,1062,850,877,612,844,861)

Table 6.2. Data on numbers of endemic plant genera in South East Asian and Oceanic countries (data from Marsh et al., 2009).

TWIG_region	Number of genera	Number of endemic genera	Number of genera standardized by area	Number of genera standardized by collecting intensity	Mean human footprint value
Andaman_Islands	453	0	943	360	34.90
Cambodia	686	0	598	922	21.95
Laos	767	4	622	1028	20.74
Myanmar	1304	10	812	1987	19.91
Nicobar_Islands	277	0	779	360	22.58
South_China_Sea	39	0	291	NA	0.62
Thailand	1605	17	1097	1959	24.60
Vietnam	1401	21	1061	1642	28.12
Borneo	1396	31	841	1704	20.27
Christmas_Island	172	0	980	NA	33.66
Cocos_Keeling_Island	62	0	438	NA	3.48
Jawa	1403	3	1307	1206	41.05
Lesser_Sunda_Islands	812	1	802	964	31.53
Malaya	1457	19	1354	1288	26.24
Maluku	898	3	1026	1009	23.88
Philippines	1406	20	1062	1467	33.79
Sulawesi	1000	5	850	1285	30.68
Sumatra	1300	7	877	1696	28.99
Bismarck_Archipelago	509	1	612	687	26.00
New_Guinea	1424	48	844	1624	16.72
Solomon_Islands	665	3	861	556	29.53

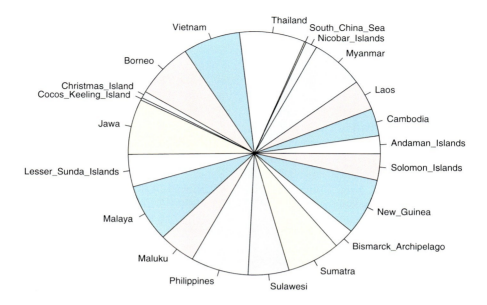

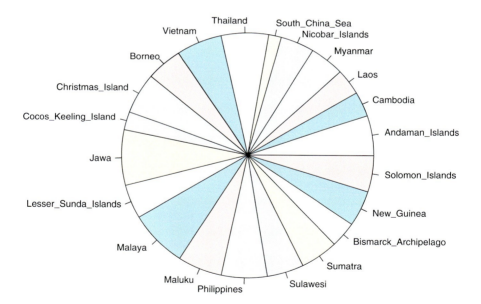

To plot pie charts of the numbers of genera and numbers standardized by area
par(mar=c(2,4.1,4.1,2.1))
par(mfrow=c(1,2))
pie(number_of_genera,labels=as.character(TWIG),cex=0.5)
pie(number_of_genera_per_area,labels=as.character(TWIG),cex=0.5)
We used the 'mar=c()' argument with the function **par** to reduce the left-hand margin of each plot from the default value 5.1 to 2.1 to avoid having too much blank space and to get the charts larger. In addition, we had to use the 'cex=' argument to reduce the font size used for the labels so as to avoid them overlapping both the graphs and each other. If you plot only a single pie chart on a page at a time, 'cex=1' is clearer.

Whilst the above functions are fine, we want you to understand that for many things you can write your own simple functions which you can then customize precisely how you want. Often this might be facilitated by installing some simple graphics library. Here we will use the package **circlize** for drawing segments of a more sophisticated pie chart of our own devising. First we must install the package from CRAN and make its functions available using **library**.
install.packages("circlize")
library(circlize)

Suppose that instead of just having just one set of proportions, we had another quantifiable variable we wanted to demonstrate. We might want to use angular spread in the pie chart for one variable and maximum radius for another. Or even both minimum and maximum radius.
Here we will take some data from Brown and Maurer (1989) on the body mass distribution of North American mammals.
num_genera<-c(2,27,70,117,113,84,75,81,52,47,26,18,7,4)
log_body_mass_base2<-c(1:14)
n<-length(num_genera)
m<-cumsum(num_genera) # **cumsum** gives the cumulative sum
x<-360*m/sum(num_genera)
x<-c(0,x) # adjacent pairs from **cumsum** are the beginning and end angles except for the first one so we append a zero
plot(NULL,xlim=c(–8,15),ylim=c(–12,12),axes=F,ann=F, type="n",xlab="",ylab="")
  # set up a blank plot area with size based on max mass (log base 2)
for(i in 1:n){
    draw.sector(start.degree=x[i],end.degree=x[i+1],rou1=0,rou2=log_body_mass_base2[i],center=c(0,0),clock.wise=FALSE,col=gray(1-(0.1+(0.9*num_genera[i]/max(num_genera)))),border="black",lwd=1,lty=0)}

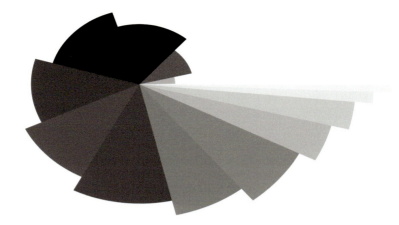

This gives us a plot in grey scale using the built-in palette 'gray'. You will have noticed '0.1+(0.9*...)'; we adjusted the scale of colours to avoid pure white. This is useful in the 'rainbow' palette because the beginning and end colours are rather similar.

For more exciting colours we suggest you try:
for(i in 1:n){
    draw.sector(start.degree=x[i],end.degree=x[i+1],rou1=0,rou2=log_body_mass_base2[i],center=c(0,0),clock.wise =FALSE,col= heat.colors(100, 0.1+(0.9*num_genera[i]/max(num_genera))), border="black",lwd=1,lty =0)}

or change 'heat.colors' to 'terrain.colors', or
for(i in 1:n){
    draw.sector(start.degree = x[i],end.degree=x[i+1],rou1= 0,rou2= log_body_mass_base2[i],center=c(0,0),clock.wise=FALSE,col= rainbow(256,0.9) [round(256*(0.1+(0.9*num_genera[i]/max(num_genera))))], border="black",lwd=1, lty=0)}

Palette 'rainbow' does not behave in quite the same way as the others. The first argument we gave '(265)' is the number of colours that we want the rainbow to be divided into; typing 'rainbow(60)' will show a smaller number of colours. The second argument (0.9) is the level of saturation. Because rainbow is a vector of colours we can access them by their index in the list.

Exercise Box 6.3

1) Use the data from R's available dataset 'InsectSprays'. Create a pie chart of total insect count for each type of spray. Use palette= to set the sector colours.
   a. Hint: To get the sums of the insects killed for each spray type you can: write a loop function, i.e. for(i in {*one value*}:{*another value*})
   b. or use the useful function **aggregate** to do it automatically, e.g.
      aggregate(count ~ spray,data=InsectSprays,sum)
   c. or use **tapply** (see Chapter 27) with **sum**, e.g.
      tapply(InsectSprays$count, InsectSprays$spray, sum)
2) Repeat using the functions from the **circlize** library.

Plotting Biological Data in Various Ways   69

> **Tool Box 6.3**
> - Use aggregate() to split data into subsets and return summary statistics for each subset.
> - Use angle= in **arrows** to set the angle of the arrowhead in degrees, 30 is default, 90 gives a T shape.
> - Use arrows() to put arrowheads on lines.
> - Use byrow=TRUE to enter elements into a matrix row by row (byrow=FALSE is the default).
> - Use code= in **arrows** to set which end(s) of the line the arrow head is placed; 3 means both.
> - Use labels=as.character() to set the sector names for a pie chart.
> - Use pie() to create a pie chart.
> - Use the **circlize** library to have more options for plotting circles, sectors, etc.

Example 9 – Fish Metacercarial Load and Box and Whisker Plots

Boxplots summarize numerical data for discrete explanatory variables in a very neat way. They can be created in R in various ways. Here we will use raw data kindly supplied by the authors from Wiriya et al.'s (2013) study of the level of trematode infection in various species of fish in Thailand. Numbers of metacercaria stages in tissue samples were recorded for various numbers of individuals of eight species of fish. The data can be entered as a table in a spreadsheet (see Chapter 31) or such a table can be created by entering relevant data (fish species, metacercaria number and tissue sample weights) as follows.

Start by assigning the numbers of metacercaria to vectors for each species of fish.
Tilapia<-5
Mystus_singaringan<-c(1,2,2,2)
Osteochilus_vittatus<-c(26,30,11,2,28,6,19,12,4,4,18,13,29,21,9,8,21,19,28)
Henicorhynchus_siamensis<-c(46,22,33,28,50,42,31,4,13,8,11,8,18)
Barbodes_gonionotus<-c(8,11,66)
Anabas_testudineus<-c(12,26)
Trichogaster_microlepis<-c(9,5)
Trichogaster_pectoralis<-c(21,12,6,3)

These can then be combined into a single vector of counts (or the counts entered as a single list)
counts<-c(Tilapia,Mystus_singaringan,Osteochilus_vittatus,
    Henicorhynchus_siamensis,Barbodes_gonionotus,Anabas_testudineus,
    Trichogaster_microlepis,Trichogaster_pectoralis)

Enter the analysed tissue sample sizes in the same order,
sample_sizes<-c(130.0,28.95,27.2,17.92,26.12,81.65,78.31,59.46,62.84,
    58.28,45.03,35.02,46.46,43.54,24.64,31.52,57.05,58.87,59.7,39.02,
    28.64,65.03,43.28,36.2,64.57,67.43,70.42,89.3,75.44,51.38,78.92,30.69,
    21.69,21.84,23.45,18.21,49.42,110,105.63,56.25,51.55,91.42,32.78,
    39.83,45.76,62.03,34.32,14.77) # in grams of tissue analysed

and then a vector ('species') to tie the names of the fish species in the same order as the data, using **rep** to replicate each name the desired number of times.
species<-c("Tilapia",rep("Mystus singaringan",4),rep("Osteochilus vittatus",
    19), rep("Henicorhynchus siamensis",13),rep("Barbodes gonionotus",3),
    rep("Anabas testudineus",2),rep("Trichogaster microlepis",2),
    rep("Trichogaster pectoralis",4))

To calculate metacercarial numbers per gram of fish tissue
counts_per_gram<-counts/sample_sizes
and we can combine these as a table using **cbind**.

The function **boxplot** provides some additional attributes. Unlike **plot** it uses the tilde character to separate the response and explanatory variables. Try first:
boxplot(counts_per_gram ~ factor(species),ylab="Metacercaria per gram")

Information Box 6.6

Hunting the tilde. Where the tilde operator character, which is very important in R, is not shown on the main keyboard, it can normally be accessed by pressing some shortcut combination of keys. Note that there are two different tildes, the one we need here is a larger one that is level with the text and is the mathematical tilde operator, and NOT the accent character that is placed, for example, over the Spanish letter ñ. Unfortunately the key combination shortcut is platform- and often country- dependent. However, both Windows and Macs have a character map option from which you can select the desired symbol. On Windows machines press the Windows key + R to open Run command box, type 'charmap' then press Enter, and the Character Map window will open. On Mac computers press the key combination Control + Command + Space bar to show the Character Viewer, and under Math symbols choose the tilde operator ~. The key combination Option + N often works too.

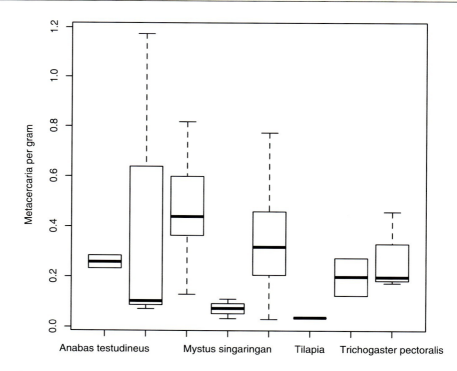

Immediately we can see a couple of issues with this particular plot – the fish species are listed in alphabetical order unlike our original list, which was aimed at comparing commercial *Tilapia* to the wild species, AND our fish names are so long that only a subset have been printed.

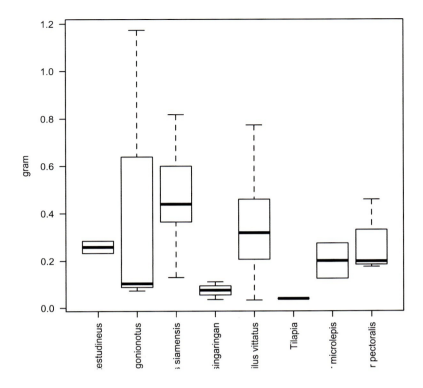

Let's deal with the fish name length first. We will make R print the names vertically by including the argument 'las=2'

boxplot(counts_per_gram ~ factor(species),ylab="Metacercaria per gram",las=2)

the names are now all there but most are too long to fit into the default bottom margin of the plot, so we must adjust the border around our plot area to increase the space at the bottom using the **par**(mar=) command. By trial and error we find that increasing the lower margin of the plot from the default 5.1 to 13 and reducing the font size to 80% we get an acceptable figure with factor (species) as the x axis labels.

par(mar=c(13,4,4,2))
boxplot(counts_per_gram ~ factor(species),ylab="Metacercaria per gram", las=2,cex=0.8)

Information Box 6.7

Boxplots present clear summaries of a lot of information. The thick horizontal line in the middle is the median of the data; the top and bottom of the box show the 75th and 25th percentiles respectively, which means that 50% of data points lie within this middle zone; the whiskers show the larger of EITHER the maximum and minimum observed values OR 1.5 × the interquartile range, the latter corresponding to approximately 95% confidence limits. If some observed values are beyond the extent of the whiskers they will be shown as separate small circles outside of the limits of the whiskers.

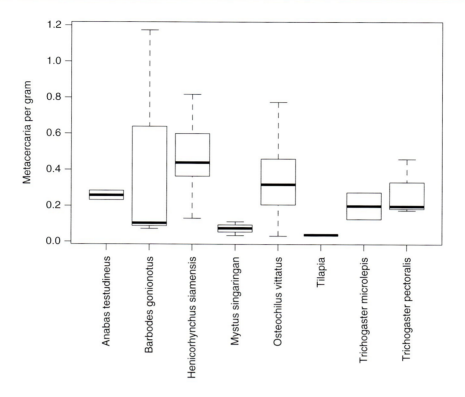

We now wish to make R present the species in the order we desire rather than in the default alphabetical order. To make interpretation clearer later on we will order the species in accordance with their mean parasite per gram count, and create a vector 'Fish' with the names in the desired order (spelled exactly the same). We use the function **tapply**, which is explained in detail in Chapter 27, to sort the vectors 'species' and 'Fish' according to the mean 'counts_per_gram'.
rownames(as.matrix(sort(tapply(counts_per_gram,species,mean))))
[1] "Tilapia" "Mystus singaringan"
"Trichogaster microlepis" "Trichogaster pectoralis"
[5] "Anabas testudineus" "Osteochilus vittatus"
"Barbodes gonionotus" "Henicorhynchus siamensis"
Fish<-rownames(as.matrix(sort(tapply(counts_per_gram,species,mean))))
and include this in the factor statement
boxplot(counts_per_gram ~ factor(species,Fish),ylab="Metacercaria per
    gram",las=2,cex=0.8)
From the boxplot we clearly see that the *Tilapia* sample has a lower metacercarial load per gram of tissue than the other species. But is this a statistically significant result? We will now go on to test this.

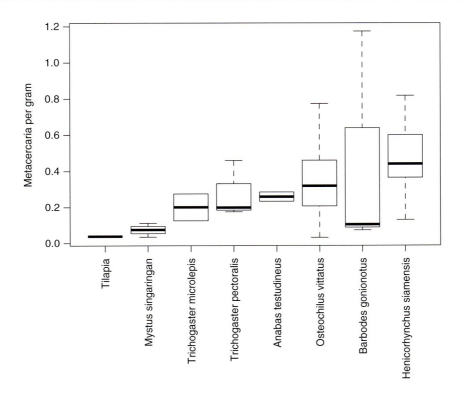

Adding Notches to a Boxplot

The function **boxplot** includes a logical argument called notch which gives a fairly reliable visual guide as to whether the medians of different sets of data will differ significantly from one another. The extent of the notch on each side of the median represents the 95% confidence interval of the median and is given by +/− 1.57 × inter-quartile range/√n – the inter-quartile distance between estimated 25th and 75th percentiles. If the notches of two boxplots do not overlap, it is very likely that their medians differ significantly at the 0.05 level, and conversely, if they do overlap, the data are unlikely to differ significantly. The resulting plot can look quite odd because if there are few samples in some categories, or the variance is high, then the ends of the notches will extend beyond the box of the boxplot. Applying the notch to the metacercaria data boxplot we write:

boxplot(counts_per_gram ~ factor(species,Fish),notch=TRUE,ylab=
    "Metacercaria per gram",las=2,cex=0.8)

```
Warning message:
In bxp(list(stats = c(0.0384615384615385,
    0.0384615384615385, 0.0384615384615385, :
some notches went outside hinges ('box'): maybe set
    notch=FALSE
```

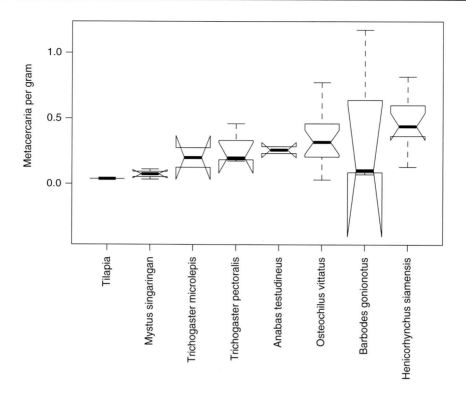

In this case the figure is indeed rather ugly, but as explained by Crawley (2007) this is intentional. The overlap between most pairs of notch ends indicates that any test applied might be invalid because of the variance/sample size values (Crawley 2007, p.157) – *viz*. most, though not all, of the notched areas overlap, and indeed, we get an error message because the notch values for some of the species extend too far.

Tukey's Honest Significant Difference Test

You cannot judge exactly whether any of these fish parasite loads differ significantly just by looking at the data. Our judgement might also be wrong because we have not made just one comparison but we have gone fishing (pun intended) making comparisons between a fairly random set of fish species. However, we can use Tukey's honest significant difference test, **TukeyHSD**, to obtain a conservative view of whether some species may be significantly more prone to heavier infections than others. **TukeyHSD** requires another function **aov** (=analysis of variance), which we will describe in more detail in Chapter 13.

```
par(mar=c(5.1, 4.1, 4.1, 2.1))
plot(TukeyHSD(aov(counts_per_gram ~ factor(species))))
```

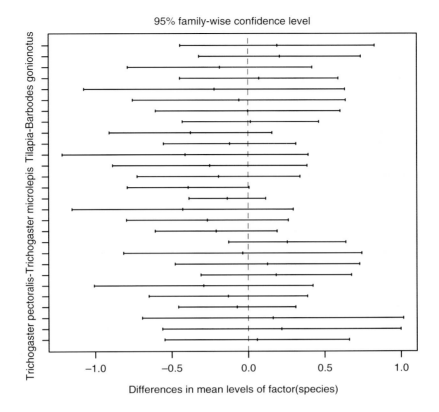

The vertical dotted line means no difference and each horizontal line is a pairwise comparison between the factors, the eight fish species. If any of the comparisons is significant (by default at the 95% confidence limit) their horizontal line will not cross the line of no difference. In this case not one pairwise comparison out of the 28 is significant. Only one is close. But which one? there is something very messy about the way that **TukeyHSD** presents the names. To make this plot more readily interpretable we can rename the fish (i.e. the factor names within 'species'). This is quite easy and is a useful thing to know how to do, but often not readily apparent. What we will choose to do is to replace each scientific name with a single letter.

The level names of species can be seen by
levels(as.factor(species))
[1] "Anabas testudineus""Barbodes gonionotus"
    "Henicorhynchus siamensis" "Mystus singaringan"
    "Osteochilus vittatus"
[6] "Tilapia" "Trichogaster microlepis" "Trichogaster
    pectoralis"
and we want to replace these with single letters 'A' to 'H' to save space. We use the preset vector of capital letters called LETTERS; LETTERS[1:8] gives 'A' 'B' 'C' 'D' 'E' 'F' 'G' 'H'
LETTERS[1:length(levels(as.factor(species)))]

we could use
species<-factor(species,labels=LETTERS[1:length(levels(as.factor(species)))])
but this will lose our species names, or we can create a new vector with each species name replaced by a letter
x<-factor(species,labels=LETTERS[1:length(levels(as.factor(species)))])
plot(TukeyHSD(aov(counts_per_gram ~ factor(x))))

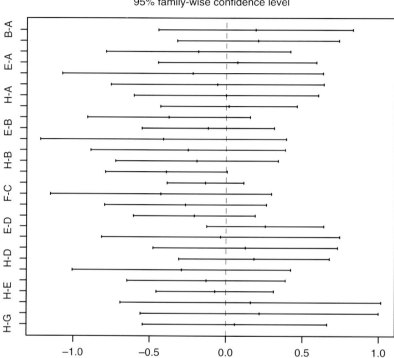

Although we still cannot read all the pairs, we can find the order by examining the object created by **TukeyHSD**
TukeyHSD(aov(counts_per_gram ~ factor(x)))[1]
$`factor(x)`

	diff	lwr	upr	p adj
B-A	0.191473212	-0.4450678	0.828014197	0.97727981
C-A	0.207751451	-0.3218827	0.737385565	0.91004896
D-A	-0.184530547	-0.7884063	0.419345254	0.97520336
E-A	0.071711105	-0.4466529	0.590075119	0.99982236
F-A	-0.220131145	-1.0741405	0.633878202	0.99065917
G-A	-0.058547097	-0.7558428	0.638748616	0.99999402
H-A	-0.001015069	-0.6048909	0.602860732	1.00000000
C-B	0.016278239	-0.4303485	0.462904939	0.99999998

```
D-B   -0.376003759   -0.9085722    0.156564638   0.34175788
E-B   -0.119762107   -0.5529646    0.313440346   0.98592908
F-B   -0.411604358   -1.2167721    0.393563377   0.72738828
G-B   -0.250020309   -0.8865613    0.386520676   0.90947725
H-B   -0.192488281   -0.7250567    0.340080116   0.93970591
D-C   -0.392281998   -0.7909761    0.006412078   0.05660124
E-C   -0.136040346   -0.3870229    0.114942192   0.66672682
F-C   -0.427882597   -1.1515006    0.295735362   0.56531484
G-C   -0.266298548   -0.7959327    0.263335566   0.74326808
H-C   -0.208766520   -0.6074606    0.189927556   0.70335602
E-D    0.256241652   -0.1273544    0.639837693   0.41106451
F-D   -0.035600598   -0.8152009    0.743999708   0.99999991
G-D    0.125983450   -0.4778924    0.729859251   0.99743495
H-D    0.183515478   -0.3095470    0.676578005   0.93021738
F-E   -0.291842251   -1.0072526    0.423568055   0.89199858
G-E   -0.130258202   -0.6486222    0.388105812   0.99197102
H-E   -0.072726174   -0.4563222    0.310869867   0.99860122
G-F    0.161584049   -0.6924253    1.015593396   0.99861926
H-F    0.219116076   -0.5604842    0.998716383   0.98452522
H-G    0.057532028   -0.5463438    0.661407828   0.99998577
```

To get a more robust idea about trematode loads, we would want to maximize comparability of samples of fish species (location, size (fish age), sex, time of year) and increase sample sizes because samples fewer than 5 (or ideally 20+) seldom give useable results. But do not despair; many studies produce far more concordant results with most or all notches ending within the limits of the boxes.

Tukey's HSD test is, as Crawley (2007) puts it, 'conservative but not ridiculously so'. Many workers have, and do, employ a very strict correction for multiple comparisons proposed by Bonferroni (1936). In Bonferroni's method the p-values obtained are multiplied by the number of comparisons that have been made. Thus if you do a single comparison and $p = 0.026$ (significant at the 0.05 level) but you make a second comparison, such as testing a second explanatory variable, the necessary p-value needed to claim significance is 0.0249999 (i.e. 2×0.0249 is < 0.05, just!). Not that unlikely, but if you have 'gone fishing' (see also Chapter 11) and tested 20 possible explanatory variables, a relationship would only be deemed significant if the p-value was $< 0.05/20 = 0.0025$. The Bonferroni correction has been accused of leading to a high level of Type II statistical errors (rejecting a true hypothesis) (Narum 2006; White et al. 2019), especially when large numbers of comparisons are being made, and is more suited for cases in which a more limited set of rather plausible explanatory variables are being tested.

Because of the possibility of an inflated rate of Type II errors, many workers employ variously less stringent methods of setting a new critical p-value. There is no correct answer, because 0.05 is an arbitrary cut-off in any case. The in-built R stats package has an automatic probability adjustment function **p.adjust** for when we have a set of p-values generated from multiple comparisons. In addition to 'bonferroni', the method argument has a number of options 'holm', 'hochberg', 'hommel', 'BH' and 'BY'; see the help page (?p.adjust) for original literature references to get some suggestions as to which adjustment method is most sensible for your data). The other parameter that has to be passed to **p.adjust** is the number of comparisons that are being made.

The 15 p-values in Narum's (2006) study, of which 12 are significant at the 0.05 level, are

Narum<-c(0.0001,0.0010,0.0062,0.0101,0.0214,0.0227,0.0273,0.0292,0.0311, 0.0323, 0.0441,0.0490,0.0573,0.1262,0.5794)

We can now try a few adjustment methods and see how many of the relationships are deemed significant at the 0.05 level after correction. Optionally, the method can be specified by 'method='.

p.adjust(Narum,"bonferroni"); length(which(p.adjust(Narum,"bonferroni")<0.05))
 [1]   0.0015   0.0150   0.0930   0.1515   0.3210   0.3405   0.4095
     0.4380 0.4665 0.4845 0.6615 0.7350 0.8595 1.0000 1.0000
 [1] 2

p.adjust(Narum,"hochberg"); length(which(p.adjust(Narum,"hochberg")<0.05))
 [1]   0.0015   0.0140   0.0806   0.1212   0.1719   0.1719   0.1719
     0.1719 0.1719 0.1719 0.1719 0.1719 0.1719 0.2524 0.5794
 [1] 2

p.adjust(Narum,"BH"); length(which(p.adjust(Narum,"BH")<0.05))
 [1] 0.00150000 0.00750000 0.03100000 0.03787500 0.04845000
     0.04845000 0.04845000 0.04845000 0.04845000 0.04845000
     0.06013636 0.06125000 0.06611538 0.13521429 0.57940000
 [1] 10

In fact, all the adjustments available with **p.adjust** gave just two significant results except the BH method.

Tool Box 6.4 ✹

- Use ~ to separate response and explanatory variables in models and boxplots.
- Use aov() to carry out analysis of variance to pass to **TukeyHSD** test.
- Use boxplot() to create a boxplot.
- Use factor() to coerce character or numeric variables to be factors.
- Use LETTERS[] to get specific or a sequence of capital letters.
- Use levels() to retrieve a list of factors.
- Use method= in **p.adjust** to set the type of method argument.
- Use n= to set the number of comparisons made.
- Use notch= in **boxplot** add a notch indicating 95% confidence limits to the median.
- Use p.adjust() to adjust the probability to take into account multiple comparisons.
- Use TukeyHSD() to plot 95% confidence overlap levels.

7 The Grammar of Graphics Family of Packages

Summary of R Packages Introduced
ggplot2
ggpubr
gridExtra

Summary of R Functions Introduced
cor.test
fortify
ggplot in "ggplot2"
ggscatter in "ggpubr"
remove.packages

In this chapter we introduce a couple of advanced graphics packages that are becoming more popular because of their customizability (**ggplot2**, **ggpubr**, **ggplotly**). The 'gg' part of their names stands for 'Grammar of Graphics'. Although in the rest of this book we will mostly use the base R plotting functions, we will occasionally use **ggplot2** functions because of their easy and extra capabilities, though using them may at first seem a bit more complicated. Grammar of Graphics packages use a layers approach and are in some ways 'intelligent', for example, with choosing colour scales for factors or continuous variables.

First we will install **ggplot2** (an updated version of the original package **ggplot**, but that still uses the function **ggplot**).
install.packages("ggplot2")
library(ggplot2)

As you'll see later your installed version of R has a number of in-built datasets and dataframes. To see the full list of installed datasets type 'data()' at the prompt. It is a large and varied selection. A more comprehensive list for several packages is available at https://vincentarelbundock.github.io/Rdatasets/datasets.html. To give a flavour we can use R's in-built dataset 'Loblolly', which gives the height, age and seed size of 84 American loblolly pine (*Pinus taeda*) trees.

head(Loblolly)

```
   height age Seed
1    4.51   3  301
15  10.89   5  301
29  28.72  10  301
43  41.74  15  301
57  52.70  20  301
71  60.92  25  301
```

The first column is the individual tree number. We will plot height versus age and colour the points based on seed weight using the R **plot** default and then using **ggplot**. The Loblolly data are already a dataframe so we can use **attach** to make the column names directly available.

attach(Loblolly)
plot(age,height,col=as.factor(Seed))

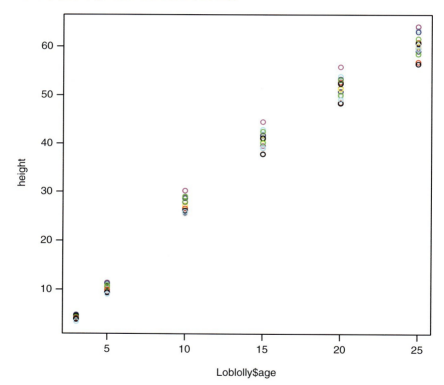

For **ggplot** we must provide the data as a dataframe, or an object that can be coerced into a dataframe using the function **fortify**.

Using **ggplot2** always comprises two or more parts. The first part is included in round brackets immediately after the **ggplot** function call defines the data and does not do the plotting. This part has the structure, ggplot({*dataset*},aes({*the variables we wish to plot*})), where **aes** stands for 'aesthetics'. Other things such as colours for plots can be specified in the **aes** function too.

The Grammar of Graphics Family of Packages

After specifying the data to be plotted we then, after a plus sign, add the type of plot, and there may be more things added in addition, each following a plus sign. There are quite a lot of plot types available in **ggplot** including **geom_point** for a scatterplot of individual points, **geom_boxplot**, **geom_histogram**, **geom_jitter**, **geom_density** and many more. Here we just simply plot the height versus age using **geom_point**. Although we specify which is the x axis and which the y axis, that is not absolutely necessary because **ggplot** by default takes the first vector to be the variable if there is more than one variable specified.

ggplot(Loblolly,aes(x=age,y=height,color=as.factor(Seed))) + geom_point()

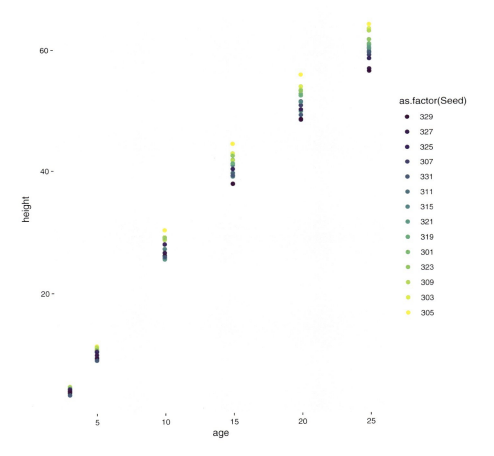

You can add many things to the plotting call, some of which are the same as in the normal **plot** function such as xlim, ylim, xlab, ylab, but the difference is that you put the values in round brackets and do not use an equals sign, e.g. specify the y axis limits as 'ylim(0,100)'. There are also differences as the plot title is not set by **main** but instead by **ggtitle**, e.g. **ggtitle**("Loblolly data").

There is a useful online cheat sheet for **ggplot** at https://rstudio.com/wp-content/uploads/2015/03/ggplot2-cheatsheet.pdf.

We have now finished with Loblolly so we should detach it to avoid later confusion with vector names

detach(Loblolly)

You may of course want to remove an installed package, either from current use or from your computer entirely.

remove.packages("ggplot2")

One limitation of **ggplot2** is that it does not let you put multiple graphs on the same plot window using par(mfrow=c(n,m)). This can be overcome using the package **gridExtra**.

The package **ggpubr** (= gg Publication Ready Plots) is based on **ggplot2** but it has a rather simpler syntax making it ideal for beginners.

install.packages("ggpubr")
library("ggpubr")

There is a useful scatterplot function in the **ggpubr** package called **ggscatter** that combines plotting the data points, calculating and adding the regression line according to whichever choice of correlation you request (Pearson, Spearman or Kendall) – but all lower case.

To demonstrate a few features and briefly introduce the syntax we will use data from Kolm *et al.* (2012) on the relationship between the proportion of ants in the diet of the fish *Corynopoma riisei* and the ornamentation of a water-flea-like lure at the apex of a projection from the operculum. In addition, we will show how to obtain and plot correlation coefficients and plot confidence intervals automatically.

x<-c(0.14990444,0.14352947,0.18572451,0.22692724,0.32947406,0.39478415,
    0.43702197,0.44205266,0.45779166,0.4772302,0.4990459,0.5148496,
    0.54387033, 0.5666838,0.61728454,0.6657855,0.72059345) # ants in fish diet

y<-c(0.074910976,0.6254242,0.3017297,0.24208963,0.14281006,
    1.0448253,0.62397754,1.1371646,1.2069718,0.8333741,1.0302496,
    0.95308137,0.70407337,0.62443894,1.1078949,0.39062706,
    1.2676938) # ornamentation of food dummy

g<-as.data.frame(cbind(x,y))
ggscatter(g,x="x",y="y",xlab="ants in diet",ylab="similarity",add="reg.line",
    conf.int=TRUE,cor.coef=TRUE,cor.method=c("spearman"))

Pearson's correlation is a parametric test, which deals with normally distributed data, whereas Spearman Rank and Kendall tests are non-parametric and are appropriate for data that are skewed or have an abnormal kurtosis (see Chapter 10). Spearman's test statistic is called *rho*, and Kendall's is called *tau* (see Appendix 3 for the equations). The test is asking whether the value of the correlation coefficient, *r*, is significantly different from what you would expect for random points. To just get the value of *r* use **cor**; to perform the statistical test use **cor.test**.

cor.test(g[,1],g[,2],method=c("spearman"))

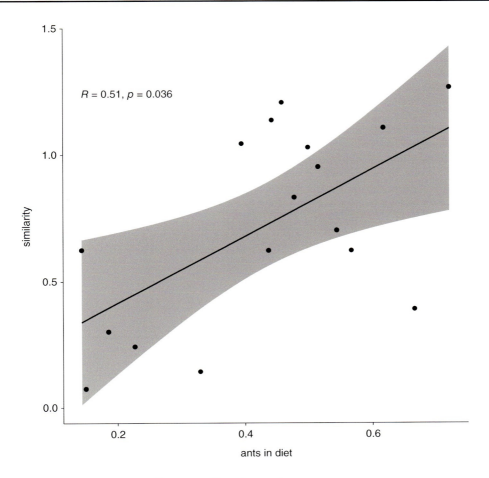

```
Spearman's rank correlation rho
data: g[, 1] and g[, 2]
S = 398, p-value = 0.0376
alternative hypothesis: true rho is not equal to 0
sample estimates:
      rho
0.512254
```

Other very neat plotting functions are included in the **ggpubr** package, e.g. **ggdensity** to produce density plots (see Chapter 20), **gghistogram**, **ggboxplot** and **ggbarplot**.

Information Box 7.1 📖

Using datasets from R package libraries. Several of the exercises we present here use datasets that are in R libraries such as **MASS** and **Stats2Data**. Having installed these packages, typing the name of a dataset will show its contents. However, to allow the data to be used in functions you need to tell R to load the dataset, using the aptly named function, **data**, e.g. **data**(Fitch) having installed the **Stats2Data** library.

Exercise Box 7.1

1) Use R's dataset 'Fitch' from the **Stat2Data**, which gives data on skull dimensions and weight for 14 species each of carnivores and primates.
2) Plot the skull length against the palate length.
3) Colour the points based on the order that the mammals belong to.
4) Add a legend.
5) Now replot the data using **ggplot**. Hints: (i) use aes in the first part of function call to specify your x and y vectors; (ii) add a geompoint() layer to tell **ggplot** to plot the points; and (iii) add geom_smooth to fit a linear model separately for each mammal order.

Tool Box 7.1

- Use add="reg.line" in **ggscatter** to include a regression line.
- Use as.data.frame(cbind()) to coerce concatenated lists into a dataframe by column.
- Use attach() to allow you to access variable (column) names of a dataframe directly.
- Use conf.int=TRUE in **ggpubr** to add shaded 95% confidence interval.
- Use cor.coef=TRUE to in **ggpubr** to add its value to the plot.
- Use cor.method=() in **ggpubr** to define the type of regression analysis ("pearson", "kendall" or "spearman").
- Use cor.test() to produce the test statistic values.
- Use data() to show a list of in-built datasets and dataframes.
- Use detach() to stop working with a particular dataframe.
- Use fortify() to coerce an object into a dataframe.
- Use geom_point() to define a geometric object for plotting points.
- Use ggplot() to plot make graphics using the **ggplot2** package.
- Use ggscatter() to create a scatterplot.
- Use head() to display the first few lines of a dataset.
- Use install.packages("{*package name*}") to download and install packages from CRAN.
- Use library() to enable R to use the functions in an installed package.
- Use mapping=aes() to specify the aesthetics of a ggplot.
- Use method= in b() to specify the correlation method.
- Use remove() or rm() to delete specific R objects.
- Use remove.packages("{*package name*}") to remove packages completely from the computer.
- Use tail() to display the last few lines of a dataset.

8 Sets and Venn Diagrams

Summary of R Libraries Introduced
    Biocmanager
    car
    ellipse
    limma
    plotrix
    VennDiagram

Summary of R Functions Introduced
    apply
    biocLite
    draw.ellipse in "plotrix"
    intersect
    readLines
    seq
    setfiff
    source
    strsplit
    structure
    union
    unlist
    VennCounts
    vennDiagram

Venn diagrams, also known as set diagrams, are commonly used to represent the overlap between sets. However, there is no in-built Venn diagram function in R so we have to use packages.

Running the following installs at the R prompt below, which downloads a package called **limma**. We have to install this slightly differently as it was recently moved from a deprecated site 'bioconductor' to CRAN.

install.packages("BiocManager")
library(limma)

and the **limma** library contains a function **vennDiagram**. A more sophisticated package is 'VennDiagram' (note the capital letters) from CRAN, but we don't have room to describe that here.

For our data we will go to the wildlife pages of three Thai National Parks (Doi Inthanon N.P., Khao Yai N.P. and Kaeng Krachan N.P.) and get their bird lists (e.g. https://www.thainationalparks.com/khao-yai-national-park/wildlife) and for ease consider only the woodpeckers (Picidae).

The downloaded woodpecker lists for the three parks are in three online data files Doi_woodpeckers.txt, KhaoYai_woodpeckers.txt and KaengKrachan_woodpeckers.txt. Use readLines to pass each of these to an appropriately named vector, such as doi, kaeng and khao.

khao<-readLines("KhaoYai_woodpeckers.txt")

```
Warning message:
In readLines("KhaoYai_woodpeckers.txt") :
incomplete final line found on 'KhaoYai_woodpeckers.txt'
```

In this instance the warning message is not important, it means that there is no carriage return after the last line in the data file, but the lines have been read just fine.

```
head(khao)
[1] "Chrysocolaptes guttacristatus, Greater flameback"
[2] "Chrysophlegma flavinucha, Greater yellownape"
[3] "Dendrocopos analis, Freckle-breasted woodpecker"
[4] "Dinopium javanense, Common flameback"
[5] "Dryocopus javensis, White-bellied woodpecker"
[6] "Gecinulus viridis, Bamboo woodpecker"
```

If we had simply grabbed the woodpecker list direct from the National Parks website and converted it to a vector, ky, using **textConnection**, you would see that all the lines of data were separated by the escape code \n, which means start a new line. Therefore, the whole result is just a single character string vector.

```
ky<-"Blythipicus pyrrhotis, Bay woodpecker
Chrysocolaptes guttacristatus, Greater flameback
Chrysophlegma flavinucha, Greater yellownape
Dendrocopos analis, Freckle-breasted woodpecker
Dendrocopos atratus, Stripe-breasted woodpecker
Dinopium javanense, Common flameback
Dryobates cathpharius, Crimson-breasted woodpecker
Dryocopus javensis, White-bellied woodpecker
Jynx torquilla, Eurasian wryneck
Picumnus innominatus, Speckled piculet
Picus chlorolophus, Lesser yellownape
Picus erythropygius, Black-headed woodpecker
Picus guerini, Black-naped woodpecker
Sasia ochracea, White-browed piculet
Yungipicus canicapillus, Grey-capped pygmy woodpecker"
ky
[1] "Blythipicus pyrrhotis, Bay woodpecker\nChrysocolaptes
    guttacristatus, Greater flameback\nChrysophlegma flavi-
    nucha, Greater yellownape\nDendrocopos analis, Freckle-
    breasted woodpecker\nDendrocopos atratus, Stripe-breasted
    woodpecker\nDinopium javanense, Common flameback\nDry-
    obates cathpharius, Crimson-breasted woodpecker\nDryo-
    copus javensis, White-bellied woodpecker\nJynx torquilla,
    Eurasian wryneck\nPicumnus innominatus, Speckled piculet\
    nPicus chlorolophus, Lesser yellownape\nPicus erythro-
    pygius, Black-headed woodpecker\nPicus guerini, Black-naped
    woodpecker\nSasia ochracea, White-browed piculet\nYun-
    gipicus canicapillus, Grey-capped pygmy woodpecker"
```

We can use the new line characters, i.e. the '\n's, to separate the species into separate elements using a combination of two functions, **strsplit** and **unlist**, where '\n' is the split character. If text had been separated by 'carriage returns' you would see '\r', or if separated by tab characters, '\t', and you could also use those as the split value. Since **strsplit** returns its results as a list we include **unlist** to obtain the separate names, otherwise the output will be nested vector containing multiple elements. The function **list** does the opposite.

khao<-unlist(strsplit(ky,"\n"))
khao
```
 [1] "Blythipicus pyrrhotis, Bay woodpecker"
     "Chrysocolaptes guttacristatus, Greater flameback"
     "Chrysophlegma flavinucha, Greater yellownape"
 [4] "Dendrocopos analis, Freckle-breasted woodpecker"
     "Dendrocopos atratus, Stripe-breasted woodpecker"
     "Dinopium javanense, Common flameback"
 [7] "Dryobates cathpharius, Crimson-breasted woodpecker"
     "Dryocopus javensis, White-bellied woodpecker" "Jynx
     torquilla, Eurasian wryneck"
[10] "Picumnus innominatus, Speckled piculet" "Picus
     chlorolophus, Lesser yellownape" "Picus erythropygius,
     Black-headed woodpecker"
[13] "Picus guerini, Black-naped woodpecker" "Sasia ochracea,
     White-browed piculet" "Yungipicus canicapillus,
     Grey-capped pygmy woodpecker"
```

We can obtain a complete list of woodpeckers by using the function **unique** to find all those elements (woodpeckers) that occur at least once in the combined list thus:

allpeckers<-unique(c(khao,doi,kaeng))
and the total number of species is
length(allpeckers)
[1] 30

OR we can treat the vectors doi, kaeng and khao as sets and use the set operations **union**, **intersect** and **setdiff**. Unfortunately, these set operations only work on pairs of sets, so must hierarchically nest further sets to achieve the desired result. The set operator **intersect** is precisely equivalent to **%in%** (see below).

allpeckers<-union(khao,union(kaeng,doi))
Which gives the same result as unique(c(khao,doi,kaeng))
in_all_three_parks<-intersect(doi,intersect(kaeng,khao))
in_all_three_parks
```
 [1] "Chrysocolaptes guttacristatus, Greater flameback"
     "Chrysophlegma flavinucha, Greater yellownape"
 [3] "Dendrocopos analis, Freckle-breasted woodpecker"
     "Dinopium javanense, Common flameback"
 [5] "Jynx torquilla, Eurasian wryneck""Picumnus
     innominatus, Speckled piculet"
 [7] "Picus chlorolophus, Lesser yellownape" "Picus guerini,
     Black-naped woodpecker"
 [9] "Sasia ochracea, White-browed piculet" "Yungipicus
     canicapillus, Grey-capped pygmy woodpecker"
```

To create Venn diagrams create a dataframe, starting by compiling a matrix with 0s and 1s for each park representing absence/presence of each woodpecker species in each of the three National Parks.

data<-matrix(NA,nrow=length(allpeckers),ncol=3)
colnames(data)<-c("kaeng","khao","doi") # assigning column names
data<-as.data.frame(data) # **as.data.frame** coerces the matrix into the dataframe structure
for(i in 1: length(allpeckers)){ # the next three lines use 'ifelse' to determine whether each woodpecker species occurs in each of the three national parks
ifelse (allpeckers[i] %in% kaeng,data$kaeng[i]<-1,data$kaeng[i]<-0)
ifelse (allpeckers[i] %in% khao,data$khao[i]<-1,data$khao[i]<-0)
ifelse (allpeckers[i] %in% doi,data$doi[i]<-1,data$doi[i]<-0)}
data

```
   kaeng khao doi
1      1    1   1
2      1    1   1
3      1    1   1
4      1    1   1
5      1    1   1
6      1    1   1
7      0    1   1
8      0    1   1
9      1    1   1
10     1    1   1
11     1    1   1
12     0    1   1
13     1    1   1
14     1    1   1
15     1    1   1
16     1    0   0
17     1    0   0
18     1    0   0
19     1    0   0
20     1    0   0
21     1    0   0
22     1    0   0
23     1    0   0
24     1    0   0
25     1    0   0
26     1    0   0
27     1    0   0
28     1    0   0
29     1    0   0
30     1    0   0
```

We can summarize our dataframe using the function **apply** to get totals of birds per park and parks per bird; **apply** applies another specified function, in this case **sum**, either to the rows of a matrix or to the columns. In this case, the same results can be obtained using two other specific functions: **rowSums** and **colSums** respectively.

Information Box 8.1 📖

The structure of the **apply** function code is **apply**(a,b,c) where

a) is the matrix or array that you want to apply a function to
b) a number, 1 stands for rows and 2 stands for columns, these specify what to apply a function to
c) the function to apply

```
apply(data,1,sum) # the 1 tells apply to work on each row
[1] 3 3 3 3 3 2 2 3 3 3 2 3 3 3 1 1 1 1 1 1 1 1 1
    1 1 1 1
apply(data,2,sum) # the 2 tells apply to work on each column
kaeng kshao   doi
   28    20    16
```

We are now ready to use the functions in the **limma** library starting with the function **vennCounts**, which creates an R object that can be processed by the plotting function **vennDiagram**.

m2<-vennCounts(data)

The **vennCounts** function adds up the numbers of species in each possible combination of parks. The parks including the species are indicated by 0s and 1s in the first three columns. 'm2' is an R object and in addition to the table includes two more rows that are used by the function **vennDiagram**

```
m2
  kaeng khao doi Counts
1     0    0   0      0
2     0    0   1      0
3     0    1   0      0
4     0    1   1      3
5     1    0   0     15
6     1    0   1      0
7     1    1   0      0
8     1    1   1     12
attr(,"class")
[1] "VennCounts"
```

vennDiagram(m2,names=c("Kaeng Krachan","Khao Yai","Doi Inthanon"),
   main= "Woodpecker species",circle.col=c("coral","paleturquoise3","bisque3"),
   lwd=3) # we livened this up a bit by colouring the circles and making
   lines thicker
mtext ("Aves: Picidae")

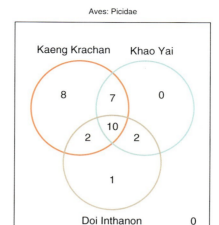

If you already know the counts in each of the sets and their intersections and do not want to create a file of presence/absence data to parse with **venn-Counts**, you can create a vennCounts object to pass to **vennDiagram** using the R function **structure** to give it the attribute 'class=VennCounts'. For example, if the parts of the Venn diagram are represented by 0s and 1s as in the vector 'm2' above, with the counts for each in the right-hand column in a vector, let's say called 'vd1', then a vennCounts object is created by

X<-structure(vd1,class="VennCounts")

Having used a package to do our Venn diagram we will now show how to write code to create one yourself. Create a blank plotting area and then draw three intersecting circles in it.

plot(NULL,xlim=c(0,2),ylim=c(0,2),axes=F,xlab="",ylab="",main="Woodpecker overlap between three Thai N.P.s")

We will define the x and y coordinates of the centres of the three circles so that they form an equilateral triangle

cx1<-0.7
cx2<-1.3
cx3<-1
cy1<-0.8
cy2<-0.8
cy3<-0.8+(((cx2-cx1)*3^0.5)/2) # the height of an equilateral triangle is $\sqrt{3}/2 \times$ its side

Use the function **seq** to create a finely divided vector (we will call it 'theta', a symbol commonly used in maths for angles) from zero to 2*pi (R uses radians not degrees to represent angles, and the circumference of a circle is 2 × pi × its radius, hence 360° is 2*pi radians). Then we set the radius and x and y coordinates of its centre.

Sets and Venn Diagrams

```
theta<-seq(0,2*pi,0.01)
radius<-0.5
x<-radius*cos(theta)  # obtain the circle's x and y coordinates using trig
    functions
y<-radius*sin(theta)
lines(x+cx1,y+cy1,lwd=3,col="blue")
lines(x+cx3,y+cy3,lwd=3,col="darkgreen")
lines(x+cx2,y+cy2,lwd=3,col="red")
text(0.7,0.22,"Kaeng Krachan N.P.",col="blue")
text(0.45,1.8,"Khao Yai N.P.",col="darkgreen")
text(cy3,0.22,"Doi Inthanon N.P.",col="red")
```

Woodpecker overlap between three Thai N.P.s

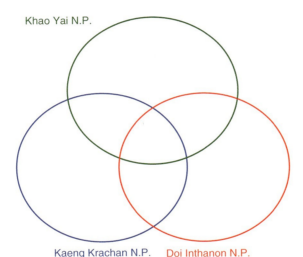

Information Box 8.2

Using symbols for circles. Here we used trigonometry to draw our circles, however, for circles, squares, rectangles, stars, thermometers and boxplots there is also a special R function **symbols** that can be used to plot them. Use ?symbols to see the options. Try:
symbols(0.7,0.8,circles=10,fg="blue",cex=20,lwd=4)
Alternatively we could use the available circle symbols (pch=1, pch=16, pch=19, pch=21) and make them as large as we want with **cex**. Using **pch** has the advantage that you just specify coordinates and size (in **plot** or **points**). The circular symbol pch=21 has the added advantage that you can control the colour of the border and fill separately, e.g. pch=21, col="red", bg="blue". The same applies to pch 22 to 27.

Now we need to find the numbers to go into the seven regions in the diagram. The first number we need to calculate is the one in the middle, the number of species occurring in all three National Parks. It would be nice to get the intersection of all three sets but R's **intersect** function only allows two vectors to be passed to it. To overcome this a simple method is to nest two or more intersect functions, e.g.

all<-intersect(intersect(kaeng,khao),doi)
text(1,1.04,length(all))

The value 1.04 is a by-eye estimate, if we want to place the number in the precise centre, the y value should be $0.8 + (0.5 + ((cx2 - cx1)*(3^{0.5})/2))/2 = 1.309808$.

Now that we know the number in the middle we can calculate those shared by pairs of National Parks but not by all three. The three values are

kk_ky_only<-length(intersect(kaeng,khao))-length(all)
kk_di_only<-length(intersect(kaeng,doi))-length(all)
di_ky_only<-length(intersect(doi,khao))-length(all)
text(1,0.65,kk_di_only)
text(0.75,1.1,kk_ky_only)
text(1.25,1.1,di_ky_only)

Finally, the numbers of woodpeckers that are unique to each park.

kk<-length(kaeng)-length(all)-kk_ky_only-kk_di_only
ky<-length(khao)-length(all)-di_ky_only-kk_ky_only

Woodpecker overlap between three Thai N.P.s

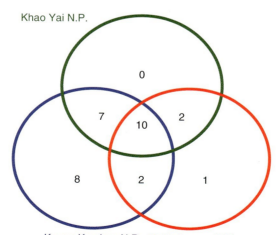

Sets and Venn Diagrams

```
di<-length(doi)-length(all)-di_ky_only-kk_di_only
text(0.6,0.65,kk)
text(1.4,0.65,di)
text(1,1.4,ky)
```

> **Information Box 8.3**
>
> *Drawing ellipses*. To draw an ellipse with the long axis either vertical or horizontal you have to divide or multiply the value of x or y in the circle drawing function by a constant. To create one at any other angle is slightly trickier. There is a CRAN package **ellipse** and ellipse-drawing is also available in the packages **car** and **plotrix**.

> **Exercise 8.1**
>
> 1) Create code to make a Venn diagram comprising four intersecting circles arranged in a square, There will be only 13 separate areas of intersection because it is not possible to create all 15 regions of overlap using circles. However it is possible if you use ellipses and draw them at angles.
> 2) Use the function **draw.ellipse** in the package CRAN **plotrix** to create a four-set Venn diagram with all 15 regions. It should look something like the figure below. To get you started create a blank plot window with xlim and ylim 0 to 10. Then use 'draw.ellipse(5,5,a=1,b=3,angle=220,deg=TRUE,border="red",col=NA)' to get a red border ellipse with no colour fill.
> 3) Fill the ellipses with partially transparent colours. Hint: you need to specify colour using the rgbt system (see Chapter 20) and include an argument 'alpha=' to specify the level of transparency (e.g. col=rgb(.9,0,0,alpha=0.3)).

Tool Box 8.1

- Use intersect(), union() and setdiff() to carryout set operations on pairs of sets.
- Use names= in **vennDiagram** to label the sets.
- Use seq() to create list of values.
- Use strsplit() to divide a string at every occurrence of a given substring.
- Use structure() to apply additional attributes to an object.
- Use unlist() to convert a list object into a vector of all its components.
- Use vennCounts() to create an object from a presence absence dataframe.
- Use vennDiagram() to plot a vennCounts object.

9 Statistics: Choosing the Right Test

Summary of R Packages Introduced
  DescTools
  ggpubr
  MASS

Summary of R Functions Introduced
  aov
  bartlett.test
  chisq.test
  cor.test
  fisher.test
  fligner.test
  ggqqplot in "ggpubr"
  glm
  glm.nb(Y ~ X) in "MASS"
  GTest in "DescTools"
  kruskal.test
  ks.test
  lm
  shapiro.test(X)
  Surv in "survival" (see Chapter 19)
  t.test
  var.test
  wilcox.test

Whilst this is not a book on statistics *per se*, one of the things R is most used for is doing statistical analyses. There are many sorts of statistical analyses and choosing the most appropriate ones can be very confusing for students. Here we will outline the basic sorts of tests and analyses. Our readers should consult additional sources before finalizing analyses and trying to publish their results. We recommend Crawley (2006, especially pp. 322–324), Dalgaard (2006), Hector (2016), McKillup (2011), McDonald (2014) and van Emden (2019), several of which are based around R. For the slightly more mathematically minded there is also Heumann *et al.* (2016)

Scientists do two principal activities. They collect data and notice patterns (e.g. correlations), and they perform experiments (or collect other data) to test their hypotheses. They may also develop hypotheses *a priori* and then carry out experiments or collect data to test whether the predictions of their hypotheses are corroborated. In testing hypotheses, the *null hypothesis* is that there is no effect, i.e. one factor does not influence the value of another. The aim of the experiment or collection of additional data is to test whether there is evidence that the null hypothesis is incorrect, i.e. there is a real association of the features.

The estimate of the probability of a given result is called its *p*-value. In biology, it is universally agreed that if the *p*-value of the observed data having resulted from there being no relationship between variables is less than one in 20 ($p < 0.05$ or $p < 5\%$) then the null hypothesis (H_0) can be rejected. In other words, it seems likely that there is an effect and we say it is significant at the 0.05 level. In high energy physics, which deals with mind-boggling numbers of observations, the standard for accepting a discovery as being real is $p < 0.0000003$. It should be emphasized that even if there is no relationship between variables, at $p = 0.05$, approximately one in 20 independent tests of the association will suggest that there is; this is a Type I error or 'false positive'. Likewise, sometimes analysing sample data will suggest that there is no significant effect when in reality there is; this is a Type II error or 'false negative'. The lower the *p*-value the more likely that the effect is real.

Emphasis on the importance of the *p*-value can lead one to lose sight of the importance of effect size. It might be that the mean number of planktonic crustaceans per 100 ml of sea water at one site is 19,000 and at another is 19,200, and such a difference might be found to be statistically significant with a sufficiently large number of independent samples BUT the difference is only about one individual per 100 and certainly no reason to think that one site is usefully more productive than another.

You should also be careful interpreting correlative evidence because even very strong correlations do not necessarily imply causality. One feature may have no influence whatsoever on the other, instead the observed correlated pattern may be due to both being driven by a third thing. To illustrate this, we will plot the number of plant genera (standardized for area) of South-east Asian countries (data from Marsh *et al.* 2009; see Chapter 4) against the human footprint values of the areas.

```
number_of_genera_per_area<-c(943,598,622,812,779,291,1097,
    1061,841,980,438,1307,802,1354,1026,1062,850,877,612,844,861)
    human_footprint<-c(34.90,21.95,20.74,19.91,22.58,0.62,24.60,28.12,20.27,
    33.66,3.48, 41.05,31.53,26.24,23.88,33.79,30.68,28.99,26.00,16.72,29.53)
plot(human_footprint,number_of_genera_per_area,xlab="Mean human
    footprint value", ylab="Number of genera standardized by
    area",pch=15)
```

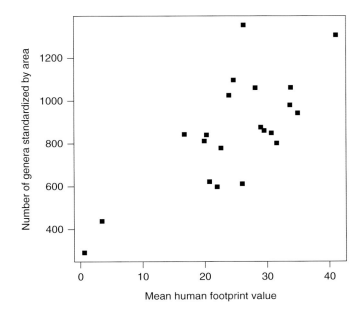

A clear case of correlation NOT causation. Both human population and its impact and the numbers of plant genera are driven by other things such as area (potential productivity and resources) and whether on islands or the mainland. More human impact does not lead to greater diversity, quite the opposite!

Explanatory and Response Variables, Experiments and Surveys

In testing hypotheses, you have two classes of variable (Crawley, 2002). You are trying to interpret what factors influence or do not influence the value of the *response variable*. The factors that you suspect influence it and measure and enter into the test or analysis are then called *explanatory variables*, i.e. they potentially explain some part of the observed variation in the *response variable*. You may have heard the response variable referred to as the dependent variable, and the explanatory variables referred to as the independent variables, however, we will stick to the terms response and explanatory in this book.

For example, you might want to determine what minerals influence fruit set in a plant. Fruit set is your response variable and the concentrations of different minerals in the soil such as potassium (K), phosphorus (P), nitrogen (N) or boron (B) are your explanatory variables.

Your data may come from experiments (especially if your plant species is small and reaches flowering stage quickly) in which case you will have more power to test the effects because you can keep all other variables constant and, in different replicates, vary the amount of just one of the given minerals, holding the others constant (presumably at some intermediate level). You would have many replicates for each treatment so that you will minimize noise due to inter-plant variation, and in a practical setting, you would randomize or stratify the layout of your experimental plants in the field or greenhouse so that you didn't put all replicates of one treatment in a shadier place, for example. This would be ideal.

On the other hand, your plant of interest might be a large tree species that takes years to reach maturity, etc. and even though it is theoretically possible to carry out an experiment, there is virtually no chance of getting the funding, land, etc., and you would be old before the first one flowered. More likely here is that you might be aware that fruit set is markedly higher in some places where the tree occurs than in others – perhaps with no obvious trend in latitude. You might therefore suspect that the variation might be affected by soil type and so you need to sample soil and measure features of its composition at many sites and at the same sites, measure fruit set (maybe mature fruit per flower or inflorescence, fruit per branch...). In this case you have no control over the soil chemistry and the variation in mineral content at each site. As with the replicates in the experimental regime, you would take replicate soil samples at each site to get a more accurate overall estimate of the site's soil chemistry, and you would measure fruit set on many trees.

Parametric versus Non-parametric Tests

Before computers became readily accessible, scientists generally were very limited in the complexity and appropriateness of the statistical tests that they could carry out and in many cases were obliged to use methods that required only relatively simple calculations. An incredibly important factor here is that the most powerful tests are based on the assumption that the departures of the data (called errors) from the best-fitted model (e.g. a slope in a regression) are normally distributed (linear models) or some other known distribution (generalized linear models). If the errors are normally distributed, or at least do not depart significantly from a normal distribution (see Chapter 30) then the appropriate statistics are called parametric – that is, they allow parameters of the normal distribution to be estimated.

This book concerns parametric tests almost entirely, because the power of modern computing and graphics allows us to examine the errors from the fitted models to determine whether they are normally distributed and independent of fitted values. If they are not, we can attempt to transform the variables such that the errors become normally distributed, after which parametric statistics can be applied. We will only normally use non-parametric tests if error normalization is really not possible (outliers we do not wish to ignore, censored data when all values above or below some limit are not recorded or are lumped together, or the data contains many zeros).

Table 9.1 presents a brief synopsis of choosing what analysis method to use; Dytham (2011) provides a far more detailed 60-couplet key to this topic. Also see below about the difference between linear models (**lm**, **aov**) and generalized linear models (**glm**). X and Y mean your data and G means groups.

Difference between Linear Models and Generalized Linear Models

Linear models fit the equation

$Y = a * X + errors$

Table 9.1. Summary of appropriate statistics and models for various types of question and data.

Question	Parametric (P) non-parametric (N)	Conditions and comments	Statistic to use	R function
All data are numeric and continuous or nearly so				
Tests of normality	P	—	Shapiro-Wilk's method (recommended)	**shapiro.test**(X)
	N	—	one sample Kolmogorov-Smirnov test (less powerful)	**ks.test**(X, "pnorm", mean=mean(X), sd=sd(X))
	—	—	visually	**ggqqplot**(X) in **ggpubr** library
Compare variances of two numeric samples	P	—	F-test based	**var.test**(X1,X2)
Homogeneity of variance	P	useful with ANOVA; best if data are normally distributed	Bartlett test	**bartlett.test**(X, {factor})
	N	not so sensitive to outliers and non-normality	Fligner-Killeen test based on medians	**fligner.test**(X1,X2) or **fligner.test**(X1 ~ X2)
Compare means of two numeric samples	P	errors normal, s.d. similar	Student's t-test, independent or paired	**t.test**(X1,X2) **t.test**(X1,X2, paired=TRUE)
	P	variances unequal	Welch's test	**t.test**(X1,X2, var.equal=FALSE)
	N	data far from normal; sample size > 14	Mann-Whitney U test, independent or paired	**wilcox.test**(X1,X2) **wilcox.test**(X1,X2, paired=TRUE)
Compare means of three or more numeric samples	P	errors can be normalized	ANOVA	**aov**(Y ~ G1, G2 ...)
	N	errors cannot be normalized; sample size > 11	Kruskal-Wallis H-test	**kruskal.test**(Y ~ G1, G2...)
Are two numeric variables correlated?	P		Pearson correlation	**cor.test**(Y, X, method="Pearson")

Continued

Table 9.1. Continued.

Question	Parametric (P) non-parametric (N)	Conditions and comments	Statistic to use	R function
	N		Spearman rank correlation	**cor.test**(Y,X, method="Spearman")
	P	scatter plots and regression analysis	linear model	**lm**(Y ~ X)
All data are counts				
Comparing proportions	N		χ^2 tests (Chi-squared tests), contingency tables	**chisq.test**()
	N	can be used when there are low expected values	G test	**GTest**() in **DescTools** library
	N	can be used when there are low expected values; gives exact probability	Fisher's exact test	**fisher.test**()
*The explanatory variables are **ALL** categorical **AND** the response variable is continuous*				
	P		Analysis of variance with explanatory variables treated as additive, interacting or nested (see Chapter 11)	**lm**(Y ~ X1 X2 X3)
*The explanatory variables include both continuous **AND** categorical variables*				
	P		Analysis of covariance (ANCOVA)	**glm**(Y ~ X1 X2 X3)
The response variable is binary (i.e. all or nothing, presence/absence, male/female)				
	P		logistic analysis (see Chapter 14)	**glm**(Y ~ X, binomial)
The response variable is a proportion (e.g. sex ratio) with unknown N				
	P		analysis of variance with arcsine transformation (see Chapter 13)	**glm**(**asin**(**sqrt**(Y)) ~ X, binomial)

Continued

Statistics: Choosing the Right Test

Table 9.1. Continued.

Question	Parametric (P) non-parametric (N)	Conditions and comments	Statistic to use	R function
	P		**glm** with binomial errors	**glm**(Y ~ X, binomial)
	P		logit transformation (see Chapter 15)	**aov**(**log**(Y/(1-Y)) ~ X)
The response variable is time of death or time of some other event				
	P		survival analysis (see Chapter 19)	**Surv**() in **survival** library
The response variable is count data				
	P		glm with Poisson errors (see Chapter 12)	**glm**(Y ~ X, poisson)
	P		log-linear regression (**lm** with the response variable logged)	lm(log(Y) ~ X)
	P		negative binomial regression	**glm.nb**(Y ~ X) in **MASS** library

Where the errors are normally (often called Gaussian) distributed. If the errors are not normally distributed we have to apply linear transformations to the variables (either the X or the Y or both) to normalize the errors. Typical transformations are taking the logarithm, the square root or a power.

Generalized linear models fit

$$f(Y) = a * X + errors$$

In generalized linear models we have a choice of several error distributions (called families) and in some cases a choice of how the Y variable is transformed; this is called the link function (f in the above equation). These **glm** families and link options are given in Table 9.2. Linear models are a special case of a glm when the link basically is just $1 \times Y$ and the errors are normal with mean zero and variance = sample standard deviation squared. In the following Table 9.2 mu means the sample mean. We will see some but not all of these used in the following chapters. In addition, negative binomial errors often used for overdispersed count data can be accommodated using the function **glm.nb** in the **MASS** library.

The quasi families fit an extra variance parameter, which means they can often deal with under- or over-dispersion. In practice they scale the variance based on Chi-squared but they complicate interpretation of the model.

Table 9.2. Error distribution families and link function options.

Family	Default link function	Possible alternative link functions (and variance)
binomial	"logit"	"probit", "cauchit", "log", "cloglog"
gaussian	"identity"	"inverse", "log"
Gamma	"inverse"	"identity", "log"
inverse.gaussian	"1/mu^2"	"inverse", "identity", "log"
Poisson	"log"	"inverse", "sqrt"
quasi	link="identity", variance="constant"	link options are: "logit", "probit", "cloglog", "log","inverse", "1/mu^2", "sqrt" variance options are: mu, mu(1-mu), mu^2, mu^3
quasibinomial	"logit"	
quasipoisson	"log"	

Our Basic Aim Is to Achieve a Near-linear QQ Plot and Even Variance

Our aim is always to get the points in the QQ plot (see Chapter 11) to form as straight a line as possible, and to try to get the variance as uncorrelated with the explanatory variables as possible (homogeneity of variance). The base R package **stats** includes the function **bartlett.test** and the **car** library includes the function **leveneTest**; both are statistical tests for homogeneity of variance. Below you will see that despite the guidelines above about how different forms of data should be analysed, following those rules does not always give a straight line QQ plot, in which case we need to try something else.

10 Commonly Used Measures and Statistical Tests

Summary of R Packages Introduced
   Deducer
   DescTools
   MASS – usually part of the installed base R
   moments
   nortest

Summary of R Functions Introduced
   ad.test in "nortest"
   binom.test
   chisq.test
   fisher.test
   GTest in "DescTools"
   hist
   ks.test
   kurtosis in "moments"
   shapiro.test
   skewness in "moments"
   t.test
   var.test
   wilcox.test

There are a number of statistical tests that are frequently used, even by non-specialists. Here we will cover tests such as Chi-squared, Fisher's exact test, Mann-Whitney U and several variations of the Student's t-test, amongst others.

Normality, Skew and Kurtosis

When we are dealing with continuous data drawn from a single population we often find a humped or bell-shaped distribution – there is a strong indication of centrality. Simple measures of such central tendency are the mean (i.e. the arithmetic average), the median (middlemost measurement) and mode (the most commonly recorded measurement). Of course sometimes we do not expect a single hump, for example, if we plot the distribution of body mass or some other measure of size, for a species, we could easily recover a bimodal distribution, often because of size differences between their sexes or annual bursts in growth.

Very often the humped distribution with large numbers of data points tends to approximate a particular curve, the normal distribution, which has enormous importance for the statistical tests we are going to concentrate upon. Critically, many powerful tests rely not on the original data themselves being normally distributed, but on the errors (deviations from our best fits) to the data, being normally distributed. When we can be confident that we can treat the errors as following a well-defined distribution (usually normal) we call the relevant statistical tests parametric – because we base them around estimated parameters of the underlying distribution of errors. If we cannot arrange or transform our data so that the errors fit to a known and parameterizable distribution, we may have to resort to a broader class of tests referred to as non-parametric. These non-parametric tests do not make assumptions based on estimates of parameters and usually rely on relative ranks of the data points.

The most common deviations from normality are skew (positive skew when the tail is extended to the right and negative when the tail is extended to the left) and narrowing or widening of the bell shape compared to a normal curve (called positive and negative kurtosis respectively). The skew of a normal distribution is 0 and its kurtosis is 3. There are no built-in R functions for calculating these but the package **moments** includes the relevant functions **skewness** and **kurtosis**.

We can always apply non-parametric tests, and there is nothing wrong with that, but if we are justified in applying parametric ones, our parametric tests are probably more powerful (more likely to indicate significant differences if there are any).

Testing Whether Proportions Agree with Null Expectations

The Chi-square test (often written as χ^2, i.e. the Greek letter chi (χ) to the power two) is used to calculate probabilities of observing numbers of observations of two or more mutually exclusive categories when you know the expected proportions from the null hypothesis.

The equation is

$$\chi^2 = \Sigma \frac{(observed - expected)^2}{expected}$$

With n – 1 degrees of freedom ($d.f. = n - 1$). Degrees of freedom is the number of independent variables in the calculation that can be changed without necessarily affecting the value of the test statistic. It is normally calculated as the sample (n) size and subtracting 1, i.e. if $n = 77$, you can change 76 of them but still obtain the same final result by adjusting the 77th to compensate. The value of chi^2 increases the greater the overall discrepancy between observed and expected values – the squaring process means all the numbers added will be positive.

For example, if a die (plural is dice) has 6 faces and is not biased, then we expect each number to come top-most on average once every 6 throws. The probability of each number is therefore 1/6 or 0.16667. If you throw it many times, say 600 times (to make calculations easier), the expected numbers of occurrences of each of the numbers 1 to 6 would be 600/6 (=100). If the observed values were 80, 105,107, 96, 90 and 122 then the value of Chi-squared will be given by

$$chi^2 = \frac{(80-100)^2}{100} + \frac{(105-100)^2}{100} + \frac{(107-100)^2}{100} + \frac{(96-100)^2}{100} + \frac{(90-100)^2}{100} + \frac{(122-100)^2}{100}$$

R will do the calculation and give the associated *p*-value using the function **chisq.test**

chisq.test(c(80,105,107,96,90,122)) # the default is that all scores are equally likely, H_0

```
Chi-squared test for given probabilities
data: c(80, 105, 107, 96, 90, 122)
X-squared = 10.74, df = 5, p-value = 0.05679
```

On this occasion, despite there appearing to be quite an excess of 6s (122), there is no significant statistical evidence at the 5% level that the die is biased.

To illustrate the Chi-squared test with a real example, we will use data from Laohapensang *et al.* (2004) on seasonal variation in hospital admissions for Buerger's disease (thromboangiitis obliterans; TAO) in northern Thailand. Buerger's disease is a subset of peripheral arterial disease (PAOD) and the number of admissions for TAO and all PAOD (including TAO) at different times of year are shown in Table 10.1.

The null hypothesis (H_0) is that the proportion of PAOD intakes with TAO does not vary with season.

TAO<-c(63,14,44)
PAOD<-c(375,235,440)
overall, TAO represents 100*121/1050
freq_TAO<-121/1050 # = 11.5% of PAOD admissions.

If the proportion of PAOD admissions that are TAO admissions is not related to season we would expect this same proportion each month and so the expected numbers of TAO each month are

expected_TAO<-freq_TAO*PAOD
expected_TAO
[1] 43.21429 27.08095 50.70476

i.e. 43, 27 and 51 patients in each season respectively. Unlike in the die throwing illustration above, we do not expect the same numbers of TAO patients in each season because our 'sample' sizes, i.e. total PAOD intake, are different.

Table 10.1. Seasonal occurrence of peripheral arterial disease (PAOD) and the subset thromboangiitis obliterans (TAO) in northern Thailand (numbers from Laohapensang *et al.*, 2004).

Months	Season	PAOD	TAO
November–February	Winter	375	63
March–May	Summer	235	14
June–October	Rainy	440	44
Totals		1050	121

Therefore we have to pass to the Chi-squared function these values too. To perform the Chi-square test with known expected values

chisq.test(TAO,p=expected_TAO,rescale.p=TRUE) # if you don't include 're-scale.p=TRUE' the probabilities must add to be precisely 1

```
Chi-squared test for given probabilities
data: observed_TAO
X-squared = 16.264, df = 2, p-value = 0.000294
```

or, since the expected values are a fixed proportion of PAOD, we could have
chisq.test(TAO,p=PAOD,rescale.p=TRUE)
or
chisq.test(TAO,p=PAOD/sum(PAOD)) # the probabilities add up precisely to 1
These all give the same result. There are two degrees of freedom ($n - 1$) because you can change any two of the observed values and still get the same overall proportion of TAO by appropriately adjusting the third – two values are free to be changed.

The Special Case of Contingency Tables

Contingency tables represent the numbers of observations for each combination of two categorical variables, though sometimes the categories may be artificially created, for example by cutting a continuous range into two at its median value using the function **median**. The simplest is the 2 × 2 contingency table with two independent variables, each with two states, and the cells of the table represent the numbers of observations in each of the four pair-wise combinations. For each cell in the table, the predicted number of observations is the proportion of total observations in the relevant row, multiplied by the proportion of total observations in the relevant column, multiplied by the total number of observations. For this reason, the number of degrees of freedom is two rather than the usual one, because you can only randomly change two of the cells out of four (d.f. = 4 – 2) without necessarily changing the value of Chi-square since you have to compensate separately for the rows and columns.

Observed values

	With X	Without X	Totals
With Y	*a*	*b*	*a + b*
Without Y	*c*	*d*	*c + d*
Totals	*a + c*	*b + d*	*a + b + c + d = n*

Expected values (under the null hypothesis)

	With X	Without X
With Y	(a + b)*(a + c)/n	(a + b)*(b*d)/n
Wthout Y	(c + d)*(a + c)/n	(c + d)*(b + d)/n

The degrees of freedom for a contingency table is (rows – 1) × (columns – 1).

Table 10.2. Infection of Korean men and women with the parasite *Clonorchis sinensis* (data from Kim *et al.*, 2009).

	Infected	Uninfected
Men	782,383	23,580,702
Women	391,841	23,828,880

An example with very large sample size is provided by Kim *et al.* (2009) from a nationwide survey of intestinal parasite infections in South Korea (Table 10.2). The human parasitic trematode, *Clonorchis sinensis*, is widespread in Asia and still actively transmitted in Korea. Here we test our null hypothesis that infection rates for men and women are equal.

We can enter the *Clonorchis* data either by creating separate vectors and binding them, e.g.

```
males<-c(782383,23580702)
females<-c(391841,23828880)
clonorchis<-rbind(males,females) # rbind sticks the rows together
clonorchis
         [,1]     [,2]
males   782383  23580702
females 391841  23828880
```

or using the function **matrix** and specifying how many rows (note that the input order in this case is by column, column 1 first, then the second column, etc.)

```
clonorchis<-matrix(c(782383,391841,23580702,23828880),nrow=2)
clonorchis
         [,1]     [,2]
males   782383  23580702
females 391841  23828880
```

To perform the Chi-squared contingency table test on these data:

```
chisq.test(clonorchis)
    Pearson's Chi-squared test with Yates' continuity
       correction
data: clonorchis
X-squared = 130775.1, df = 1, p-value < 2.2e-16
```

With such a huge sample size it is not surprising that the difference in infection rates between men and women is so highly significant ($p < 2.2 \times 10^{-16}$ or $p < 0.00000000000000022$) so we reject our null hypothesis. Men have approximately double the infection rate of women.

Hardy-Weinberg Equilibrium

Another common biological application of the Chi-square test is to determine if gene allele frequencies in a population are in Hardy-Weinberg equilibrium. If there are two alleles of a given gene locus in a population of a sexually reproducing species, we can predict the genotypes in the next generation if we know the

> **Information Box 10.1**
>
> *Odds ratio and relative risk.* Chi-squared is a statistical test of association and not a measure. Particularly in the medical world you will often see the use of two measures of the association, the odds ratio (OR) and relative risk (RR). Using the same *a, b, c, d* as in the Chi-square table above then OR is defined as
>
> $$OR = \frac{\text{odds with treatment 1}}{\text{odds with treatment 2}} = \frac{a/b}{c/d} = \frac{a \times d}{b \times c}$$
>
> Typically the rows represent different treatments or exposures, and the columns the outcome (live vs die, better vs worse). The OR then indicates the extent to which the treatment affects the outcome, with a value of 1 meaning no effect.
> The RR value is given by
>
> $$RR = \frac{a/(a+b)}{c/(c+d)} = \frac{a(c+d)}{c(a+b)}$$
>
> as with OR, an RR of 1 indicates no treatment effect.
> References: Holcomb *et al.* (2001); Porta (2014)

> **Exercise Box 10.1**
>
> 1) Use the dataset faithful in the **MASS** library. Plot the data (plot and/or histogram). Choosing an appropriate statistical analysis determine whether the duration of the eruption of the geyser Old Faithful is correlated with the time elapsed between eruptions. Hint: you can split both vectors at the median.
> 2) Write code to calculate the Odds Ratio and the Relative Risk for the sexes with the *Clonorchis* data.

frequencies of the two alleles – call the alleles a and b and their respective frequencies p and q. because there are only two alleles, their frequencies must add to one.

$p + q = 1$

Then, if all the assumptions of the Hardy-Weinberg equilibrium are satisfied the frequencies of the homozygotes and heterozygotes are

p^2, $2pq$ and q^2

We can plot these and at the same time introduce a new graphical argument.
```
p<-seq(0,1,0.001)
plot(NULL,xlim=c(0,1),ylim=c(0,1),xlab="p",ylab="Frequency")
lines(p,p^2,col="red",lwd=2)
lines(p,(1-p)^2,col="blue",lwd=2)
lines(p,2*p*(1-p),col="purple",lwd=2)
text(0.2,0.95,expression('q'^2*' Propn(aa)'),col="blue") # use expression
     and place text to be superscripted between ^and*
text(0.8,0.95,"q^2 Prop(AA)",col="red")
```

```
text(0.5,0.68,"2pq Prop(Aa)",col="purple")
    arrows(0.2,0.93,0.13,0.7569,code=2,col="blue",length=0.1) # 0.7569 is
        our x value^2
arrows(0.8,0.93,0.87,0.7569,code=2,col="red",length=0.1)
arrows(0.5,0.66,0.5,0.5,code=2,col="purple",length=0.1)
```

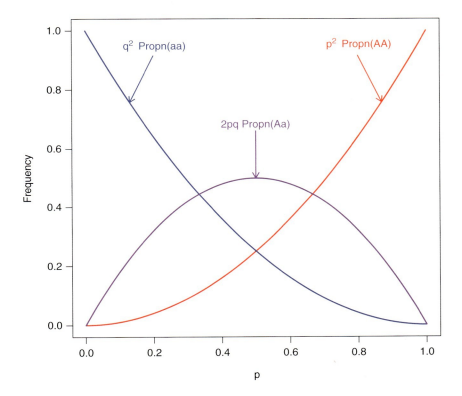

The assumptions of Hardy-Weinberg are: no genetic mutation, random mating, no gene flow into or out of the population, infinite population size(!) and no selection. Most of these are almost certainly violated in any real population. A significant result indicates that the alleles in the population are not at Hardy-Weinberg equilibrium and significantly violate at least one of the above assumptions in some way.

As a simple classic example we will use the data of the famous ecological geneticist Professor E.B. Ford of Oxford University. Ford and co-workers studied a population of the scarlet tiger moth, *Panaxia dominula* (Lepidoptera: Erebidae: Arctiinae) over many years (see Manly, 1985, pp. 220–221). Three morphotypes called dominula, medionigra and bimacula can be recognized in the field and these correspond to three genotypes (AA, Aa and aa respectively). The numbers of individuals of the three types in the year 1957 for example were:

1469 white spotted (AA)
138 medium number of spots (Aa)
5 few spots (aa)

From these numbers we can calculate the frequencies of the two alleles. Each AA individual has two copies of A and each Aa individual has one copy of A The total alleles (2 per individual) is 2*(1469 + 138 + 5)

obs<-c(1469,138,5)
Tot<-2*(1469+138+5)
fA<-(2*1469+138)/Tot

ditto for the a allele

fa<-(2*5+138)/Tot

fa+fA # to check that the allele frequencies do add up to 1
[1] 1

our expected proportions are therefore:

expected<-c(fA^2,2*fA*fa,fa^2) # these are expected phenotype proportions

We can now test whether the observed moth allele frequencies were in Hardy-Weinberg equilibrium; we use 'p=' to pass expected phenotype frequencies

chisq.test(obs,p=expected)
```
Chi-squared test for given probabilities
data: obs
X-squared = 0.83095, df = 2, p-value = 0.66
```

As would appear to be the case by eye, the observed allele frequencies do not differ significantly ($p > 0.05$) from null expectations; the moth colour genotypes are in Hardy-Weinberg equilibrium.

Information Box 10.2

Care with Chi-square proportions. Be very careful that you distinguish passing expected proportions from observed numbers to **chisq.test**. If you carry out chisq.test(obs,expected) the test will treat the expecteds as real observations and you will get $p = 0.1991$. It is true that in this example the p-value is also not significant, but it is closer to significance than the correct value. It is easy to imagine that with slightly different observed values you might wrongly get a significant result.

Alternatives to the Chi-squared Test under Some Circumstances

Instead of the Chi-squared test, some researchers use a closely related alternative based on likelihoods called the G-test (see Appendix 3 for equation). The G-test is slightly more accurate, but it is not included in the R base package. It is available in the **DescTools** library and as the function **likelihood.test** in the **Deducer** package.

install.packages("DescTools")
library(DescTools)

Applying the G-test to the *Clonorchis* data that we analysed above with **chisq.test** we get a very similar answer. Generally when sample sizes are large (> 1000) the G-test is probably better. The G-test is also more flexible and can be used for testing more complicated hypotheses (see McDonald, 2014, pp. 68–76).

```
GTest(clonorchis)
Log likelihood ratio (G-test) test of independence without
    correction
data: clonorchis
G = 133280, X-squared df = 1, p-value < 2.2e-16
```

Fisher's exact test – a note of caution for small expected values

When the expected number of observations in one or more cells of a contingency table falls below five, the results of Chi-squared tests are unreliable. Fisher's exact test should then be used to calculate the exact probability of the observed values. To perform this test use the function **fisher.test** in exactly the same way as we used **chisq.test** above. If you use the function **chisq.test** on a contingency table with a low expected value you will get something like

```
Warning message:
In chisq.test(rbind(a, b)) : Chi-squared approximation may be
    incorrect
```

Testing Whether Two Means Are Significantly Different

The t-test (or Student's t test) is widely employed. There are three basic ways it can be used:

- The mean of a single sample can be tested against some otherwise known value.
- The means of two independent samples can be compared.
- As above but with paired data.

See Appendix 3 for the relevant equations.

The basic assumptions of the t-test are that the values of the sample or samples are normally distributed AND when two samples are being compared, they have the same variances. Student's original t-test still gives quite accurate probabilities even if the two variances differ providing that sample sizes are the same or nearly the same. A variant of Student's t-test by Welch (Welch's t-test) does not require equal variances irrespective of sample sizes. Importantly, the function **t.test** in R automatically does Welch's t-test. Further, for 'moderately large' samples, commonly stated as being greater than 40 values, the probability given by the t-test is relatively accurate even if the samples are not normally distributed.

Single-sample t-test

In a single sample (or one sample) t-test you are asking whether the mean of a set of observations differs from an already known value (see Appendix 2). This is very seldom applicable to biology because there are very few things for which you know what the mean ought to be.

Two-sample t-test

In R we need two vectors of numerical data or they can be columns within a table. Here we will use data from Sombatboon's (2014) study of captive bullfrogs, comparing the weights of males and females during the winter. Note that weight is a continuous measure.

frog_males<-c(395,390,410,390,395,455,400,370,460,490)
frog_females<-c(335,360,315,335,405,265,315,390,405,380)
t.test(frog_males,frog_females)
```
Welch Two Sample t-test
data: frog_males and frog_females
t = 3.4287, df = 17.546, p-value = 0.003085
alternative hypothesis: true difference in means is not
    equal to 0
95 percent confidence interval:
25.09774 104.90226
sample estimates:
mean of x mean of y
415.5 350.5
```
We see that the t statistic is 3.43 with 17.54 degrees of freedom, which indicates that, at that time of year, males are significantly heavier than females, and we can see this visually by plotting the data

boxplot(cbind(frog_males,frog_females),col="greenyellow")

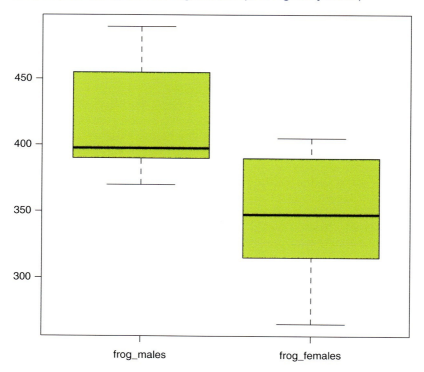

Paired t-test

Sometimes samples are paired. Some examples might be:

- In medical studies they may be before and after treatment results for a given patient.
- In agriculture the pair of values could be fruit weight on fertilized and unfertilized plants with each pair coming from a different farm.
- In ecology they could be beetles trapped in pitfall traps with or without detergent in the water, with each pair of pitfalls closely spaced and so likely to be sampling approximately the same density of beetles, but with different pairs of traps widely spaced.

In each case, the amount of experimental, individual, environmental variation within a pair of samples is likely to be markedly less than between pairs where other factors may predominate. Therefore, if data are paired, using a paired t-test eliminates the effect of inter-pair variation.

Here as an example we use data on the consistency with which Akha herbalists in northern Thailand and in southern China report the use of herbs for treating nine types of medical disorder (data from Inta *et al.* 2008).

```
Chinese<-c(0.875,0.781,0.75,0.677,0.655,0.5,0.4,0.333,0.167)
Thai<-c(0.818,0.435,0,0.771,0.829,0.5,0.677,0.429,0.6)
herbalists<-rbind(Chinese,Thai)
t.test(Chinese,Thai,paired=TRUE)
     Paired t-test
data: Chinese and Thai
t = 0.074582, df = 8, p-value = 0.9424
alternative hypothesis: true difference in means is not
   equal to 0
95 percent confidence interval:
-0.2626227 0.2801782
sample estimates:
mean of the differences
            0.008777778
```

The result is non-significant ($p = 0.94$) showing that there is no difference between Chinese and Thai herbalists in the consistency with which various plant treatments are recommended for ailments.

Testing Whether Three or More Means Differ from One Another

This is an analysis of variance problem which uses the function **aov** (see Chapter 13). The idea is that we have samples from three or more populations and our null hypothesis is that all the populations have the same mean. You could perform many pairwise t-tests but **aov** takes care of everything so you only have to do one analysis. The assumptions of an ANOVA are that the samples are normally distributed, independent of one another and have similar variance. Since we are only considering one explanatory factor with three or more levels, then this is a one-way ANOVA.

Comparing Two Variances

We use the F-test to determine whether the variances of two samples differ significantly from one another. The relevant R function is **var.test**. Using the same data as above we can use this to test whether there is any significance in the variance of the herbalist consistency scores

var.test(Chinese,Thai)
```
F test to compare two variances
data: Chinese and Thai
F = 0.80373, num df = 8, denom df = 8, p-value = 0.7648
alternative hypothesis: true ratio of variances is not
    equal to 1
95 percent confidence interval:
0.1812953 3.5631407
sample estimates:
ratio of variances
         0.8037293
```
and again, there is no significant difference.

Non-normally Distributed Data with Small Sample Sizes – Mann-Whitney U test

We will use data from Offenberg et al. (2004) on leaf damage to mangrove trees caused by leaf beetles and crabs (see Chapter 12 for further analysis of their dataset). Percentage leaf damage caused by beetles, which was always rather low, was measured for 10 mangrove trees that had no resident tree ants, and 18 trees which had at last one nest.

without_ants_beetle_damage <-c(0.44,0.49,0.59,0.70,0.73,0.94,1.03,1.09, 1.42,2.10)

with_ants_beetle_damage<-c(0.19,0.39,0.03,0.04,0.19,0.36,0.42,0.20,0.21, 0.09,0.05,0.16,0.07,0.05,0.06,0.06,0.01,0.14)

boxplot(without_ants_beetle_damage,with_ants_beetle_damage,ylab= "Percent leaf damage",names=c("without","with"),col="olivedrab")

text(1.5,-0.45,"Presence of tree ant nest(s)",cex=1.1,xpd=NA)

xpd=NA is included because we want to write a text label outside of the plotting area

Commonly Used Measures and Statistical Tests

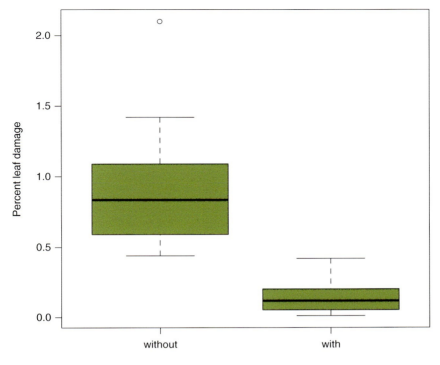

To explore whether the data are normally distributed, first we will plot histograms of the two sets of measures using the function **hist**. It should be emphasized that histograms are tricky things and what you get graphically depends greatly on the number of bins you sort your data into and on the exact break points chosen. It is worth mentioning another histogram function **truehist** in the MASS package that provides a number of extra options.

par(mfrow=c(2,2))
hist(without_ants_beetle_damage)
hist(with_ants_beetle_damage)

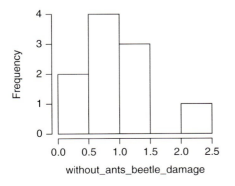

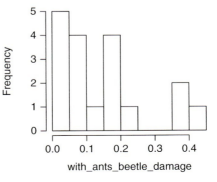

Extent of leaf damage on trees without ant nests is close to normal, certainly a humped distribution with a hint of skew, though the sample size (10) is rather too small to be certain. Leaf damage on trees with ant nests is very decidedly non-normal – it may be a highly skewed normal distribution but our data aren't sufficient to tell. There is also a more formal R function, the Shapiro-Wilk normality test, which is carried out using the function **shapiro.test**.
shapiro.test (with_ants_beetle_damage)

```
    Shapiro-Wilk normality test

data: with_ants_beetle_damage
W = 0.8583, p-value = 0.01148
```

And indeed when we apply it to the 'with ant nests' data we see that the probability of the data having been drawn from a normal distribution is < 0.05, so we can also statistically reject the idea that they are normally distributed. With small samples like these, however, by-eye judgement may be all you need.

Another widely used (and more versatile but less sensitive test is the Kolmogorov-Smirnov test, which compares the sample's cumulative distribution with a user-specified one, in the case of testing for normality we include the argument **pnorm**. Given a large sample size it is sensitive to differences in mean, variance, skew and kurtosis.
ks.test(with_ants_beetle_damage,"pnorm")

```
    One-sample Kolmogorov-Smirnov test

data: with_ants_beetle_damage
D = 0.50399, p-value = 0.0002137
alternative hypothesis: two-sided

Warning message:
In  ks.test  (with_ants_beetle_damage,  "pnorm")  :  ties
    should not be present for the Kolmogorov-Smirnov test
```

You will note that while both tests reject the null hypothesis of normality at the 0.05 level, they differ markedly in the estimated *p*-value. There are other tests for normality available in R packages such as the Anderson-Darling test (**ad.test**) in the **nortest** (normality test) package, and these sometimes give different results because of differing sensitivities to extreme values and outliers – Shapiro test is probably adequate for most biological situations.
install.packages("nortest")
library(nortest)
ad.test(with_ants_beetle_damage)

```
    Anderson-Darling normality test
data: with_ants_beetle_damage
A = 0.96314, p-value = 0.01164
```

and here we can see that it agrees closely with the Shapiro test.

Non-parametric Two-sample Tests

The most familiar, non-parametric alternative to the t-test is the Mann-Whitney U test (also called Wilcoxon rank-sum test; they are identical) and this is implemented in R as **wilcox.test**(y,x) – use '?wilcox.test' to see details in R.

It is strongly recommended to use **t.test** as long as your data satisfy the above criteria. We can use this test if our sample sizes are less than 40 and all we want to know is whether the presence of ant nests has a significant effect on leaf damage. Applying this to the weaver ant data:

wilcox.test(without_ants_beetle_damage,with_ants_beetle_damage)

```
Wilcoxon rank sum test with continuity correction

data: without_ants_beetle_damage and with_ants_beetle_damage
W = 180, p-value = 1.763e-05
alternative hypothesis: true location shift is not equal
   to 0
Warning message:
In wilcox.test.default(without_ants_beetle_damage, with_
   ants_beetle_damage) : cannot compute exact p-value
   with ties
```

The test result is clear, there is highly significantly less leaf damage due to leaf beetles on trees that support tree ant colonies. The **wilcox.test**(x,y) performs the Mann-Whitney U test if both x and y are numeric; technically you can specify this by adding 'paired=FALSE' (cf. Wilcoxon Signed Rank Test below); **wilcox.test**(x~y) performs the Mann-Whitney U test if x is numeric and y is a binary factor; **wilcox.test**(x,y,paired=TRUE) performs the Wilcoxon Signed Rank Test if both x and y are numeric.

If there is more than one factor level, the **kruskal.test**(x~y) performs the Kruskal-Wallis rank sum test, sometimes called one-way ANOVA on ranks, if x is numeric and y is a factor.

The Kolmogorov-Smirnov test that we demonstrated above for comparing a distribution against a prediction of normality can also be used to compare the distributions of two samples, and given large sample sizes, can detect differences not just in mean but also in skewness and kurtosis.

```
    Two-sample Kolmogorov-Smirnov test
data: without_ants_beetle_damage and with_ants_beetle_
   damage
D = 1, p-value = 5.215e-06
alternative hypothesis: two-sided

Warning message:
In ks.test(without_ants_beetle_damage, with_ants_beetle_
   damage) : cannot compute exact p-value with ties
```

Binomial Test

The function **binom.test** gives the exact probability of observing a given number of occurrences for a given sample size and an optional probability value, the default probability is $p = 0.5$, i.e. like the probability of getting a head from the toss of a fair coin. As an example, Verme and Ozoga (1981) examined the effect of latency between oestrus and insemination in captive white-tailed deer (*Odocoileus virginianus*) (Table 10.3). Taking the first row, we use **binom.test**

Table 10.3. Latency between oestrus and insemination in captive white-tailed deer (*Odocoileus virginianus*) (numbers from Clutton-Brock and Iason, 1986).

Hours after oestrus	Sample size	Number male fawns	*p*-value
13–24	28	4	0.00018
25–36	31	12	0.281
37–48	40	25	?
49–96	26	21	?

to calculate the probability of getting four male fawns out of 28 given a 50/50 expectation for mammalian sex-determining (XX, XY chromosome) genetics.

binom.test(4,28)

```
    Exact binomial test

data: 4 and 28
number of successes = 4, number of trials = 28, p-value
    = 0.00018
alternative hypothesis: true probability of success is
    not equal to 0.5
95 percent confidence interval:
0.04033563 0.32665267
sample estimates:
probability of success
             0.1428571
```

Exercise Box 10.2

1) Calculate the last two *p*-values and think of some possible biological mechanisms and explanations (see Clutton-Brock and Iason, 1986, for discussion on mammalian sex ratios).

Tool Box 10.1

- Use Boxplot() to create a boxplot.
- Use "pnorm" in **ks.test** to set the user-defined distribution to normal.
- Use ad.test() in **nortest** to test for normality using the Anderson-Darling test.
- Use chisq.test() to calculate *p* values using Chi-squared.
- Use fisher.test() to calculate exact probabilities for small sample sizes.
- Use GTest() or likelihood.test() to calculate likelihood of expected values of observed data.
- Use hist() or truehist() to plot histograms.
- Use Kruskal.test() to carry out a Kruskal-Wallis one way ANOVA.
- Use ks.test() to test for user-defined distribution using the Kolmogorov-Smirnov test or to compare two distributions.
- Use p= in **chisq.test** to set expected probabilities.
- Use paired=TRUE in t.test to calculate the paired t-test values.
- Use rescale.p=TRUE in **chisq.test** to rescale probabilities so they add to 1.
- Use shapiro.test() to test for normal distributions with the Shapiro-Wilk test.
- Use t.test() to calculate the Welch two-sample t-test.
- Use var.test() to compare variances using the F test.
- Use wilcox.test() to carry out a Mann-Whitney U test.

11 Regression and Correlation Analyses

Summary of R Packages Introduced	exp
diptest	ggqqplot in "ggpubr"
	lm
Summary of R Functions Introduced	nls
abline	predict
abs	pt
AIC	qt
anova	summary
dip in "diptest"	

Correlation and regression analyses are used to test whether, and to what degree, variation in one continuous variable is related to variation in another continuous variable. In correlation analysis we have no control over either variable, they are just data collected, and indeed, even if two variables are strongly correlated, they may not be influencing one another but simply both being affected by a third which perhaps we did not measure. The initial assumption of the analysis is that the values of both variables are drawn from a normal distribution.

In regression analysis we are controlling one of the variables seeing whether changing its value affects the other. The variable we are controlling is the explanatory variable (sometimes called the treatment) and the other is the response variable. As we are controlling the explanatory variables they are probably going to be set at specified values or set increments and are therefore not normally distributed. There may be more than one explanatory variable. If all the explanatory variables are categorical then we call the regression an ANOVA.

In both cases we use the function **lm** (= linear model) to carry out the analysis. The function **lm** finds the slope (a) and intercept (b) of the line through the data points that minimizes the sum of the squares of the residuals from that line, along the y axis to each data point.

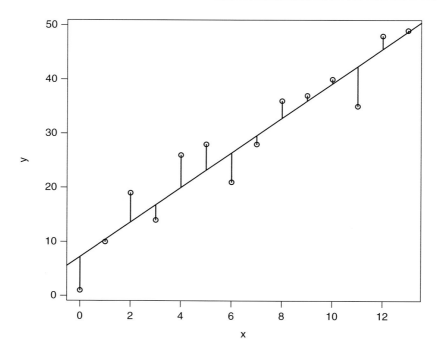

Linear versus Non-linear Regression

In statistics, regression usually is called linear regression if you can fit a straight line to the points, for example:

$$y = a + bx$$

or with multiple explanatory variables (x) all their relationships with the response variable (y) are linear

$$y = a + bx_0 + cx_1 + dx_2 + ex_3$$

The equations are also called linear even if some of the explanatory variables are raised to powers, e.g.

$$y = a + bx_0 + cx_1 + dx_2^2$$

that is because such equations are 'linear in their parameters', i.e. the effect of each parameter on y does not change with x. Thus, polynomial regression (see Chapter 11) is a linear regression – power functions are transformable to linear by taking roots. However, if an equation cannot take the form above, it is non-linear. In non-linear equations, the right-hand side is not a sum of linear functions of x (or a power of x). Non-linear equations cannot be linearized by transforming one or both of the variables, whereas linear equations can.

Biologists work with many non-linear equations, which are common in population models. For example, exponential growth functions are based on all individuals having the same reproductive rate irrespective of population size and give J-shaped curves. In contrast, logistic equations give S- (or reversed S) shaped curves perhaps because after an initial exponential increase, *per capita* growth rate starts to diminish as the population increases further, finally reaching an asymptote when the population reaches the carrying capacity. Similar non-linear equations turn up in many other areas too, such as survivorship. Exponential functions ranging from simple ones with two parameters to fit (*a* and *b*)

$$y = ae^{bx}$$

or with three parameters to fit (*a*, *b* and *c*),

$$y = a - be^{-cx}$$

We can see that the first of these is non-linear in its parameters thus: let $a = 1$ and $x = 1$, and consider only the parameter *b*, if $b = 0$, $y = 1$, if $b = 1$, $y = e = 2.718282$, and if $b = 2$, $y = e^2 = 7.389056$. These are not a linear progression with equal increments of *b*, and there is no transformation of *y*, or *x* or of both that will make the results so. See Appendix 3 for further examples of commonly used non-linear equations.

In all cases, the more parameters that have to be fitted, the larger the number of data points you need in order for them to be estimated. This will lead us on to model simplification where we ask whether including another parameter that explains a smaller amount of the variation provides a significantly better explanation than a simpler model.

Log-log Plot Example Correlation of Numbers of Species with Area

The shape of the species–area curve has been much discussed. It commonly follows a power relationship (Preston, 1962) that results in a straight line when plotted on a log-log scale

$$\log(species) = a^* \log(area) + b$$

though exponential and logistic relationships have also been advocated (see Scheiner, 2003). Island faunas have long been popular for the study of species–area relationships because their faunas are discrete, and indeed, the field has been so popular that it is widely known as 'island biogeography' (MacArthur and Wilson, 1967), although this is not quite the whole story for islands (Lomolino, 2000) with the smallest islands behaving somewhat differently. Here we will analyse the numbers of species of butterflies found on various Caribbean islands that vary considerably in land surface area (data abstracted from Scott, 1972).

```
CB<-read.csv("CarribTot.csv")
par(mfrow=c(2,2))
plot(CB)
plot(log(CB,10),xlab="log(Area)",ylab="log(Total species)")
mod_CB<-lm(log(CB$Tot_butterflies,10) ~ log(CB$Area_square_km,10))
summary(mod_CB)
```

```
Call:
lm(formula = log(CB$Tot_butterflies, 10) ~ log(CB$Area_square_km,
    10))
Residuals:
    Min       1Q   Median       3Q      Max
-0.08911 -0.07749  0.02635  0.03855  0.10454
Coefficients:
                            Estimate Std. Error t value
Pr(>|t|)
(Intercept)                  0.96546    0.06717   14.37
1.79e-08 ***
log(CB$Area_square_km, 10)   0.25189    0.01959   12.86
5.72e-08 ***
---
Signif. codes:  0 '***' 0.001 '**' 0.01 '*' 0.05 '.' 0.1 ' ' 1
Residual standard error: 0.0683 on 11 degrees of freedom
Multiple R-squared: 0.9376,    Adjusted R-squared: 0.9319
F-statistic: 165.2 on 1 and 11 DF,  p-value: 5.717e-08
abline(mod_CB,col="blue",lwd=2)
```

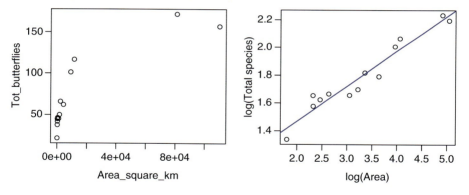

Obviously the relationship is highly significant, as well as there being a significant intercept. The equation describing the relationship is therefore:

$$\log_{10}(species) = 0.25189 * \log_{10}(area) + 0.96546$$

which implies that when $\log_{10}(area) = 0$ (i.e. the area is km²) we would expect almost exactly one species of butterfly.

Information Box 11.1

Default logarithm base in R. In R the default **log** function returns the natural logarithm, i.e. log to the base e ($e = 2.71828$ approx.). To obtain logarithms in base 10, use **log**(x,10) or in base 2 **log**(x,2), etc. However, just for bases 2 and 10, there are actually in-built specific functions **log2()** and **log10()**.

Despite this all looking very good, it would probably be a good idea at this stage to see what the model is actually predicting. The function **simulate** generates random sets of predicted response variable values for the x axis values and we

can plot these over our original data to get a visual estimate of how well the model fits our data.
a<-simulate(mod_CB,10)
plot(CB,pch=15)
for(i in 1:10) points(cbind(CB[1],10^a[i]),pch=3,col="red") # because 'modelt' was based on log10 we have to back convert by taking 10 to the power of the model values

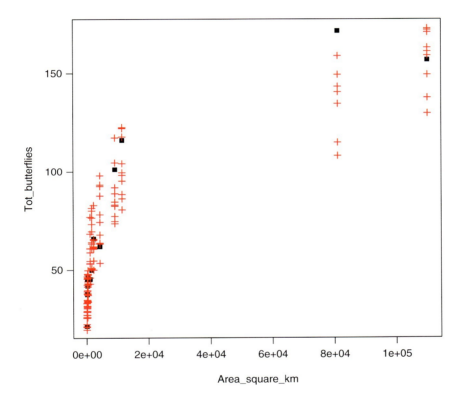

So how good do you think our 'modelt' is?

Linearizing Data with No Known Underlying Model

For a second example we will use the data of Offenberg et al. (2004) on the protection from herbivory afforded to mangroves (*Rhizophora mucronata*) by weaver ants (*Oecophylla smaragdina*) in Thailand. The numbers of weaver ant nests per tree is the explanatory variable, and two measures made for foliage damage, that caused by leaf beetles (Chrysomelidae) and that caused by crabs.
nests<-c(0,0,0,0,0,0,0,0,0,0,1,1,2,2,2,2,2,4,5,5,7,7,9,11,15,20,24,30)
total_damage_percent<-c(3.72,3.11,2.15,1.40,1.12,0.94,0.55,0.59,0.73,0.83, 0.19,0.44,0.19,0.64,0.66,0.68,1.03,0.19,0.20,0.13,0.10,0.59,0.09, 0.05,0.06,0.06,0.04,0.14)

```
leafbeetle_damage_percent<-c(0.44,0.49,0.59,0.70,0.73,0.94,1.03,
     1.09,1.42,2.10,0.19, 0.39,0.03,0.04,0.19,0.36,0.42,0.20,0.21,0.09,
     0.05,0.16,0.07,0.05,0.06,0.06,0.01,0.14)
plot(nests,total_damage_percent,xlab="Number of weaver ant nests",ylab=
     "Percentage leaf damage",pch=1)
points(nests,leafbeetle_damage_percent,pch=4)
```

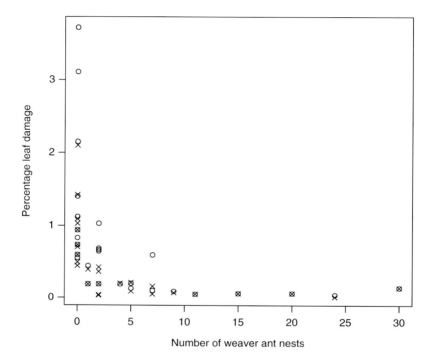

It is immediately obvious that the x and y values are not linearly related and before any regression analysis can be carried out, one or both axes will have to be transformed. This also demonstrates another common problem: the variance is related to the magnitude of the explanatory variable.

First we'll run a model and look at a number of diagnostic plots by plotting the model. The one we will be most interested in is the QQ plot (labelled in R as the normal Q-Q plot). To check normality of errors in R after the first application of a linear model, you plot the model, the model itself being an R object. This will produce four plots of which two are particularly important. R estimates the mean and variance of the errors and then plots these values against the observed ones in a QQ plot (sometimes Q-Q). If the residuals are normally distributed, the line of points will be close to straight and we can believe the stats. But if the QQ plot shows distinct curvature (concave, convex or sigmoid), the results of the statistical test will be invalid. Obviously, our data are our data so we cannot change them, but we are allowed to transform them in consistent ways to make the QQ plot acceptably straight (that's a matter of judgement and may take experience). Common ways of transforming one of the variables include rooting, logging and squaring. As long as the resulting QQ plot is acceptably straight, these transformations only improve the accuracy of your statistical inference. If you have a seriously concave QQ plot with your raw data, log the explanatory variable. If it is still quite concave, log the log. If it still isn't acceptably straight, log the log of the log. This is all OK; any combination of linear transformations is OK.

Exercise Box 11.1

1) Using the dataset 'Fitch' in the **Stat2Data** library, manipulate the size and/or weight data until you can carry out a linear regression analysis on the weight to skull length data for the two orders of mammals. Hint: to access the 'Fitch' data use **data**(Fitch).
2) Can you manipulate the weight data as a single dataset, or do you need to separate it into orders? Does it make a difference to the result? Why?

Errant Points and Leverage

The fourth of the **plot**(model) graphs shows Cook's distances of all the sample points. This index combines the amount of departure of each point from the best-fit regression 'multiplied by' the distance the point is away from the median of points. In regression analysis, deviation from the trend by points towards the extreme of the distribution have undue weight. Such points are labelled in the R output. As the great physicist Richard Feynman said, do not trust the extreme points on a graph. In his case the reason could easily be that if you could have measured more extreme values easily, you would have done so, and therefore the most extreme point in your data would likely be where the equipment was at its limits. A similar rationale applies to biology. If say, you are examining some physiological process as a function of body mass or latitude, there is good chance that those individuals or populations very close to the end of the range may behave differently.

If you have some strong outliers in the Cook's distance plot you should re-examine them. It is always possible that something like a typo has happened during data entry (maybe typing a 0 instead of a 9) or the study subject was sick at the time (a very low weight mouse might be ill or dying and therefore showing other odd physiological or behavioural features). In any case you would be well advised to carry out separate analyses with and without these influential points included. If you are lucky, the p-values and slopes will be hardly changed, but with small datasets it is always possible that including or excluding such influential points might make the difference between reporting a significant or a N.S. result. In the latter case you need to think carefully and maybe collect more independent data.

modelA<-lm(leafbeetle_damage_percent ~ nests)

If we just try to plot the model ('plot(modelA') you will be asked to hit the return key until you get the QQ or other plots (pressing the escape key terminates this). Therefore we will plot all four graphs in the same window.

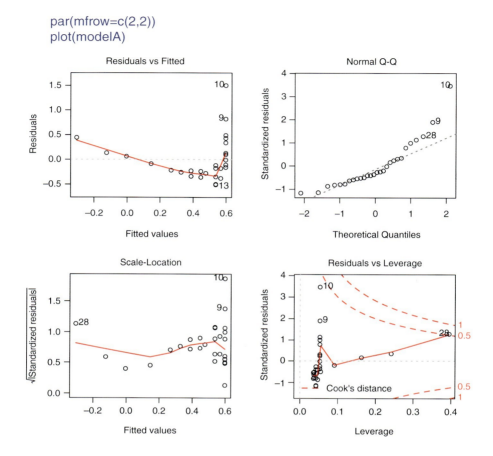

```
par(mfrow=c(2,2))
plot(modelA)
```

These four plots represent important features of the model.

- Top left shows how the residuals (differences between observed data points and the best-fit ANOVA model) are related to the fitted values – we want to see no pattern increasing or decreasing (or possibly a strongly arced pattern) because that would mean that variance was not constant. There is no pattern so that is OK.
- Top right tests whether the residuals are normally distributed plotting the observed values against theoretical values obtained from the best-fit model. We want to see a straight line. Here we have a bit of a problem because although most points are a good fit to a straight line, the data on line 10, and to a lesser extent on line 18, are some way off the line. More on this below with the logit transformation method.
- Bottom left shows the relationship between √(standardized residuals) and fitted values, we don't want to see any strong relationship – and there isn't one.
- Bottom right shows the Cook's distance of each data point and is important for identifying outliers. Cook's distance combines the distance of points

from the best-fit line (residual) and how far they are from the middle of the explanatory variable which indicates how much leverage they have on the overall fit. In this case no data points are highlighted and no points are close to the red dashed 'danger' lines.

The QQ relationship from 'modelA' is very strongly concave (in the sense that biologists use the term). In such cases the first transformation to try is logging the x variable, in this case the number of nests. Note that we have had to add 1 to the logarithm of number of nests because the logarithm of zero is –infinity and R cannot handle infinity.

modelB<-lm(leafbeetle_damage_percent ~ log(nests+1))
plot(modelB)

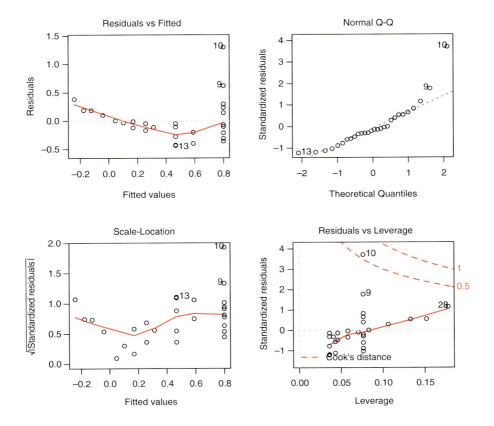

Not much better. The authors then achieved a better fit by cube-rooting (^(1/3)) the response variable though the result is still not perfect – in such cases it may be that there are two different things going on, maybe something dramatically different occurs when ant nests are present versus absent. Anyway, we will plot their model and the regression lines.

```
plot(log(nests+1),total_damage_percent^0.333,xlab="ln(weaver ant
    nests+1)",ylab= "Percentage leaf damage ^0.33",pch=1,col="red")
points(log(nests+1),leafbeetle_damage_percent^0.333,pch=4,col="green")
modelC<-lm(total_damage_percent^(1/3) ~ log(nests + 1))
summary(modelC)
```

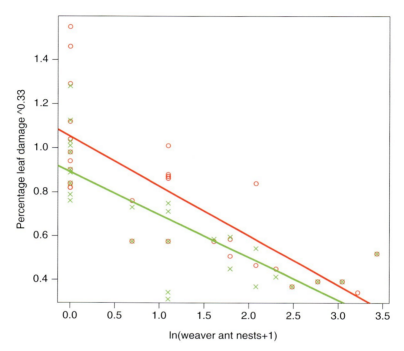

```
Call:
lm(formula = total_damage_percent^(1/3) ~ log(nests + 1))
Residuals:
     Min        1Q    Median        3Q       Max
-0.32081  -0.12548  -0.04836   0.06879   0.49707
Coefficients:
                Estimate Std. Error t value  Pr(>|t|)
(Intercept)      1.05239    0.05539   19.00   < 2e-16 ***
log(nests + 1)  -0.22605    0.03379   -6.69  4.25e-07 ***
---
Signif. codes:  0 '***' 0.001 '**' 0.01 '*' 0.05 '.' 0.1
    ' ' 1
Residual standard error: 0.2004 on 26 degrees of freedom
Multiple R-squared: 0.6326,   Adjusted R-squared: 0.6184
F-statistic: 44.76 on 1 and 26 DF,  p-value: 4.253e-07
modelT<-lm(leafbeetle_damage_percent^(1/3) ~ log(nests+1))
summary(modelT)
```

```
Call:
lm(formula = leafbeetle_damage_percent^(1/3) ~ log(nests
    + 1))
Residuals:
Min 1Q Median 3Q Max
-0.36646 -0.09599 0.00186 0.07582 0.38916
Coefficients:
              Estimate Std. Error t  value    Pr(>|t|)
(Intercept)    0.89142  0.04554   19.573   < 2e-16 ***
log(nests + 1) -0.19500 0.02778   -7.019   1.88e-07 ***
---
Signif. codes:  0 '***' 0.001 '**' 0.01 '*' 0.05 '.' 0.1
    ' ' 1
Residual standard error: 0.1648 on 26 degrees of
    freedom
Multiple R-squared: 0.6545, Adjusted R-squared: 0.6412
F-statistic: 49.26 on 1 and 26 DF, p-value: 1.878e-07
```

We will plot the regression lines resulting from each model using the function **abline**, which can be used either with the '**lm**' equation or the R object containing the model thus

abline(lm(total_damage_percent^(1/3) ~ log(nests+1)),lwd=2,col="red")
abline(modelT,lwd=2,col="green")

QQ Model Plot from the *car* Library

Here we have been using the model summary feature of R to examine QQ plots. An alternative is available in the **car** package with the function **qqPlot**. Using the ant data from Kolm *et al.* data again (reload x and y and create 'g' as in Chapter 7, p. 121) we create a linear model using **lm**.

modelAnts<-lm(g$y ~ g$x)
install.packages("car")
library(car)
qqPlot(modelAnts)

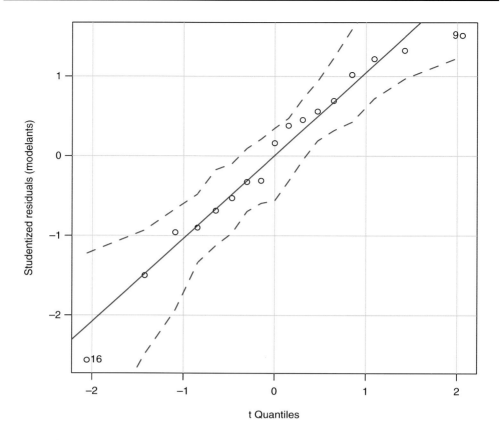

The area between dashed lines shows the default 95% confidence interval. Although the points deviate somewhat from a straight line towards the ends, the deviation is not significant at the $p = 0.05$ level, and importantly, no point is outside of the 95% confidence interval. That the majority of points are good fit to a straight line and most of the deviation involves two points near the extremes suggests that these might perhaps reflect some observation or experimental problems.

Comparing Regression Slopes and Intercepts Using t-test

We often want to know whether an explanatory variable has the same effect in two or more situations, such as on different species or on samples from different localities. For didactic purposes, we will show the workings of the calculation with a fairly obvious example. The data are from Garzón and Schweigmann (2015), who investigated whether the effect of temperature on the larval development time of the mosquito, *Ochlerotatus albifasciatus*, differed between a population from Buenos Aires, which has a temperate climate, and from Sarmiento Valley far to the south where the weather is much colder.

```
Ochlero<-read.table("culicid_DT.txt",header=TRUE)
head(Ochlero)
  localities  Temperature_C  ln.Development_time.
1 Sarmiento              11              3.521450
2 Sarmiento              11              3.344547
3 Sarmiento              11              3.306459
4 Sarmiento              11              3.268804
5 Sarmiento              11              3.217727
6 Sarmiento              11              3.185641
with(Ochlero,plot(Temperature_C,ln.Development_time.,col=localities,
    pch=as.numeric(localities)+7)) # using with to avoid typing "Ochlero"
    every time
```

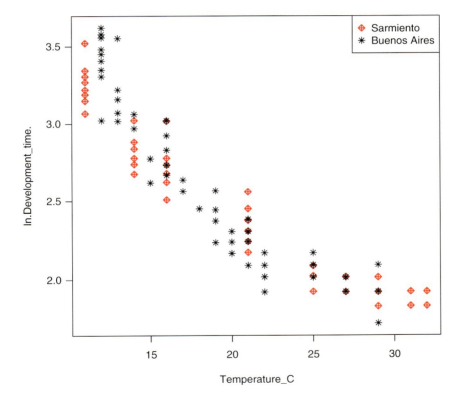

Visual inspection shows that even though the original article presented the natural logarithm of development time, the overall scatter of points is not linear, therefore we need to transform one of the variables to linearize them and normalize the errors before we can compare regressions. The standard approach for this shaped curve is to try logging or taking the square root of the x axis. We tried that and the result was improved but not perfect, so we took the logarithm of the log and plotted the diagnostic plots together. The top row is the **lm** of the raw data, the lower row is the **lm** of the data with log(log(Temperature_C)).

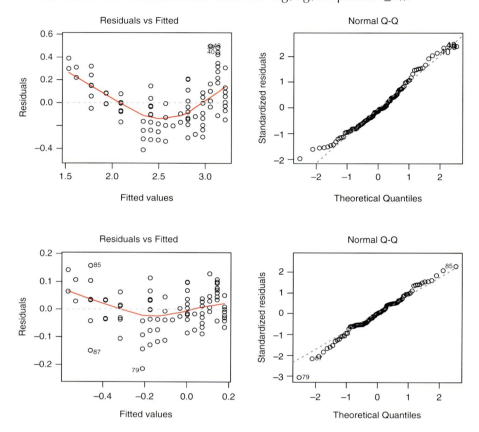

Now we are much happier with the overall model fit, we can now calculate the regressions separately for the two localities and test the significance of the difference using t or Z with the following equation:

$$Z = \frac{slope1 - slope2}{\sqrt{(standard\ error\ 1)^2 + (standard\ error\ 2)^2}}$$

with $n1 + n2 - 4$ degrees of freedom.

Information Box 11.2

The meaning of Z. The number of standard deviations away from the mean is denoted by Z. It is similar to t except that it implies we know the population mean, or our sample size is greater than about 30 in which case our sample standard deviation, s, can be accepted as being sufficiently close to the population value, σ, that we can use Z.

All the values we need to calculate the t statistic are in the summaries of the lm models for each site. We can extract the data points for Sarmiento and Buenos Aires separately by using **subset**
Sar<-subset(Ochlero,localities=="Sarmiento")
BAires<-subset(Ochlero,localities=="BuenosAires")
and then use **lm** to get the regression slopes and intercepts and their standard errors.
regrSar<-lm(log(log(Sar$ln.Development_time.)) ~ Sar$Temperature_C)
regrBA<-lm(log(log(BAires$ln.Development_time.)) ~ BAires$Temperature_C)
We could copy the values by hand from summary({*model name*}) but we can also extract them by their indices from the summary. So to get the regression slopes for Sarmiento and Buenos Aires
summary(regrSar)[[4]][2]; summary(regrBA)[[4]][2]
    [1] -0.03191494
    [1] -0.04066265
and for their standard errors
summary(regrSar)[[4]][4]; summary(regrBA)[[4]][4]
    [1] 0.001289394

Information Box 11.3

Indices of values in model summaries. When extracting values from summaries or models, etc. it is important to check the precise position in each case. In our example, the model fits both intercept and slope so the **summary**(model)[[4]] has eight elements. If we had forced the regression through the origin there would only be the four slope elements.

---

    [1] 0.002017866
Putting these into the formula, using **abs** to ignore signs, we get
Z<-abs(summary(regrSar)[[4]][2]-summary(regrBA)[[4]][2])/sqrt(summary
    (regrSar)[[4]][4]^2+summary(regrBA)[[4]][4]^2)
Z
    [1] 3.653031
with 83 (nrow(Ochlera) – 4) degrees of freedom. We can also get this by adding the degrees of freedom given in the summaries (36 + 47).

Now we need to know is whether this t-value is a significant and to do this we use **qt** with following structure: **qt**({*critical value of probability*},{*degrees of freedom*}) (see Chapter 30 on R's in-built distribution functions). The next important question is whether we had any prior reason to expect that the slope for one locality might be greater than for the other – we don't and indeed that is nearly always the case – and therefore we use a two-tailed test, which means we are interested in whether our value of t lies within the 2.5% zone on either side (2 × 0.025 = 0.05, the cut-off for significance), so we are looking within 97.5% or 0.975.
qt(0.975,83)
    [1] 1.98896
As our value of Z is larger than the critical level we conclude that the critical value for $p = 0.05$. In fact if you try **qt**(0.9995,83) you will see that the prob-

ability of slopes being drawn from the same distribution is less than 1 in 1000. Therefore we can conclude that the development time of the cold-adapted mosquito population is significantly faster at lower temperatures compared to the Buenos Aires insects.

Non-linear Regression

In non-linear regression analysis we use the function **nls** instead of **lm**, and within **nls** we specify the non-linear equation we want to fit PLUS we usually need to specify some estimated starting values for the parameters. R will then search around these best guesses of ours to find the values that minimize the squared errors.

Now we will examine just such data, the growth of the snakehead fish, *Channa striata*, in lotic habitats (Jutagate *et al.*, 2013)

age<-c(6.96,6.96,7.94,5.89,7.88,10.90,13.83,14.89,14.84,16.87,17.94,
    20.81,22.93, 24.86,25.83)
length<-c(12.78,13.54,14.12,28.15,26.93,18.70,27.46,28.22,38.0,29.28,
    28.72,35.39, 36.74,48.15,38.28)
plot(age,length,xlim=c(0,25),ylim=c(0,50),ylab="Length (cm)",xlab="Age (months)")

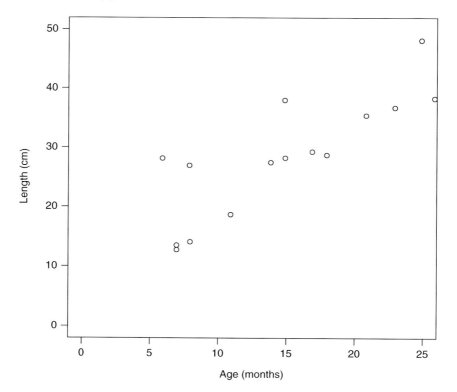

In this case we know that the curve must pass through the origin because on this scale, the length at time of hatching is effectively zero. An appropriate non-linear equation for asymptotic growth is

$$y = c\left(1 - e^{-ax}\right)$$

Where x is age, y is size, c is some upper limit of growth and a is the constant describing the shape of the curve. The function **exp** raises e, the base of natural logarithms, to the power passed to it.

Looking at our graph we might guess that c is somewhere around 100 cm (i.e. where the graph appears to be asymptoting to) and a is about 0.1 (because it doesn't seem to be very curved and x and y values are approximately the same order of magnitude).

```
modelG<-nls(length ~ C*(1-exp(-a*age)),start=c(C=100,a=0.5))
summary(modelG)
Formula: length ~ C * (1 - exp(-a * age))
Parameters:
    Estimate    Std. Error    t value    Pr(>|t|)
C   53.06428    14.85729      3.572      0.00341 **
a   0.05603     0.02644       2.119      0.05390 .
---
Signif. codes:  0 '***' 0.001 '**' 0.01 '*' 0.05 '.' 0.1 ' ' 1
Residual standard error: 6.241 on 13 degrees of freedom
Number of iterations to convergence: 7
Achieved convergence tolerance: 1.572e-06
```

The best fit with these data has final size c significantly larger than zero at 53 cm though markedly less than our guess of 100, but the curvature is only marginal ($p < 0.1$). *Channa striata* can grow up to a metre in length, but the present data suggest that the fish in the study might not attain that IF the model equation used is appropriate, which it might not be. In fact, using an equation with an asymptotic size for fish growth, and only two fitted parameters, might not be appropriate, but with so few data points and such variance, fitting a more complex model would not be justifiable. To view the shape of the curve create a finely divided vector of x axis values, xv, using **seq**

```
xv<-seq(0,25,0.1)
```

This generates a sequence from 0 to 25 in increments of 0.1, it is shorthand for seq(from=0, to=25, by=0.1); seq() can also take length as an argument so this is the same as seq(0,25,length=251).

Then we pass the model and x-values to the function **predict**, which then generates the values of the response variable according to the specified model.

```
yv<-predict(modelG,list(age=xv)); lines(xv,yv,lwd=2,col="blue")
```

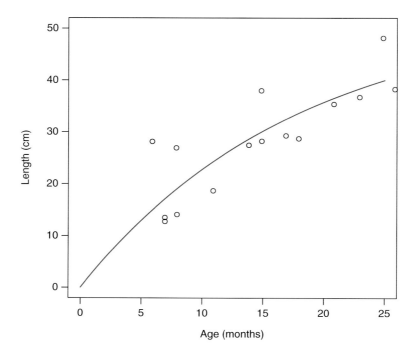

Despite being way off, our guestimate parameters seem to have been near enough, but using **nls** can easily go wrong, because it is often very difficult for us to make reasonably intelligent guesses about the values of these parameters, especially when there are several. You can always play around just plotting the function with parameters that you think might be OK until you get a reasonably shaped curve. We are fortunate that R includes some special self-starting (SS) functions that make reasonable initial estimates of the parameters for ten of the commonly used equations including:

SSasymp (asymptotic regression without intercept)
SSasympOff (asymptotic regression with intercept)
SSasympOrig (asymptotic regression with fit forced through origin, i.e. x = y = 0)
SSlogis (three parameter logistic regression)
SSgompertz and **SSweibull** for Gompertz and Weibull growth models, respectively.

In these cases the model function takes the form:

model<-nls(response_variable ~ SSfunction(explanatory_variable, a,b,c…))

If the fish growth dataset had had rather more data points and/or the data were better-behaved (less variance) and we had good reason to expect an asymptotic size, we would have fitted a Weibull model, but that has to fit four parameters and we only have 15 data points for *Channa* so that would be just under four points per parameter – simply not doable.

Multiple Regression

When there is more than one continuous explanatory variable that we have a suspicion may affect a continuous response variable, we use multiple regression. Multiple regression analyses are carried out like linear regressions except that there is more than one explanatory variable. We are asking whether each explanatory variable is significantly related to the response variable and (if so) how much of the variation in the response variable (y) is explained by each explanatory variable. As mentioned before, if you have sample sizes in the millions, you might well find that many explanatory variables are significantly associated with the response variable, but it is almost never the case that you will be interested in anything whose effect size is less than 1% or even 10% – physicists with huge sample sizes may be because even such small effects might mean that their models are not fully accurate. The focus on *p*-value rather misses the point that effect size is what is important.

A simple regression model takes the form

lm(y ~ x)

and a multiple regression with more than one explanatory variable (x1, x2, x3, etc.)

lm(y ~ x1 + x2 + x3…xn)

The situation can be more complicated, however, because the separate x variables may not all be independent. Non-independence of these explanatory variables is called an *interaction*. Specifically, that means the effect of one explanatory variable depends on the value of a second one.

Here is a simple example: we want to know whether cholesterol in your diet is good for you or bad for you? Well the obvious analysis would look at some measure of 'good for you' as a function of dietary cholesterol intake. Easy – not quite. Most of us will be aware that many older people are at a high risk of developing heart disease and that this could be exacerbated by a high cholesterol intake. Depending on the demography (age distribution) of your subjects, a simple linear regression of 'good for you' ~ 'cholesterol' may indicate no significant relationship. But if age was also included as an explanatory variable along with 'cholesterol', we might obtain a very significant result, 'cholesterol' might be good for you when you are young and growing, but bad for you when you are old and liable to develop heart disease if you eat too many fatty things. Depending on your sample you might find that overall, neither cholesterol nor age had a significant effect on 'good for you' BUT that the interaction 'cholesterol'*'age' was highly significant.

Interaction terms are specified in R's regression models in a variety of ways, some of which are beyond the scope of this book (see Crawley, 2007, pp. 329–330). If we have just two explanatory variables and we want to look at the relationship of the response variable to each of them plus their interaction, we use (note the * sign) lm(y ~ x1*x2)

That means that the amount of variance in y will be partitioned among three explanatory variables, x1, x2 and x1*x2. Again, here we can see the importance of sample size. Suppose we have four possible explanatory variables, x1, x2, x3 and x4. In a model without interactions

lm(y ~ x1 + x2 + x3 + x4)

R only has to estimate four values – a multiplier for x1 ... x4 and check whether they are significantly correlated with the response variable. But if we want to assess whether any interaction terms between any of our four explanatory variables and y are important via

lm(y ~ x*1 + x*2 + x*3 + x4)

then R checks all of the following combinations.

x1	x2	x3	x4	x1*x2
x1*x3	x1*x4	x2*x3	x2*x4	x3*x4
x1*x2*x3	x1*x2*x4	x2*x3*x4	x1*x3*x4	x1*x2*x3*x4

That means that 15 separate fit parameters and each of their probabilities will need to be calculated from the data. With sample sizes in the hundreds and thousands or more, this is quite feasible.

For the above reason, if you don't have thousands of data points, you should not 'throw in' all possible explanatory variables (phase of the moon, how many cigarettes you smoked before coming into the lab) because you do not have any reasonable belief that they would have an effect on your response variable (see Chapter 6: **TukeyHSD**, Bonferroni correction, etc.). In conclusion, using an interaction expression between explanatory variables '*' is only likely to be possible for small number of explanatory variables, and you should think first of all, which ones might be expected to influence the effect of the other.

Pairwise Plots of Explanatory Variables to Visually Inspect Interactions

It can be helpful to look at a number of pairwise comparisons if you only have only a few factors that you already suspect may be informative. The function **pairs** automatically plots each pair of vectors against one another. We can use the 'ChickWeight' dataset in R's base package.

pairs(ChickWeight, panel=panel.smooth,lwd=3)   # panel=panel.smooth fits a
    line to the data

Regression and Correlation Analyses

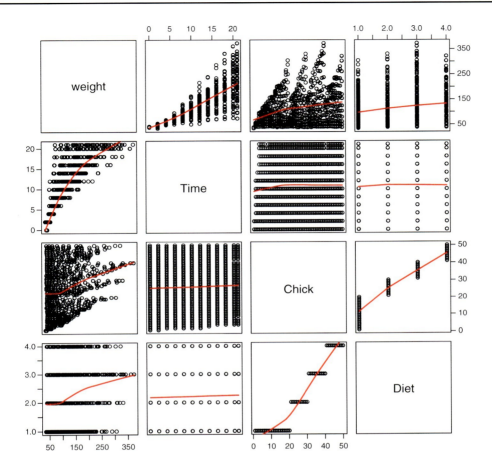

The block of plots shows each pairwise plot. The top row plots height (on the *y* axis) against age (top middle plot) and against seed weight (top right). The second row plots age (on the *y* axis) against height (middle left plot) and against seed weight (middle right plot), and so on. The two diagonal fits show a strong correlation between height and age and only a weak trend between height and seed (top right).

Information Box 11.4

Potential problem with using T & F rather than TRUE & FALSE. If 'T' (or 'F') have been used as names for vectors R will give the following error message when used in a logical test, e.g.
`Error in !header : invalid argument type`
If you get this message check that the issue is what we suspect
class(T)
`"character"`
so we can either (best practice) write 'header=TRUE' and/or reset the class of 'T' to logical
class(T)="logical"
However, this may only work for the given R session. Be warned, do not use 'T' or 'F' as variable names.

Polynomial Regression and Model Simplification

Finally in this chapter we will look at polynomial regression in which we see how much of the variation in the response variable is explained by an explanatory variable (x), by x^2, x^3, etc. There are not that many occasions where this might be relevant in biology, but possible examples might include physiological features such as respiration that could be affected differently according to surface area (i.e. size^2) or volume (size^3). In ecology, some aspects of species richness might scale with powers of linear dimensions.

Here we will look at oxygen consumption by an Antarctic krill (Crustacea: Mysidae) in relation to wet weight using data from Van Ngan *et al.* (1997) on oxygen consumption rate versus krill wet weight.

knor<-read.table("krill.txt",header=TRUE)
head(knor)
```
    wwt_grams    ulO2_min
1   0.07984726   6.954309
2   0.08838326   12.794991
3   0.08716658   14.757405
4   0.08920311   16.801575
5   0.09170859   18.638888
6   0.09604174   20.147910
```
First we will look at two plots, the simple plot of O_2 consumption versus wet weight, and the log-log plot as presented by Van Ngan *et al.* (1997)
par(mfrow=c(2,2))
plot(knor$wwt_grams,knor$ulO2_min,ylab="Oxygen consumption (ul/min)", xlab="Wet weight(grams)")
plot(log10(knor$wwt_grams),log10(knor$ulO2_min),ylab="log(Oxygen consumption (ul/min))",xlab="log(Wet weight(grams))")

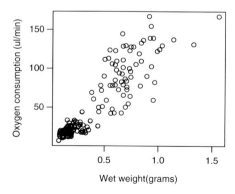
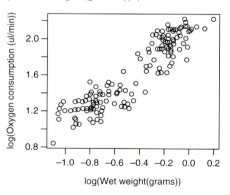

The left-hand plot clearly shows a massive increase in variance with increasing wet weight (heteroscedasticity). Logging both axes (right-hand figure) gets rid of this so is clearly appropriate, though there seems to be a bit of a deficit of individuals with middling weights (i.e. wet weight is bimodal). Anyway, we cannot help that in this case so we will proceed and apply a simple linear model to the data.
par(mfrow=c(2,2))

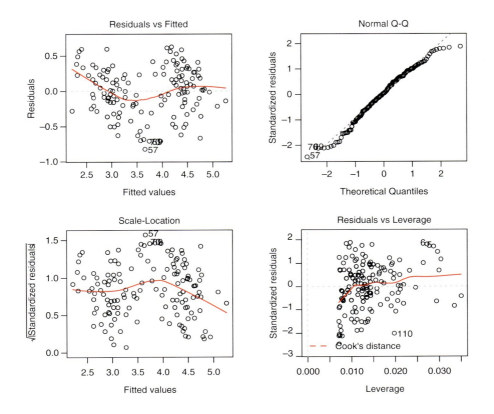

```
model1<-lm(log(knor$ulO2_min) ~ log(knor$wwt_grams))
plot(model1)
```
We can see that the vast majority of points in the QQ plot form an essentially straight line, approximately 10 at the left-hand end deviate one way and four at the right-hand end deviate the other, though none by very much, and no points lie outside the Cook's distance 95% confidence limits. This pattern indicates some under dispersion in the residuals but this is probably acceptable and so we will proceed. If this were for a publication we might try deleting the more extreme outliers and seeing whether that made any difference to our results. We suspect that it would not change anything much in this case.
```
summary(model1)
Call:
lm(formula = log(knor$ulO2_min) ~ poly(log(knor$wwt_grams,
    3)))
Residuals:
     Min       1Q   Median       3Q      Max
-0.82340 -0.20866  0.01216  0.23879  0.62782
Coefficients:
```

```
                        Estimate Std.  Error t
 value     Pr(>|t|)
(Intercept)                    3.71643    0.02825
131.56       <2e-16 ***
poly(log(knor$wwt_grams, 3))  9.22200    0.33544
 27.49       <2e-16 ***
---
Signif. codes: 0 '***' 0.001 '**' 0.01 '*' 0.05 '.' 0.1 ' ' 1
Residual standard error: 0.3354 on 139 degrees of
   freedom
Multiple R-squared: 0.8447, Adjusted R-squared: 0.8435
F-statistic: 755.8 on 1 and 139 DF, p-value: < 2.2e-16
```

There is obviously a highly significant, approximately linear, positive correlation between the logged variables. But look more closely, are those doubly logged points fitting well to a straight line? Without even plotting the model fit line it appears to the eye that the points form a slight concave curve. Further, we have an *a priori* expectation that functions such as metabolic rate and oxygen consumption might scale according to some more complicated measure of body size, maybe surface area or even linear dimension might have an effect. Therefore we will construct a polynomial regression model. We have no *a priori* expectation that oxygen consumption should scale with some higher order than power three so we will limit our investigation accordingly. What we want to determine is whether oxygen consumption is explained significantly better if we include a squared, and/or a squared and cubed term in our model.

There are two ways of calling for a polynomial regression, either by specifying the higher order terms individually, i.e.

model3I<-lm(log(knor$ulO2_min) ~ knor$wwt_grams+I(knor$wwt_grams^2) + I(knor$wwt_grams^3))

Or using the argument 'poly'.

modelpoly3<-lm(log(knor$ulO2_min) ~ poly(log(knor$wwt_grams),3))

The resulting models are not the same and we can see this by examining the model summaries:

summary(model3I)

```
Call:
lm(formula=log(knor$ulO2_min)~knor$wwt_grams+I(knor$wwt_
   grams^2) +
I(knor$wwt_grams^3))
Residuals:
    Min         1Q   Median         3Q      Max
-0.73140   -0.18058  0.00491   0.24138  0.63513
Coefficients:
            Estimate Std.  Error t    value Pr(>|t|)
(Intercept)            2.0947   0.1144  18.310  < 2e-16 ***
knor$wwt_grams         5.4907   0.7746   7.089 6.53e-11 ***
```

```
I(knor$wwt_grams^2)   -3.4258    1.2751   -2.687   0.00811 **
I(knor$wwt_grams^3)    0.6955    0.5756    1.208   0.22901
---
Signif. codes:  0 '***' 0.001 '**' 0.01 '*' 0.05 '.' 0.1 ' ' 1
Residual standard error: 0.3164 on 137 degrees of
   freedom
Multiple R-squared: 0.8638, Adjusted R-squared: 0.8608
F-statistic: 289.5 on 3 and 137 DF, p-value: < 2.2e-16
summary(modelpoly3) # the model with orthogonal variables
Call:
lm(formula = log(knor$ulO2_min) ~ poly(log(knor$wwt_grams), 3))
Residuals:
    Min       1Q   Median       3Q      Max
-0.79567  -0.17519  0.02326  0.24278  0.66977
Coefficients:
                              Estimate Std.    Error
t value      Pr(>|t|)
(Intercept)                    3.71643    0.02678
 138.761       < 2e-16 ***
poly(log(knor$wwt_grams), 3)1  9.22200    0.31803
  28.997       < 2e-16 ***
poly(log(knor$wwt_grams), 3)2  0.88492    0.31803
   2.783       0.00616 **
poly(log(knor$wwt_grams), 3)3 -1.00048    0.31803
  -3.146       0.00203 **
---
Signif. codes:  0 '***' 0.001 '**' 0.01 '*' 0.05 '.' 0.1 ' ' 1
Residual standard error: 0.318 on 137 degrees of freedom
Multiple R-squared: 0.8624, Adjusted R-squared: 0.8594
F-statistic: 286.2 on 3 and 137 DF, p-value: < 2.2e-16
```

Both of our above models indicate significant higher order effects but what is behind this difference? The answer is that specifying the individual power terms in the model using 'I(x^2)' etc. does not allow for the almost inevitable correlation between the different power terms, whereas using 'poly' makes each power term orthogonal (i.e. independent) from each of the others. So looking at it, there probably is significant autocorrelation between power terms and so we should use the formula including the 'poly' argument. This model indicates that there are significant independent effects of linear, squared (=quadratic) and cubed functions of krill wet weight.

Model Simplification

Our aim in statistical modelling is to end up with the simplest model that does a good job of explaining the variance in our response variable – we call this the minimum adequate model (see Crawley, 2007, p. 327). When we carry out a multiple regression with a range of what you think are potentially important explanatory variables, and interactions between them (a maximal model), we are likely to find that some of the explanatory variables are actually not important, and therefore we should discard them and re-run the model. The next step is to remove the least significant of the remaining terms then re-run the model again and compare the new result statistically with the previous one using the function **anova**, (i.e. **anova**({first model},{simplified model}). If deleting the term did not cause a significant difference to the fit of the model, leave it out. If it made a significant difference put it back in and move to the next least significant term. Repeat this process until all the remaining terms are significant and there you have it, your minimum adequate model. We do this because if we add more explanatory terms, even if they are completely random, we will always be able to explain at least a tiny bit more of the variance. What is important is that if the extra variables are random then their addition to the model should not improve the fit of the model significantly so they can be deleted. Other approaches use Akaike's information criterion (see below).

With the krill data above what we want to know is whether the model with three terms is a significant improvement over the model with just two terms. The first thing we should look for is which of the variables have the most similar effects. In our example, this isn't too complicated – the estimates of the square and cubic terms are closest but they are not that close. So, the question in this case is whether a model including both squared and cubed terms explains significantly more of the variation in oxygen consumption than either alone. To test this in the first instance we must run a second model with only the squared term included and compare it with the more complicated three-term model using **anova**.

```
model2<-lm(log(knor$ulO2_min) ~ poly(log(knor$wwt_grams),2))
anova(model2,modelpoly3)
Analysis of Variance Table
Model 1: log(knor$ulO2_min) ~ poly(log(knor$wwt_grams), 2)
Model 2: log(knor$ulO2_min) ~ poly(log(knor$wwt_grams), 3)
  Res.Df    RSS  Df Sum of Sq      F   Pr(>F)
1    138 14.857
2    137 13.857   1    1.001  9.8966 0.002031 **
---
Signif. codes:  0 '***' 0.001 '**' 0.01 '*' 0.05 '.' 0.1 ' ' 1
```

The ANOVA is highly significant, so we should not simplify the model, and indeed there are both significant square and cubic terms. Had the **anova** been non-significant we should have combined the pair of explanatory variables and progressively combined non-significantly different terms until all explanatory terms were significant and there were no non-significant ones. That is our minimally explanatory model.

As our models differ in the number of explanatory variables, we can also use Akaike's information criterion, which is a measure of a model's fit. When we add more parameters (i.e. more explanatory variables) we will get a better fit. As Crawley (2007) put it, 'There is always going to be a trade-off between the goodness of fit and the number of parameters required by parsimony'.

In R, Akaike's information criterion is calculated using the function **AIC**. We prefer the model with the lowest AIC.

AIC(model2,model3I)
```
          df         AIC
model2    4          90.85449
model3I   5          83.02003
```
As with the ANOVA comparison, because the AIC value of model3I is lower than that of model2, we prefer it even though the model has a larger number of parameters.

Information Box 11.5

Akaike's Information Criterion (AIC). This useful criterion was first published by Akaike (1971) in a rather obscure place so it did not get widely used until much more recently. The actual value of AIC is not very meaningful, its value comes in comparing different models. What it does is to reward models that produce a high goodness-of-fit score but penalize them for having more parameters. Formally it is defined as:

$AIC = 2 * \{number\ of\ parameters\} - 2 * \log_e(\{likelihood\ of\ model\})$

The more parameters the larger the AIC so the less we prefer it.

Exercise Box 11.2

1) Use the function **dip** in the package **diptest** to test whether the krill wet weight distribution really is bimodal rather than unimodal.
2) Use the data from R's available data set 'trees'. Choosing an appropriate statistical analysis determine whether the height of black cherry trees is correlated with their girth.

> **Tool Box 11.1** 🛠
>
> - Use −1 in a model to force regression lines through the origin.
> - Use abline() to plot the regression line on the graph.
> - Use anova() to compare the fit of two models.
> - Use exp() to raise *e* to the power of something.
> - Use ggqqplot() to plot quantile values against each other.
> - Use lm() to create a linear model or regression.
> - Use nls() to fit a non-linear least squares regression model.
> - Use log() to calculate natural logarithm.
> - Use log10() or log(x, 10) to calculate log base 10.
> - Use log2() or log(x, 2) to calculate log base 2.
> - Use predict() to generate the values of the response variable.
> - Use qt(critical value of *p*, degrees of freedom) to decide if something is significant.
> - Use seq(a,b,c) to generate a sequence from a to b in increments of c.
> - Use SS... self-starting function to obtain reasonable starting parameter values for **nls**; a list of common ones can be found on p. 136.

12 Count Data as Response Variable

Summary of R Packages Introduced	Summary of R Functions Introduced
vcd	goodfit in "vcd"

We are devoting a short chapter specifically to count data for three reasons: (i) they are common in ecological studies (e.g. clutch sizes, numbers of fledglings from a nest, numbers of seeds per pod...); (ii) they are simple to collect and are therefore often the data collected by students (e.g. numbers of beetles in a pitfall trap, number of pollinator visits to flowers...); and (iii) they pose numerous issues that linear models with their normal error structure cannot deal with.

One obvious difference between models based on continuous variables and ones based on count data is the predicted values. Continuous variable models will predict continuous values of your response variable, but being count data, your response variable can only be integer values. Whether this is important depends a lot on your question and whether the count data are sufficiently finely divided that this doesn't really matter in practice even though it is a technical violation.

You will see many published papers that deal with count data in exactly the same way as continuous variables (though usually not in major tier 1 journals, but nevertheless). We will show here how analysing data that way can lead to very wrong estimates and often to the wrong statistical conclusions. Analysis of count data is a tricky area in the sense that it depends to a large extent on the range of values. You cannot have negative counts, a bird may lay a clutch of 1, 2, 3, ... 10 eggs but cannot lay –1 or –2. Therefore, the clutch sizes cannot be normally distributed, and we say that they are left-censored, with values less than one (or sometimes less than zero) missing. Not only are such count data censored but very often they will be highly skewed. Depending on the type of data there might be many zeros and that will always be problematic for analyses. In contrast to the bird egg clutches, if we were to be investigating the numbers of eggs laid by marine turtles, the numbers of fish in a shoal or harem size in elephant seals, each of which are count data, the mean values are likely to be far from zero and often in the hundreds and there will be no ones or twos. In such cases there is no problem with censoring and the distributions may appear to be close to normal, albeit that non-integer values are not allowed.

When events are rare you will always have low means with many zeros; this is the nature of the Poisson distribution in which the mean and the variance are equal (see Chapter 31). A classic example might be the numbers of galls made by gall wasps (Cynipidae) on oak leaves. If the gall wasps are uncommon with respect to oak leaves and if they oviposit independently, any leaves on a tree may have no galls, some with one, a few with two, fewer with three, etc. When there are very many zeros it becomes impossible to normalize the errors or to fit them meaningfully to any distribution-based model and, in these cases, we may have to resort to Monte Carlo type statistics (see Chapter 16).

Usually the correct way of dealing with count data is to use a Poisson model, which we apply with a **glm** specifying family="poisson", but see below for caveats. Using Poisson errors is of course, only likely to be valid if the error structure is close to a Poisson distribution—we show below one way of checking this. However, errors with count data often deviate from a pure Poisson structure, and somewhat more relaxed approaches include quasipoisson errors (see Chapter 15) and negative binomial regression, both of which have an additional fit parameter which allows for some overdispersion of the data (i.e. data in which the variance is greater than the mean). The difference between family=poisson and family=quasipoisson is in how the variance is related to the mean, linearly or as a quadratic function. For a detailed comparison of these see Verhoef and Boveng (2007). Shmueli et al. (2004) have recently considered a generalization of the Poisson distribution that can exhibit both under- and overdispersion relative to the Poisson distribution. They use the name COM–Poisson because the distribution was proposed originally by Conway and Maxwell (1962).

We will examine two studies with the response variable being counts, starting with one that nearly fits the ideals of a Poisson distribution well, the other less so. Another study with count data as the response variable in an ANCOVA will be examined in Chapter 14.

Example 1 – Fledgling Numbers in Relation to Clutch Initiation Date

The data, from Shustack and Rodewald (2011), are on the northern cardinal bird, *Cardinalis cardinalis*, and were collected to test the hypothesis that birds that start their clutches later may suffer higher pre-fledging offspring mortality. For simplicity, clutch initiation dates are given in days starting with the earliest clutch. First we read the data and plot number of fledglings against clutch initiation date.

```
d<-read.csv("fledge_number.csv")
par(mfrow=c(2,2))
plot(d,xlab="Clutch initiation date",ylab="Number of fledglings",pch=17,
    col="blue")
```
Without having to do any model inspection it is clear that there is a negative trend in the variance of fledgling number with date. It is also clear that there are a substantial number of zeros. Also, by eye, there appears to be a negative trend in the mean, but is it significant? Therefore we will plot the frequencies of number of fledged chicks, which we can easily do using the function **table** (note the [,2] meaning the second column)
```
freq<-table(d[,2])
freq
 0  1  2  3  4  5  6  7
19 27 35 27 18  8  4  3
barplot(freq,xlab="Number of fledglings",ylab="Frequency")
```

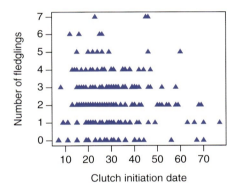
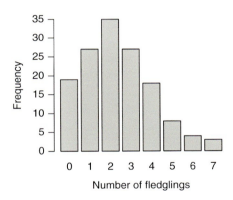

The distribution is modal at two fledglings, clearly left censored below zero and is broadly similar to some of the Poisson distribution plots shown in Chapter 31. The next step is therefore to see how closely the number of fledglings conforms to a Poisson distribution. We know that in Poisson distributions the mean is equal to the variance; the two are usually called the parameter lambda (λ).
```
mean(d[,2])
[1] 2.375887
var(d[,2])
[1] 2.8077
```
The mean and variance are not identical but they are pretty similar; the data are slightly overdispersed because the variance is larger than the mean. Very seldom do we find almost exactly equal means and variances in these sorts of real data. Before going on to conduct a glm with Poisson errors we will use a package **vcd** that contains a function **goodfit** that will allow us more precisely

to compare the observed frequencies with optimal Poisson fits. **goodfit** has two fit options, maximum likelihood ("ML"), which we will use, and minimum Chi-squared ("MinChisq"). Its **plot** function also has two display options: 'type="standing"' and 'type="hanging"'. We will show both but, unfortunately, they cannot be shown in a single window, and we have pasted them side-by-side to save space. They just show the same thing but in different ways.

install.packages("vcd")
library(vcd)
gf<-goodfit(freq,type="poisson",method="ML")
plot(gf,type="standing",xlab="Number of fledged chicks",ylab="Frequency", scale="raw")
plot(gf,type="hanging",xlab="Number of fledged chicks",ylab="Frequency", scale="raw")

The red line is the predicted Poisson distribution based on the best-fit parameters. Both plots show that overall the fit obtained by goodfit is very good but our data have a slight excess of zeros and also of sevens (that is the overdispersion. If you ask for a summary of the model (**summary**(*gf*)) we see that our data show nowhere near a significant departure from Poisson expectations, so we can proceed with using a **glm** with Poisson errors.

model_card<-glm(d$fledged ~ d$date,family="poisson")
summary(model_card)

```
Call:
glm(formula = d$fledged ~ d$date, family = "poisson")

Deviance Residuals:
    Min       1Q   Median       3Q      Max
-2.3580  -0.9724  -0.1523   0.5443   2.5980
Coefficients:
             Estimate Std. Error z value Pr(>|z|)
(Intercept)  1.067142   0.129818   8.220   <2e-16 ***
d$date      -0.006386   0.003808  -1.677   0.0936 .
---
Signif. codes:  0 '***' 0.001 '**' 0.01 '*' 0.05 '.' 0.1 ' ' 1

(Dispersion parameter for poisson family taken to be 1)

Null deviance: 191.33  on 140  degrees of freedom
Residual deviance: 188.44  on 139  degrees of freedom
AIC: 531.38

Number of Fisher Scoring iterations: 5
```

Count Data as Response Variable

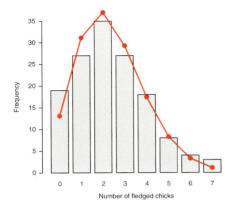
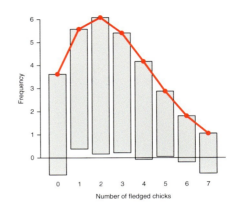

The analysis shows that the weak negative trend of fledgling success with later clutch initiation dates is not significant ($p = 0.0936$).

Exercise Box 12.1

1) Replot the cardinal fledging data, and add the regression line in a different colour.
2) Compensate for the overdispersion by running a glm with quasipoisson errors and examine the model and its summary. What would you now conclude?

Example 2 – Pollinator Flower Visits in *Passiflora* in Relation to Flower Size

The data are the number of visits by potential pollinators of a Neotropical vine, *Passiflora speciosa* (Longo and Fischer, 2006). First load and plot the points.

```
PV<-read.csv("Passiflora_visits.csv",header=TRUE)
head(PV)
  Flower_diam_mm Total_visitors
1           8.80              5
2           9.61              7
3           9.51              1
4          10.20              7
5          10.11             10
6          10.40              1
plot(PV,pch=16,col="red")
```

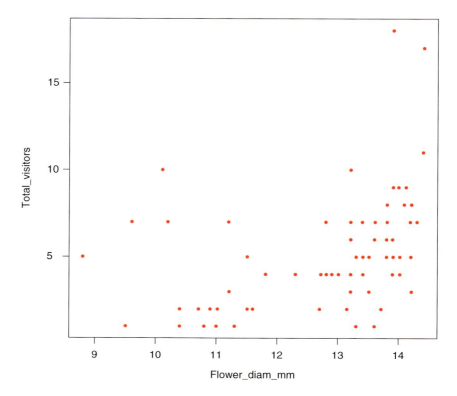

The data show very obvious problems for **lm**: there are quite a lot of low counts (ones and twos), and the variance appears to increase markedly towards the right. There are no zeros in this dataset because only flower visits were recorded, so the data are zero censored. Do these data fit a Poisson distribution?
mean(PV$Total_visitors); var(PV$Total_visitors)
[1] 5.203125
[1] 11.6565
Unlike the previous example the count variable shows marked overdispersion, the variance is more than twice the mean, and therefore this needs to be taken into account. A glm with Poisson errors cannot be used because the assumed error structure is violated. To deal with this, we will use quasipoisson errors, which fits an additional parameter to account for the extra variance.
modelQP<-glm(PV$Total_visitors ~ PV$Flower_diam_mm,family="quasipoisson")
summary(modelQP)

```
Call:
glm(formula  =  PV$Total_visitors  ~  PV$Flower_diam_mm,
    family = "quasipoisson")

Deviance Residuals:
    Min       1Q   Median       3Q      Max
-2.5189  -1.0082  -0.3613   0.5333   3.7730
```

```
Coefficients:
               Estimate Std.  Error t value Pr(>|t|)
(Intercept)       -0.96832 0.81289  -1.191   0.23812
PV$Flower_diam_mm  0.20242 0.06176   3.278   0.00172 **
---
Signif. codes: 0 '***' 0.001 '**' 0.01 '*' 0.05 '.' 0.1 ' ' 1
(Dispersion parameter for quasipoisson family taken to
   be 1.939608)
    Null deviance: 130.80 on 63 degrees of freedom
Residual deviance: 107.46 on 62 degrees of freedom
AIC: NA
Number of Fisher Scoring iterations: 5
```

modellm<-lm(PV$Total_visitors ~ PV$Flower_diam_mm)
summary(modellm)

```
Call:
glm(formula = PV$Total_visitors ~ PV$Flower_diam_mm,
    family = "poisson")

Deviance Residuals:
    Min      1Q   Median      3Q      Max
-2.5189 -1.0082  -0.3613  0.5333   3.7730

Coefficients:
               Estimate Std.  Error z value Pr(>|z|)
(Intercept)       -0.96832 0.58368  -1.659   0.0971 .
PV$Flower_diam_mm  0.20242 0.04434   4.565   5e-06 ***
---

Signif. codes: 0 '***' 0.001 '**' 0.01 '*' 0.05 '.' 0.1 ' ' 1

(Dispersion parameter for poisson family taken to be 1)

    Null deviance: 130.80 on 63 degrees of freedom
Residual deviance: 107.46 on 62 degrees of freedom
AIC: 323.6

Number of Fisher Scoring iterations: 5
```

If we plot 'modelQP' we see that the points in the QQ plot deviate a little above the straight line at the right, indicating a bit of right skew, i.e. if you plot a histogram of number of flower visitors, there are a few more high counts than expected (hence right skew). Inspecting the plot, it is easy to spot these values (data points 5, 49 and 50 being particularly noticeable). The amount of skew is nevertheless quite small and is probably acceptable. Since the errant points are at both ends of the x axis no simple transformation such as logging or square-rooting the x axis will remove the skew, and therefore only deleting these extremes would give a straight QQ plot. We can reasonably conclude that there is a small but significant, positive effect of flower size on insect visitors.

Information Box 12.1

Plotting text at a precise angle relative to a plotted line. To prettify graphs you might want to print text in parallel to a plotted line. The answer to Exercise 12.2 includes code for doing it with that example. It is often easiest to try a few guestimates, but to get the angle precise you need to calculate the x and y limits of the line using par("usr") and make an adjustment using the ratio of width to height of the whole plot window using par("pin")[2]/par("pin")[1].

Exercise Box 12.2

1) Replot the *Passiflora* pollinator data and add regression lines (in different colours) from simple (inappropriate) **lm**, and more appropriate **glm** with Poisson errors and the final **glm** with quasipoisson errors. Hint: use **abline**.
2) Add text labels above each line at the same angle and same colours specifying which model was used. Hint: use srt= to specify the angle of the text in degrees; use guestimates for angles.

Tool Box 12.1

- Use glm() with family=poisson to perform regression or correlation with a Poisson distributed response count variable.
- Use glm() with family=quasipoisson to perform regression or correlation with a count response variable showing marked under- or overdispersion.
- Use goodfit() to visualize how well data fit a Poisson, binomial or negative binomial distribution.
- Use table() to get a table of frequencies of particular numbers, characters or factors.

13 Analysis of Variance (ANOVA)

Summary of R Functions Introduced
    aov
    asin
    is.ordered
    sqrt

Analysis of variance is used to analyse the differences between group means in a sample, when the response variable is numeric (real numbers) and the explanatory variable(s) are all categorical. Each explanatory variable may have two or more factor levels, but if there is only one explanatory variable and it has only two factor levels, you should use Student's t-test (**t.test** in Chapter 10) – the result will be identical.

Basically an ANOVA fits an intercept and slopes for one or more of the categorical explanatory variables. We usually perform an ANOVA using the linear model function **lm**, or the more specific function **aov**, but there is a special function **oneway.test** when there is only a single explanatory variable. For a one-way ANOVA the non-parametric equivalent (if variance assumptions are not met) is the **kruskal.test**, which we met in Chapter 10.

Example 1 – A One-way ANOVA, the InsectSprays Dataset

We have already met the in-built dataset 'InsectSprays' in Chapter 6, which presents data on numbers of insects in a field following six different spray treatments with 12 replicates each; data from Beall (1942). Here we will use ANOVA to determine whether the treatments affect the numbers of remaining insects.

```
data<-InsectSprays
oneway.test(InsectSprays$count ~ InsectSprays$spray)
One-way analysis of means (not assuming equal variances)
data:  InsectSprays$count and InsectSprays$spray
F = 36.065, num df = 5.000, denom df = 30.043, p-value =
   7.999e-12
```

The spray treatment had a highly significant effect on numbers of insects surviving. That is all there is to it if that is all we want to know. But to look at the effects in more detail we should use **lm**.

modelSprays<-lm(InsectSprays$count ~ InsectSprays$spray)
summary(modelSprays)

```
Call:
lm(formula = InsectSprays$count ~ InsectSprays$spray)
Residuals:
    Min      1Q  Median      3Q     Max
 -8.333  -1.958  -0.500   1.667   9.333
Coefficients:
                    Estimate Std.  Error  t value  Pr(>|t|)
(Intercept)          14.5000       1.1322  12.807   < 2e-16 ***
InsectSprays$sprayB   0.8333       1.6011   0.520     0.604
InsectSprays$sprayC -12.4167       1.6011  -7.755  7.27e-11 ***
InsectSprays$sprayD  -9.5833       1.6011  -5.985  9.82e-08 ***
InsectSprays$sprayE -11.0000       1.6011  -6.870  2.75e-09 ***
InsectSprays$sprayF   2.1667       1.6011   1.353     0.181
---
Signif. codes:  0 '***' 0.001 '**' 0.01 '*' 0.05 '.' 0.1 ' ' 1
Residual standard error: 3.922 on 66 degrees of freedom
Multiple R-squared: 0.7244,    Adjusted R-squared: 0.7036
F-statistic: 34.7 on 5 and 66 DF,  p-value: < 2.2e-16
```

That is much more information. The first row shows the mean of the counts for spray 'A', and it is highly significantly different from zero. The subsequent rows show which sprays are significantly different from the intercept. We can look at the individual means easily using **tapply** (see Chapter 27).

sprayMeans<-tapply(InsectSprays$count,InsectSprays$spray,mean)
sprayMeans

```
        A         B         C         D         E         F
14.500000 15.333333  2.083333  4.916667  3.500000 16.666667
```

In agreement with the p-values, the non-significant ones have means closest to that of spray A. Finding out which treatments have the greatest effect on the results is considerably more complicated and beyond the scope of this book, and we refer you to Crawley (2011, pp. 370–380). However, since we are performing multiple comparisons a useful first step would be to use Tukey's HSD test to the ANOVA model (see Chapter 6). For this we need to use the **aov** function.

aovSprays<- aov(InsectSprays$count ~ InsectSprays$spray)
TukeyHSD(aovSprays)

```
Tukey multiple comparisons of means
95% family-wise confidence level
```

```
Fit: aov(formula = InsectSprays$count ~ InsectSprays$spray)
$`InsectSprays$spray`
            diff        lwr         upr     p adj
B-A    0.8333333  -3.866075    5.532742 0.9951810
C-A  -12.4166667 -17.116075   -7.717258 0.0000000
D-A   -9.5833333 -14.282742   -4.883925 0.0000014
E-A  -11.0000000 -15.699409   -6.300591 0.0000000
F-A    2.1666667  -2.532742    6.866075 0.7542147
C-B  -13.2500000 -17.949409   -8.550591 0.0000000
D-B  -10.4166667 -15.116075   -5.717258 0.0000002
E-B  -11.8333333 -16.532742   -7.133925 0.0000000
F-B    1.3333333  -3.366075    6.032742 0.9603075
D-C    2.8333333  -1.866075    7.532742 0.4920707
E-C    1.4166667  -3.282742    6.116075 0.9488669
F-C   14.5833333   9.883925   19.282742 0.0000000
E-D   -1.4166667  -6.116075    3.282742 0.9488669
F-D   11.7500000   7.050591   16.449409 0.0000000
F-E   13.1666667   8.467258   17.866075 0.0000000
```

Conveniently this time, the 'InsectSprays' dataset has nice short factor names. Looking at the adjusted *p*-values we can indeed see that there are many significant differences between pairs of spray treatments.

Example 2 – ANOVA with Proportion Data as Response Variable Using Arcsine Transformation

In our first example we will explore Tunjai and Elliott's (2012) data on the relationship between various seed traits and establishment success of 19 tree species in Thailand. This sort of data allows several types of analysis, though there are also several caveats with this example: the relatively small number of species investigated limits its power, and the appearance of the data (see below) is strongly influenced by a single data point. Further, as with all cross-species comparisons there is a potential problem of phylogenetic non-independence, which is that the apparent importance of a trait may be overstated because several closely related species share the trait and have similar successes but because of other features, not necessarily the seed trait. Given these issues do not be surprised if we fail to get a very clear answer to our questions in this example.

The first thing to note here is that the response variable, the establishment success, is measured as a percentage (a proportion × 100). This has important consequences for data analysis, because the errors from a fitted model cannot be normally distributed since they are constrained to be 0 minimum and 100% at maximum, and the variance will not be constant – this is called heteroscedasticity.

Here we will use two traditional and valid ways to try to overcome this: one the so-called arcsine transformation (which technically is the arcsine of the square root of the proportion), and the other the logit transformation. A good discussion on the analysis of these sorts of data is provided by Kieschnick and McCullough (2003).

Arcsine transformation, also called angular transformation, re-ranges proportion data from minus infinity to plus infinity just like the range of the normal distribution. Although this method is widely recommended (Sokal and Rohlf, 1995; Gotelli and Ellison, 2004), it is not without its problems, and is not really recommended nowadays as there are better alternatives (see Warton and Hui's, 2011 amusingly titled paper).

We will explore data on plant attributes and establishment from Tunjai and Elliott (2012) (Table 13.1). We have saved the above table from Excel as a text file, 'Tunjai_regeneration.txt', so we can now access it using **read.table**.

tj<-read.table("Tunjai_regeneration.txt",header=TRUE)
head(tj)
```
Species             Regeneration_guild
Size_categorical Shape Coat Coat_mm X.MC establishment
1 Archidendron_clyperia Pioneer_exclusion
                  I    R    Tn   0.07    M          ME
2 Artocarpus_dadah Late-successional_non-dominants
                  I    R    M    0.13    M          ME
3 Callerya_atropurpurea Late-successional_non-dominants
                  L    R    Tk   1.34    M          HE
4 Cinnamomum_iners Late-successional_non-dominants
                  I    R    Tn   0.08    L          ME
5 Diospyros_oblonga Late-successional_non-dominants
                  I    O    M    0.31    L          HE
6 Diospyros_pilosanthera Late-successional_non-dominants
                  I    O    M    0.29    L          ME
  establishment.
6     11.25
2     36.25
3     41.57
4     21.25
5     43.13
6     17.50
```

Note that the name of the last column, our *response variable*, has been changed in the save-read process from 'establishment%' to 'establishment.'. This was because % has a special meaning in R. This is just an inconvenience and we will change it to something easier.

colnames(tj)[9]<-"PercentEstab"

We will transform the response variable by converting to a proportion and applying the trigonometric arcsine ($\sin^{-1}$) to the square root of data, referred to as arcsine transformation. This should then give us constant variance and normalized errors. The R function for arcsine is **asin**, thus:

arcsine_establishment<-asin(sqrt(tj$PercentEstab/100)) # division by 100 converts percentages to proportions and we will bind the proportions onto the right-hand end of our dataframe

Analysis of Variance (ANOVA)

Table 13.1. Seed traits and establishment success for 19 tree species (from Tunjai and Elliott, 2012).

Species	Regeneration_guild	Size_categ.	Shape	Coat	Coat_mm	% MC	Establishment[1]	Establishment%
Archidendron_clyperia	Pioneer_exclusion	I	R	Tn	0.07	M	ME	11.25
Artocarpus_dadah	Late_successional_non_dominants	I	R	M	0.13	M	ME	36.25
Callerya_atropurpurea	Late_successional_non_dominants	L	R	Tk	1.34	M	HE	41.57
Cinnamomum_iners	Late_successional_non_dominants	I	R	Tn	0.08	L	ME	21.25
Diospyros_oblonga	Late_successional_non_dominants	I	O	M	0.31	L	HE	43.13
Diospyros_pilosanthera	Late_successional_non_dominants	I	O	M	0.29	L	ME	17.5
Garcinia_cowa	Late_successional_subcanopy	I	O	M	NA	M	EF	NA
Garcinia_hombroniana	Late_successional_subcanopy	I	O	Tk	0.52	M	ME	15.65
Garcinia_merguensis	Late_successional_subcanopy	I	O	M	NA	L	EF	NA
Lepisanthes_rubiginosa	Late_successional_subcanopy	I	R	M	0.15	H	LE	0.63
Litsea_grandis	Late_successional_subcanopy	I	R	M	0.13	M	ME	18.13
Microcos_paniculata	Pioneer_exclusion	I	R	Tk	1.04	L	ME	24.38
Morinda_elliptica	Pioneer_exclusion	S	F	M	0.13	M	ME	3.13
Pajanelia_longifolia	Pioneer_exclusion	S	F	Tn	NA	H	EF	NA
Palaquium_obovatum	Late_successional_subcanopy	I	O	M	0.43	M	ME	28.75
Peltophorum_pterocarpum	Pioneer_exclusion	S	F	M	0.13	L	LE	4.38
Scolopia_spinosa	Late_successional_subcanopy	S	R	Tn	0.09	M	LE	10
Sandoricum_koetjape	Late_successional_non_dominants	I	R	Tk	0.34	M	LE	5.63
Vitex_pinnata	Late_successional_subcanopy	I	R	Tk	0.57	L	ME	25.63

[1]EF = establishment failure, none of the seedlings survived; LE = low establishment (<10%); ME = moderate establishment (10–40%); HE = high establishment (>40).

tj<-cbind(tj,arcsine_establishment) # the vector name will automatically be assigned as the column name.

The dataset 'tj' includes a three-level categorical variable describing features of the tree's seeds, a biological feature that plausibly could affect establishment success.

levels(tj$Size_categ) # seed size (intermediate, large and small)
[1] "I" "L" "S"
levels(tj$Coat) # seed coat thickness (medium, thick and thin)
[1] "M" "Tk" "Tn"
levels(tj$Shape) # shape (flat, oval and round)
[1] "F" "O" "R"

First we will examine just seed size (tj$Size_categ). Always plot your data first as it will show you if you have made any serious mistakes, such as missing out a decimal point. Note that R automatically recognizes the data type and produces a box and whisker plot.

boxplot(tj$arcsine_establishment ~ tj$Size_categ)

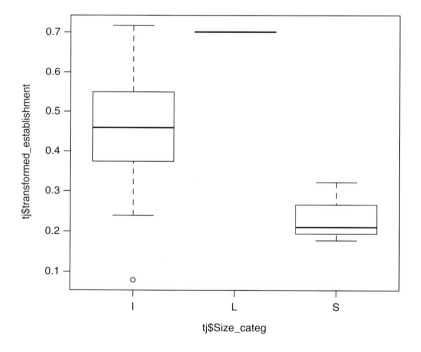

Immediately we can see that the order along the x axis is illogical. In fact, in all cases as our table is presented, plotting the response variable against each explanatory variable will produced a rather unhelpful plot, because the levels in each vector are unordered – R does not know that 'intermediate' comes between 'small' and 'large' for example. To view the data properly we need to order the levels in the first two factors – it is not immediately obvious to us that flat, oval and round really is a logical order on a continuous shape spectrum but seed size and coat thickness certainly are. We can show that they are NOT ordered at present by

is.ordered(tj$Size_categ) # commands that start with 'is.' are asking whether or not the following vector has the given property, in this case, is it ordered?

[1] FALSE

So now we need to put each of them in order
tj$Size_categ<-ordered(tj$Size_categ,levels=c("S","I","L"))
tj$Coat<-ordered(tj$Coat,levels=c("Tn","M","Tk"))
Check it worked OK
is.ordered(tj$Size_categ)
An alternative is to use
tj$Size_categ<-factor(tj$Size_categ,levels=c("S","I","L"),ordered=TRUE)
tj$Size_categ
[1] I I L I I I I I I I I S S I S S I I
is.ordered(tj$Size_categ)
[1] TRUE

The ordering worked OK so now we can create a plot
boxplot(tj$arcsine_establishment ~ tj$Size_categ)

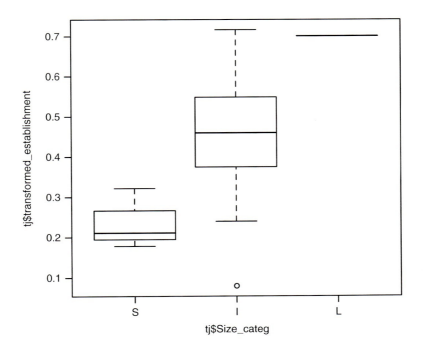

This graph certainly shows a progression in the data from low establishment success for trees with small seeds to greater success with increasing size. But is it significant? Our opinion might be strongly influenced by the single data point for large seeds. We need to perform an analysis of variance (ANOVA) using the function **aov**

modelA<-aov(tj$arcsine_establishment ~ tj$Size_categ)

We have now performed an ANOVA but before we can trust the statistical result we need to check whether the data agree with the assumptions of a parametric analysis (e.g. errors are normally distributed, show no major trends and there are no very high leverage outliers) by examining the four plots in the R object (modelA) we have just created:

par(mfrow=c(2,2)) # two rows and two columns of graphs to see all four plots at the same time rather than having to step through them one by one
plot(modelA)

You will get an error message
Warning messages:
1: not plotting observations with leverage one: 3
2: not plotting observations with leverage one: 3

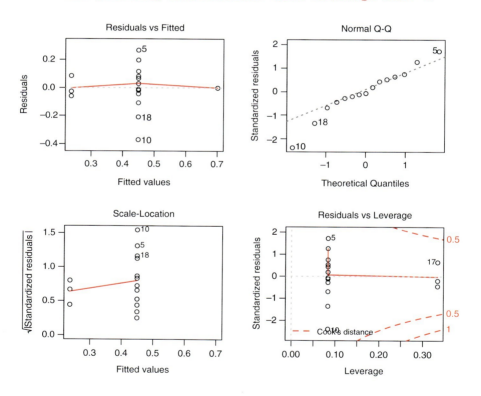

Note that in the two lower plots only two columns of data points have been plotted; the single large seed size one is omitted. Do not worry, this is because it is a categorical value and potentially could have enormous leverage. However, the departures from a straight line in the QQ plot do indicate some overdispersion, i.e. the tails of the error distribution are extended beyond a

theoretical normal one. For the moment we will ignore this and look at the results of the ANOVA.
summary(modelA)
```
                Df Sum Sq  Mean Sq  F value  Pr(>F)
tj$Size_categ    2 0.18945 0.094726  3.6047  0.05682 .
Residuals       13 0.34162 0.026278
---
Signif. codes:  0 '***' 0.001 '**' 0.01 '*' 0.05 '.' 0.1 ' ' 1
3 observations deleted due to missingness
```
The *p*-value (probability of the relationship resulting from chance alone) is 0.05682, i.e. not significant, but note that the three species that completely failed to establish were treated as NA.

Suppose we had not transformed the percentage data.
modelB<-aov(tj$PercentEstab ~ tj$Size_categ)
summary(modelB)
```
                Df Sum Sq Mean Sq  F value Pr(>F)
tj$Size_categ    2 1062.5  531.25   4.2209 0.03868 *
Residuals       13 1636.2  125.86
---
Signif. codes:  0 '***' 0.001 '**' 0.01 '*' 0.05 '.' 0.1 ' ' 1
3 observations deleted due to missingnesssummary(modelB)
                Df Sum Sq Mean Sq  F value Pr(>F)
tj$Size_categ    2 1062.5   531.25  4.2209 0.03868 *
Residuals       13 1636.2   125.86
---
Signif. codes:  0 '***' 0.001 '**' 0.01 '*' 0.05 '.' 0.1 ' ' 1
3 observations deleted due to missingness
```
If you plot modelB you will see that the errors are far more overdispersed. But we should still be worried about the overdispersion in the model with arcsine transformation, especially because the *p*-value is not highly significant so a slightly inappropriate model could give a misleading answer. You can try a transformation of the data to try to get rid of the overdispersion, for example, sqrt(tj$arcsine_establishment). If you do this the QQ plot is better still except for one outlier, and the *p*-value even further from significance.

Example 3 – Analysis with Proportion Data as Response Variable Using Logit Transformation

Now we will analyse the same data using logit transformation as recommended by Warton and Hui (2011).

The equation for logit transformation is

$$logit = log\left(\frac{p}{1-p}\right)$$

Where *p* is the proportion (percentage divided by 100). See Crawley (2011, pp. 572–573) for the derivation.

> **Information Box 13.1**
>
> *Logit with zero or 1 values in response variable.* The logarithm of zero is minus infinity, which means that if you have any zeros in your proportion data you cannot use the simple logit transformation. Further if the proportion is 1 the transformation results in division by zero. The standard way to deal with this issue is to add a small value, ε, to the zeros. Warton and Hui (2011) recommend setting this value to the smallest non-zero value, but clearly that loses the distinction. Others recommend adding 0.5 × the smallest non-zero value to the zeros and subtracting ε from the 1s. This also might be problematic if ε is greater than the difference between 1 and the largest non-1 value, so we would recommend subtracting half the difference between 1 and the largest non-1 value from the ones. Importantly there is no fixed rule and Warton and Hui (2011) encourage users to experiment with different values of ε.

tj$logit_establishment<-with(tj, log((PercentEstab/100)/(1-(PercentEstab/100))))
The above code adds a new column 'logit_establishment' to the dataset, and we re-run the ANOVA using this:
modelC<-aov(tj$logit_establishment ~ tj$Size_categ)
plot(modelC,which=2) # the which argument allows us to select which plots to view

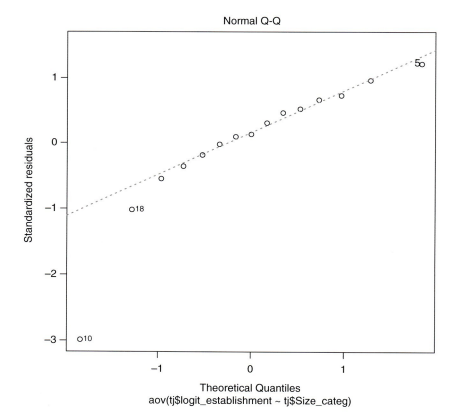

The model has a markedly straighter QQ plot (though there is one outlier that could have enormous influence on our result) so it should provide a better explanation of the data.
summary(modelC)

```
               Df   Sum Sq   Mean Sq   F value   Pr(>F)
tj$Size_categ   2    6.015    3.007     2.13     0.158
Residuals      13   18.356    1.412
3 observations deleted due to missingness
```

Again the *p*-value is not significant so we can reliably conclude that there is no effect of seed size on regeneration, given these data. In summary, we would have concluded erroneously that seed size was a significant factor in determining the degree of establishment success among those species that did sometimes become established! We should re-run the model without the outlier just in case.

Exercise Box 13.1

1) Re-analyse the seed establishment data excluding the outlier point.
2) Re-analyse using logit transformation the seed establishment data including the three species that completely failed to establish (*G. cowa*, *G. merguensis* and *P. longifolia*). Hint: use which(is.na(tj$PercentEstab)). Examine the model. Is it good?
3) Analyse the extended dataset without transformation, and compare the model with the transformed data. Which is a better model?
4) What do you conclude about the potential influence of seed traits on establishment?

Tool Box 13.1 ✕

- Use aov() to perform analysis of variance.
- Use asin() to calculate the arcsine ($\sin^{-1}$) of a number.
- Use is.na() to determine which components of a vector are NAs.
- Use is.ordered() to display whether levels of are ordered.
- Use levels() to see categories of an attribute of a variable.
- Use ordered(filename, levels=c()) to order levels.
- Use sqrt() to square root a number.
- Use which= to specify which summary plot from a model to display.

14 Analysis of Covariance (ANCOVA)

Analysis of covariance or ANCOVA is a combination of ANOVA and regression. It tests the effects of a mix of continuous and categorical variables on a continuous response variable.

Example 1 – Growth of Tagged Gobies

Malone *et al.* (1999) investigated the effects of two types of tagging (acrylic paint and subcutaneous microtags) on the growth of the coral reef goby, *Coryphopterus glaucofraenum*, in the British Virgin Islands and included initial size as a continuous explanatory variable. We will use ANCOVA to determine how initial size and tagging type influence growth of adult fish, and as we always should, first we plot the data and then examine the model. Note that the interaction term '*' in the model expression is just a shorthand for {*explanatory variable 1 + explanatory variable 2 + explanatory variable 1: explanatory variable 2*}; the two formats will give identical results.

```
Goby_growth<-read.csv("Chapter_14_Malone_gobies.csv")
with(Goby_growth,plot(Initial_length.mm,Growth.mm,pch=c(rep(1,22),
    rep(19,24)),cex=3,col=c(rep("black",22),rep("grey50",24))))
legend("topright",c("Visual implant","Acrylic paint"),pch=c(1,19),bty="n",
    col=c("black","grey50"),cex=1.2)
```

Analysis of Covariance (ANCOVA)

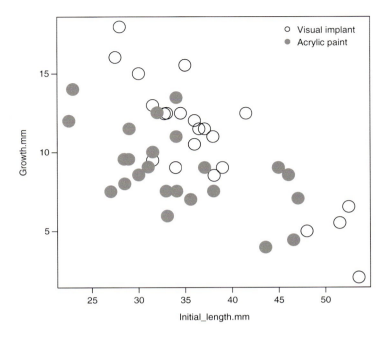

```
par(mfrow=c(2,2))
goby_model<-lm(Goby_growth$Growth.mm~Goby_growth$Treatment*
    Goby_growth$Initial_length.mm)
plot(goby_model)
```

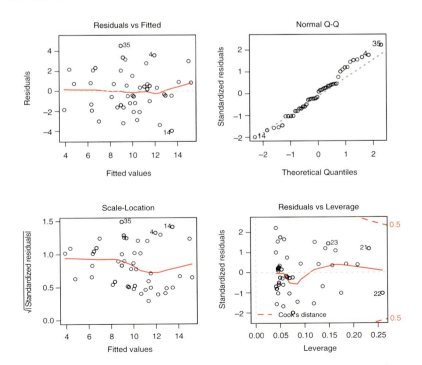

On this occasion the model looks reasonably acceptable despite the slight disjunction of the upper nine points in the QQ plot, so we will examine its implications.
summary(goby_model)
```
Call:
lm(formula = Goby_growth$Growth.mm ~ Goby_growth$Treatment *
Goby_growth$Initial_length.mm)
Residuals:
    Min      1Q  Median      3Q     Max
-3.9933 -1.3862 -0.0813  0.8599  4.5258
Coefficients:
                                     Estimate Std.   Error
t value  Pr(>|t|)
(Intercept)                          16.00725 2.12672
   7.527  2.59e-09 ***
Goby_growth$TreatmentVisual_implant  11.17647 3.13042
   3.570  0.000909 ***
Goby_growth$Initial_length.mm        -0.20708 0.06107
  -3.391  0.001529 **
Goby_growth$TreatmentVisual_implant:
   Goby_growth$Initial_length.mm     -0.22761 0.08567
  -2.657  0.011101 *
---
Signif. codes: 0 '***' 0.001 '**' 0.01 '*' 0.05 '.' 0.1 ' ' 1
Residual standard error: 2.082 on 42 degrees of freedom
Multiple R-squared: 0.6372,   Adjusted R-squared: 0.6112
F-statistic: 24.58 on 3 and 42 DF,  p-value: 2.415e-09
```
From the summary we see that growth is highly significantly negatively correlated with initial length ($p < 0.005$), and that treatment also has a highly significant effect on growth ($p > 0.001$). Further the interaction term is significant meaning that treatment affects the relationship between growth and initial length (i.e. the slopes differ significantly).

Example 2 – Fitting through the Origin and Count Data as Response Variable

For many types of data, we can get a feel for what we should expect if we imagine (calculate) what we would find at the extremes. For example, if we measure some difference from normal and apply a treatment that goes from nothing to some maximum then we know that at zero treatment, there must be zero effect. In other words, our plotted relationship must necessarily pass through the origin (0,0). When we know this, we can force the line of best fit to pass through the origin by including '–1' in the model call.

Here we will analyse data from Srimuang et al. (2010) on the number of pollinaria removed by pollinators from inflorescences of two Sirindhornia orchid

Analysis of Covariance (ANCOVA)

species (*S. monophylla* and *S. mirabillis*) in relation to the number of flowers in the inflorescence (also count data) and the orchid species (categorical) using ANCOVA with Poisson errors. We include '–1' in the model call because it is obvious that if there are no flowers, there can be no pollinaria removed.

```
sir<-read.csv("CHAP14_Sirindhornia_count.csv")
modelPoiss<-glm(sir$pollinaria ~ sir$flowers*sir$species-1,family="poisson")
summary(modelPoiss)
Call:
glm(formula = data$pollinaria ~ data$flowers * data$species - 1,
    family = "poisson")
Deviance Residuals:
     Min       1Q    Median       3Q      Max
-2.08415 -0.64772 -0.07994  0.63934  2.81307
Coefficients:
                                     Estimate    Std. Error
z value  Pr(>|z|)
data$flowers                         0.035197    0.006794
  5.180   2.21e-07 ***
data$speciesmirabilis                1.236014    0.195548
  6.321   2.60e-10 ***
data$speciesmonophylla               1.510332    0.288695
  5.232   1.68e-07 ***
data$flowers:data$speciesmonophylla  0.032097    0.019378
  1.656     0.0976 .
---
Signif. codes:  0 '***' 0.001 '**' 0.01 '*' 0.05 '.' 0.1 ' ' 1
(Dispersion parameter for poisson family taken to be 1)
Null deviance: 786.112  on 29  degrees of freedom
Residual deviance:  37.344  on 25  degrees of freedom
AIC: 157.66
```

The results are very clear, the number of pollinaria removed is highly significantly related to the number of flowers in the inflorescence and there are highly significant effects of species but there is no significant difference between the numbers of flowers per inflorescence between the species.

Exercise Box 14.1

1) Investigate the difference in the model if we had not forced the fit to pass through the origin.
2) The most appropriate way to analyse the *Sirindhornia* data might not be so obvious. Plot the data and reanalyse them as if the response variable were continuous and plot both that model and the one above using **glm**. What do you conclude?
3) Plot both models in the same plotting window to facilitate comparison. Which do you think is better?
4) Use Akaike's information criterion to compare the models. Were you right?

Tool Box 14.1 ✜

- Use anova() to show a table of analysis of variance tables for one or more fitted models.
- Use as.factor() to treat variables as categorical factors.
- Use as.numeric() to treat variables as numbers.
- Use family=binomial in **glm** to set the parameter for family as binomial (or another specified error structure).
- Use glm() to carry out a generalized linear model.
- Use table() to summarize variables.

15 More Generalized Linear Modelling

Summary of R Libraries Introduced
   aod
   MASS

Summary of R Functions Introduced
   colMeans
   colSums
   dose.p in "MASS"
   jitter
   rowMeans
   rowSums
   split

This is a huge topic and we will only give the basics; for further knowledge and understanding you should consult books such as Crawley (2012) and Dunn and Smyth (2018). We employ generalized linear modelling using the function **glm** (pronounced glim) when we know that variances are not constant with one or more explanatory variables and/or we know that the errors cannot be normally distributed, for example, they may be binary data, or count data where negative values are impossible, or proportions which are constrained between 0 and 1. A **glm** seeks to determine how much of the variation in the response variable can be explained by each explanatory variable, and whether such relationships are statistically significant. The data for generalized linear models take the form of a continuous response variable and a combination of continuous and discrete explanatory variables.

Model Inspection

For the resulting probabilities to be valid the main criterion to be met is that the residuals from the best-fitted relationship are normally distributed. If they are not, then one or more of the variables can be transformed in a linear fashion until that criterion is met. Of course, there is always noise in real data, but what we need to achieve is as near to normal a distribution of residuals as possible with no obvious trends in relation to the response variable. In this chapter, we take the reader through a number of ways of altering the data to achieve a normal distribution.

Binary Response Variable with One Continuous Explanatory Variable

Example 1 – Logistic regression of gall former predation

Sometimes data are binary, typically presence or absence of something, or alive or dead. Obviously binary data are not normally distributed and errors from the observed points cannot be either. To test whether such data show a significant trend with some other variable, we use a logistic regression. Data on the presence/absence of an Indian braconid wasp, *Bracon garugaphaga*, in galls formed by its host psyllid on the plant *Garuga pinnata* are given in Table 15.1.

In this case we cannot load the data into R directly using **read.table** (**textConnection**({"*cut & paste the table here*"}),header=TRUE) because the column headers contain spaces and we will get an error message such as
Error in scan(file = file, what = what, sep = sep, quote = quote, dec = dec, :
line 1 did not have 16 elements
We could delete the header row and read the table with 'header=FALSE', we could manually edit the column heading is the table or we can type in the vectors as below.
psyllids<-c(0:9)
no_Bracon<-c(5,139,153,109,70,28,20,2,3,1)
Bracon<-c(63,57,61,36,20,9,3,0,0,0)
The proportion containing a *Bracon* can be calculated by dividing by the total Bracon + no_Bracon
PropBracon<-Bracon/(Bracon+no_Bracon)
plot(psyllids,PropBracon,pch=15,xlab="Number of psyllids in gall",ylab= "Proportion of galls containing a Bracon")

Table 15.1. Relationship between numbers of gall-forming psyllids and number of entomophytophagous parasitic *Bracon* wasp larvae in galls (based on the raw data used to construct Fig. 5 in Ranjith *et al.*, 2015).

Number of psyllids in gall	Number galls with *Bracon* absent	Number of galls with *Bracon* present
0	5	63
1	139	57
2	153	61
3	109	36
4	70	20
5	28	9
6	20	3
7	2	0
8	3	0
9	1	0

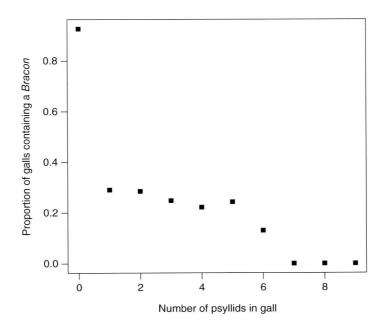

Because these data are proportions, the errors cannot be normally distributed (they are bounded by zero at bottom and 1 at top) so we should use a **glm** with binomial errors, but further we cannot do the logistic regression straight on the proportions, we need a vector containing the numbers of galls with and without *Bracon* separately
data<-cbind(Bracon,no_Bracon)
model_galls<-glm(data~psyllids,family=binomial) **# we can omit "family="**
summary(model_galls)

```
Call:
glm(formula = data ~ psyllids, family = binomial)

Deviance Residuals:
    Min       1Q   Median       3Q      Max
-4.0171  -0.4778  -0.0484   1.0680   7.1137

Coefficients:
            Estimate Std.Error z value Pr(>|z|)
(Intercept)  0.14286   0.14348   0.996    0.319
psyllids    -0.42500   0.06147  -6.914 4.71e-12 ***
---
Signif. codes:  0 '***' 0.001 '**' 0.01 '*' 0.05 '.' 0.1 ' ' 1

(Dispersion parameter for binomial family taken to be 1)

    Null deviance: 131.651  on 9  degrees of freedom
Residual deviance:  75.423  on 8  degrees of freedom
AIC: 110.35

Number of Fisher Scoring iterations: 4
```

The relationship is highly significant $p = 4.71\mathrm{e}^{-12}$. To plot the regression line on our figure we use **predict** and pass to it a vector of x values (xv) spanning the range of observed x values with many fine divisions so the curve will appear smooth.

xv<-seq(0,9,0.01)
yv<-predict(model_galls,list(psyllids =xv),type="response")
lines(xv,yv,lwd=2,col=4)

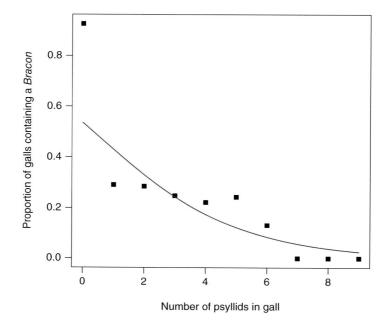

Other than showing proportions we might want to plot the relationship in visually informative ways. We want to distinguish how many galls contained or did not contain a *Bracon* larva. We could use **text** to write the number either instead of or above a point, we could adjust the size of the symbol displayed (varying the value of the **cex** argument according to the number), we could draw circles of different sizes using the equation of a circle or we can use **jitter**, which adds a small, specifiable amount of noise to the data points so that several coincident points are now randomly shifted so they will appear distinct. For fun we will explore the latter since you have already seen how to change symbol sizes and put text on figures. We will create a two-column vector from our data. Every row represents a gall with the first column the number of psyllids found and second column the absence (0) or presence (1) of a *Bracon* in the gall. This will give exactly the same raw data as were collected in the field notebook. The process will use a nested loop.
nPsyllid<-NULL
Bracon_status<-NULL

```
for(i in 1:10){
    np<-i-1 # because the first row in data corresponds to zero psyllids
    if(data[i,1]>=1){
        for(j in 1:data[i,1]){
            nPsyllid<-c(nPsyllid,np)
            Bracon_status<-c(Bracon_status,0)}}
    if(data[i,2]>=1){
        for(j in 1:data[i,2]){
            nPsyllid<-c(nPsyllid,np)
            Bracon_status<-c(Bracon_status,1)
            }} # end if statement and j loop
} # end i loop
xpandedData<-cbind(nPsyllid,Bracon_status)
```

This bit of code is made easier to follow by indenting for loops and **if** statements, which is generally good programming practice. The **if** statements are there so that nothing is added on to either the vectors 'nPsyllid' or 'Bracon_status' if there were no galls with a given combination, otherwise by just starting the 'for(j in…' loops a value will be added to the vectors. Now we can plot this, also adding a customizable amount of random noise using the function **jitter** to each of the points.

```
plot(jitter(xpandedData[,1],1,0.15),jitter(xpandedData[,2],1,0.03),
    ylab="Bracon in gall",xlab="Total number of psyllids in gall",main="",
    pch=20,cex=0.3,axes=F)
axis(1,c(0:9),at=c(0:9)); axis(2,c("absent","present"),at=c(0,1))
```

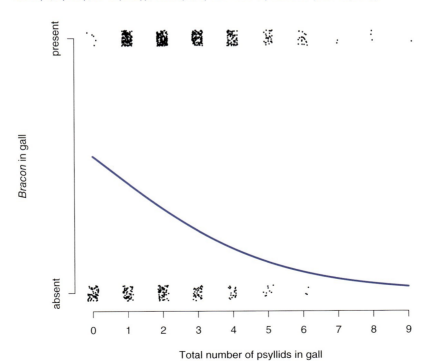

The second and third arguments passed to **jitter** are not necessary, but they allow you to customize the spread; type '?jitter' at the prompt for more details. We think the effect is quite nice.

When we present data graphically we also need to write informative figure legends and these are often a good place to give the details of the results of the statistical test.

Here is an example of what we might write for the *Bracon* gall data. The relationship between number of psyllids and the presence/absence of a *Bracon* larva is incredibly highly significant and R gives it as an exponent '$p < 5e^{-12}$', which means 5×10^{-12} or $p < 0.000000000005$.

Fig XX. Logistic regression of presence or absence of *Bracon* in galls (n=779) versus number of psyllids present, showing highly significant negative relationship (GLM with binomial errors: null deviance = 976.24 on 778 degrees of freedom, residual deviance = 920.01 on 777 degrees of freedom, AIC = 924.01, p<5e-12).

LD50s

Another important aspect of logistic regression is in the estimation of lethal dosages (e.g. LD50). For those that do this sort of research there is a useful function **dose.p** in the **MASS** library that, given a logistic model, will give estimates of LD at whatever level is desired. **dose.p** takes the name of the fitted model, an argument 'cf=', which will normally be c(1,2) if the model contains just one explanatory variable, and 'p={a vector of probabilities that you want estimates for}'. The output calls the estimates 'Dose' but can be other things depending upon what you are analysing.

Exercise Box 15.1

1) The dataset 'menarche' in the **MASS** library gives the ages of menarche for a sample of Polish girls in 1965. Fit and plot a logistic regression model through the proportion of girls having reached menarche at a given age. Hints: (i) the response variable will be **cbind** (Menarche,Total-Menarch); and (ii) use binomial errors with probit link.
2) Use **dose.p** to estimate the best estimate of ages when 10%, 50% and 90% of girls will have reached menarche.

Information Box 15.1

Logit versus probit link function. Logit and probit are both based on cumulative distributions: logistic and standard normal, respectively. In practice, which one you use usually makes very little difference and choice between the two often depends on the field, probit being more popular in the biomedical sciences.

Example 2 – Pollinator counts – showing importance of deviance

For count data, we must use a generalized linear model because linear models may predict negative counts, which would be impossible. Further, count data often follow or approximate a Poisson distribution in which the variance increases with the mean so the residual deviances in the top left of the set of four diagnostic plots will not be flat. Count data are not the same as proportion data because we only know the number of instances that fall in any category, not the numbers that do not. In the following worked example, we know the number of bees that visited a flower, but we do not know how many did not!

O'Hanlon et al. (2014) described a simple experiment with the flower-mimicking Malaysian orchid mantis, *Hymenopus coronatus*, which closely resemble flowers of *Asystasia intrusa* among which they often live, and their resemblance is thought to attract and deceive pollinating insects, which they then devour. The mantid also occurs in southern Thailand around Krabi. The authors counted pollinator visits to three sticks placed in various locations, one with an orchid mantis, one with a flower and a plain control. We can enter the data as follows

```
mantid<-c(2,8,10,12,3,5,2,0,15,6,2,2,1,3,7,16,5,17,9,8,3,8,4,6,9,4,
    10,10,17,14,32,6,12)
flower<-c(4,2,4,7,3,5,4,2,8,10,1,12,0,1,6,11,4,9,9,10,2,4,8,4,4,7,4,8,
    4,5,28,5,5)
control<-c(0,0,0,0,2,2,0,0,0,4,0,0,0,0,0,1,0,1,0,0,0,0,0,0,1,0,1,0,
    1,0,2,0,0)
m<-as.data.frame(cbind(as.numeric(mantid),as.numeric(flower),
    as.numeric(control)))
colnames(m)<-c("Total.Mantid","Total.Flower","Total.Control")
head(m)
  Total.Mantid   Total.Flower   Total.Control
1            2              4               0
2            8              2               0
3           10              4               0
4           12              7               0
5            3              3               2
6            5              5               2
```

The data are also available as an online text file and can be read using
`m<-read.table("mantid.txt",header=TRUE)[,2:4]`

We can extract column and row totals and means using **colSums**, **rowSums**, **colMeans** and **rowMeans** functions, e.g.

```
colMeans(m[,1:3])
Total.Mantid   Total.Flower   Total.Control
   8.1212121      6.0606061       0.4545455
rowSums(m)
 [1]  6 10 14 19  8 12  6  2 23 20  3 14  1  4 13 28  9 27 18 18  5
     12 12 10 14 11 15 18 22 19 62 11 17
```

to get these, we could alternatively use **apply**, one of a family of R base functions thus

apply(m,1,sum) # the 1 means that the function **sum** will be applied to the rows
```
[1]  6 10 14 19  8 12  6  2 23 20  3 14  1  4 13 28  9 27 18 18  5
    12 12 10 14 11 15 18 22 19 62 11 17
```
apply(m,2,sum) # the 2 applies it to the columns
```
Total.Mantid   Total.Flower   Total.Control
         268            200              15
```
We can see that there were on average more pollinator visitors to the mantids than to the flowers and both received far more than the controls, as expected. We should plot the data to see how they look, using the histogram function **hist** and we will specify that we want bins of width 1 using a vector of break points. Also we will add a vertical line on each to show where the mean number of visitors is, using the function **abline** with 'v=' to indicate we need a vertical line; to plot an horizontal line you use 'h='

par(mfrow=c(2,3)) # Creating a plot area so we can put three histograms side-by- side and not have them too tall
par(mar=c(2,2,2,2)) # setting the margin around each plot to 2, reducing the border around the plots so they look neater
brks<-seq(1:34)-1.5 # creates a sequence from –0.5 to 32.5, which spans the full range of data with breaks always between other integer counts
hist(m$Total.Mantid,breaks=brks,main="Visits to mantid")
abline(v=mean(m$Total.Mantid),col="blue",lwd=2,lty=2)
hist(m$Total.Flower,breaks=brks,main="Visits to flowers")
abline(v=mean(m$Total.Flower),col="blue",lwd=2,lty=2)
hist(m$Total.Control,breaks=brks,main="Visits to controls")
abline(v=mean(m$Total.Control),col="blue",lwd=2,lty=2)

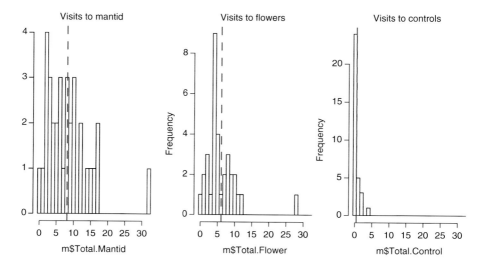

More Generalized Linear Modelling

However, to determine whether the type of item presented (mantid, flower, nothing (=control)) influences the number of visits we need to put the type of model in one column of a dataframe and the numbers of visits in another. Because it is abundantly clear that the control sticks received virtually no visitors we will concentrate only on the numbers of potential pollinators visiting the mantid and real flower.

```
reps<-nrow(m)
type_list<-c(rep("Mantid",reps),rep("Flower",reps))
visitors<-c(m$Total.Mantid,m$Total.Flower)
# new.data<-as.data.frame(cbind(type_list,visitors))
model1<-glm(visitors ~ type_list,family=poisson)
par(mfrow=c(2,2))
plot(model1)
hat values (leverages) are all = 0.03030303
    and there are no factor predictors; no plot no. 5
```

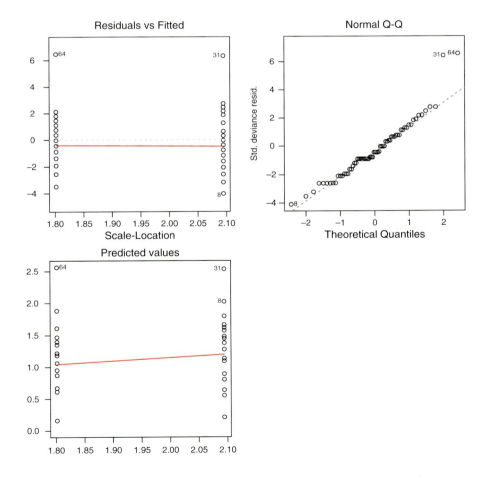

In this example, there is no fourth plot as all the points are coincident. With the exception of two outliers (data lines 31 and 64), which we can see on the histograms and QQ plots above, the data are behaving well, so we will look at the result of the model, though we should also perform the analysis with the two outlier points excluded. If both results are virtually the same we have no need to worry. However, if they differ we should be more cautious and probably need to collect more data.

```
summary(model1)
Call:
glm(formula = visitors ~ type_list, family = poisson)

Deviance Residuals:
    Min       1Q   Median       3Q      Max
-4.0302  -1.5487  -0.4444   0.7505   6.4671

Coefficients:
                 Estimate Std. Error z value Pr(>|z|)
(Intercept)       1.80181    0.07071  25.481  < 2e-16 ***
type_listMantid   0.29267    0.09344   3.132  0.00174 **
---
Signif. codes:  0 '***' 0.001 '**' 0.01 '*' 0.05 '.' 0.1 ' ' 1

(Dispersion parameter for poisson family taken to be 1)

    Null deviance: 263.30  on 65  degrees of freedom
Residual deviance: 253.38  on 64  degrees of freedom
AIC: 487.91

Number of Fisher Scoring iterations: 5
```

The results seem to show that the mean number of pollinator visits to the flower mimicking mantis (8.12) is significantly higher than the number of visits to the real flowers (6.06). However, there is another thing we must look at in the output of the **glm**, which we see by typing the model name, which is how much of the observed variation the model explains.

```
model1
Call:  glm(formula = visitors ~ type_list, family = poisson)

Coefficients:
    (Intercept)  type_listMantid
         1.8018           0.2927

Degrees of Freedom: 65 Total (i.e. Null);  64 Residual
Null Deviance:      263.3
Residual Deviance:  253.4    AIC: 487.9
```

It is important to look at two numbers in the output – the Null Deviance and the Residual Deviance – both are measures of how badly the model fits. If the model is a good fit (i.e. it explains most of the variation) then both should be

small relative to their respective degrees of freedom. In this particular case, if the number of pollinator visitors was very well explained by whether it was a flower or a mantid then the Null Deviance will be small, and there would be little variation left unexplained. On the contrary, a large part of the variation is left unexplained by the fitted model. Further note that the residual deviance of 253.4 (i.e. unexplained variation) is a lot larger than the residual degrees of freedom. This means that we cannot trust this probability because the data are overdispersed. This means that there are probably other important factors that influenced the numbers of pollinator visits which were not measured. Here, only whether it was a flower or a mantid was recorded. In the worst case, everything that was measured might actually have been unimportant (Crawley, 2008). One way of addressing overdispersion is to use quasipoisson errors rather than Poisson errors, and we do this in a new **glm** using family=quasipoisson
model2<-glm(visitors ~ type_list,family=quasipoisson)# plot(model2)
summary(model2)

```
Call
glm(formula = visitors ~ type_list, family = quasipoisson)

Deviance Residuals:
    Min       1Q   Median       3Q      Max
-4.0302  -1.5487  -0.4444   0.7505   6.4671

Coefficients:
                  Estimate Std. Error t value Pr(>|t|)
(Intercept)         1.8018     0.1516  11.886   <2e-16 ***
type_listMantid     0.2927     0.2003   1.461    0.149
---
Signif. codes:  0 '***' 0.001 '**' 0.01 '*' 0.05 '.' 0.1 ' ' 1

(Dispersion parameter for quasipoisson family taken to
   be 4.595668)

Null deviance: 263.30  on 65  degrees of freedom
Residual deviance: 253.38  on 64  degrees of freedom
AIC: NA

Number of Fisher Scoring iterations: 5
```

Applying quasipoisson errors to take into account the overdispersion now gives a different result, i.e. the number of visits to the mantid and flowers are not significantly different from one another. Where might the overdispersion result come from? Well, the data are in fact paired, with each triplet of tests being carried out at different times and places. If we run a paired Mann-Whitney U test on the mantid and flower visitor numbers (remember the raw data are skewed

rather than normal so we cannot use a parametric t-test directly) we do indeed find a significant difference between mantid and flower visitors:
wilcox.test(mantid,flower,paired=T)
```
Wilcoxon signed rank test with continuity correction
data: mantid and flower
V = 349, p-value = 0.01672
alternative hypothesis: true location shift is not equal to 0
```
We get two warning messages because of ties and zeros, but *p*-value is quite significant. Therefore, in this case the overdispersion that led to the use of **glm** with quasipoisson errors was correctly saying that the data themselves (not taking into account any pairing of visitor numbers) do not warrant the conclusion that mantids receive more pollinator visits than real flowers. The overdispersion was probably due to the effects of site, time and mantid type, and the paired nature of the original study allows for that to be factored out. Indeed, it seems that the flower-mimicking mantid is a more effective attractor of pollinators than the real flowers used.

The CRAN package **aod** implements several methods for analysing overdispersed data.

Example 3 – Proportion data with N known

The snails dataset in the **MASS** library includes data on number of deaths, sample sizes (conveniently all 20) and several environmental treatments for two species of snail. We will look at the treatments temperature and humidity and make separate boxplots for each snail species.
library(MASS)
data(snails)
spA<-snails[1:48,]; spB<-snails[48:96,]
par(mfrow=c(2,2))
boxplot(spA$Deaths~spA$Temp,main="Species A",xlab="Temperature")
boxplot(spA$Deaths~spA$Rel.Hum,main="Species A",xlab="Relative Humidity")
boxplot(spB$Deaths~spB$Temp,main="Species B",xlab="Temperature")
boxplot(spB$Deaths~spB$Rel.Hum,main="Species B",xlab="Relative Humidity")

More Generalized Linear Modelling

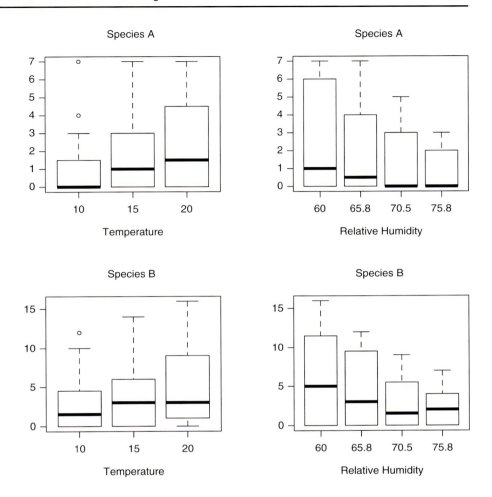

For both species there are consistent trends increased mortality higher temperature and decreased mortality with higher humidity. But notice the y axis scales are very different between species, so we really do not need to test if there is a species effect. Therefore we will nest the analyses within snail species using the '/' sign so we will get separate results for each species.

The dataset only gives numbers of deaths so we will add a 'column' for numbers dead or alive and we can pass these pairs of numbers to the **glm**.
snails$deadORalive<-cbind(snails$Deaths,20-snails$Deaths)

Information Box 15.2 📖

*Using **split***. In the snails example above we specified which rows were species A or species B. A useful alternative, especially when there are more than two categories is to use the function **split** thus,
spp<-split(snails,snails$Species)
and then we can access each species separately as spp$A and spp$B.

```
head(snails$deadORalive)
     [,1] [,2]
[1,]    0   20
[2,]    0   20
[3,]    0   20
[4,]    0   20
[5,]    0   20
[6,]    0   20
m1<-glm(deadORalive ~ Species/(Temp+Rel.Hum),family=binomial,
    data=snails)
```

Plotting the model shows considerable irregularity in the QQ plot and strong heteroscedasticity so we will also run a model with quasibinomial errors. Comparing their summaries is very informative about taking the problem with variance into account.

```
m2<-glm(deadORalive~Species/(Temp+Rel.Hum),family=quasibinomial,
    data=snails)
summary(m1)
Call:
glm(formula = deadORalive ~ Species/(Temp + Rel.Hum),
    family = binomial,
    data = snails)

Deviance Residuals:
   Min      1Q   Median      3Q     Max
-4.445  -1.980  -1.085   1.457   3.753

Coefficients:
                   Estimate Std. Error z value Pr(>|z|)
(Intercept)         1.47497    1.44869   1.018 0.308613
SpeciesB            1.87636    1.76834   1.061 0.288652
SpeciesA:Temp       0.08313    0.02952   2.816 0.004856 **
SpeciesB:Temp       0.06753    0.02049   3.297 0.000979 ***
SpeciesA:Rel.Hum   -0.07675    0.02084  -3.683 0.000230 ***
SpeciesB:Rel.Hum   -0.08583    0.01465  -5.859 4.65e-09 ***
---
Signif. codes:  0 '***' 0.001 '**' 0.01 '*' 0.05 '.' 0.1 ' ' 1

(Dispersion parameter for binomial family taken to be 1)

Null deviance: 539.72  on 95  degrees of freedom
Residual deviance: 416.82  on 90  degrees of freedom
AIC: 587.69

Number of Fisher Scoring iterations: 5

summary(m2)
```

```
Call:
glm(formula = deadORalive ~ Species/(Temp + Rel.Hum),
    family = quasibinomial,
data = snails)

Deviance Residuals:
   Min      1Q   Median      3Q      Max
-4.445  -1.980  -1.085   1.457   3.753

Coefficients:
                   Estimate Std. Error t value Pr(>|t|)
(Intercept)         1.47497    2.88755   0.511  0.61074
SpeciesB            1.87636    3.52467   0.532  0.59580
SpeciesA:Temp       0.08313    0.05883   1.413  0.16110
SpeciesB:Temp       0.06753    0.04083   1.654  0.10163
SpeciesA:Rel.Hum   -0.07675    0.04153  -1.848  0.06792 .
SpeciesB:Rel.Hum   -0.08583    0.02920  -2.940  0.00418 **
---
Signif. codes: 0 '***' 0.001 '**' 0.01 '*' 0.05 '.' 0.1 ' ' 1

(Dispersion parameter for quasibinomial family taken to
    be 3.972894)

    Null deviance: 539.72  on 95  degrees of freedom
Residual deviance: 416.82  on 90  degrees of freedom
AIC: NA

Number of Fisher Scoring iterations: 5
```

The first thing to note is that when the heteroscedasticity is corrected for using the quasi version, we lose several apparently highly significant effects. We conclude that despite the same visual trends, only in the case of species B is there statistical support for an effect of humidity on survival, though the result for species A is suggestive as one might expect for snails.

Exercise Box 15.2

1) Re-run the analyses (quasibinomial errors) with species and other variables as separate factors. What do the summaries indicate?

Tool Box 15.1

- Use apply(filename, 1, sum) to calculate the sum of rows.
- Use apply(filename, 2, mean) to calculate the mean of columns.
- Use colMeans() to calculate the means of columns.
- Use colSums() to calculate the sums of columns.
- Use family= to specify the error distribution for a glm.
- Use glm() to calculate the amount of variance explained by the variables.
- Use h= in **abline** to set a line as horizontal.
- Use jitter() in plot to separate coincident points so that they appear distinct.
- Use mar= in par to set the margins.
- Use rowMeans() to calculate the means of rows.
- Use rowSums() to calculate the sum of rows.
- Use split() to separate values in a vector according to a supplied list of factors.
- Use v= in **abline** to set a line as vertical.

16 Monte Carlo Tests and Randomization

Summary of R Libraries Introduced	Summary of R Functions Introduced
picante	paste
	round

In this chapter we explore the statistical approach that gets its name from the Mediterranean city famous for its casinos. It involves randomizing the observed numbers many times and comparing the randomized results with the original observed data. It is easy to go wrong when what you are trying to simulate is complicated. Here we will stick to straightforward examples. We will also show how randomization can be used in experimental design and sampling.

If the measure of the original data being considered lies within those of 95% of the results of randomizations, we can assume that the original data probably resulted from chance alone. On the other hand, if the observed metric lies within the 5% areas at either extreme (2.5% on each tail) then we have statistical evidence that the observations are reflecting some degree of positive or negative correlation (clustering).

The function **sample** randomly selects elements from a list. Random numbers in R can therefore be generated using **sample** from a list. If the list is from 0 to 1, you will get a random binary number, if it is from 0 to 9, a random decimal number etc.

Random Number Generator Code

You can write an R script that simulates **sample**. The following uses a commonly employed algorithm that includes a large prime number (M), a seed value and another large number (A). Here we can generate a sequence of 10,000 prime numbers between 0 and 9 and use **hist** to plot their relative frequencies.

Random<-NULL # this will become our vector of 10,000 random numbers
seed<-9997 # less than 10,000 and each seed will lead to a different random number sequence
A<-16877 # this number is 7^5

```
M<-39119 # a large prime number; 2147483647 = 2^31-1 is often used
for(i in 1:10000){
    seed<-M*A*seed %% M/M # %% means the modulus, i.e. the re-
        mainder after division; M*A*seed is likely too large for the function
        round to be used; %/% gives the integer part of a division
    s<-as.character(seed)
    Random<-c(Random,as.numeric(substr(s,nchar(s),nchar(s))))  # gets
        the last digit of seed
} # end i loop
hist(Random,breaks=c(0:10)-0.5) # plots the frequencies of values in bins
    from minus 0.5 to 9.5; 0 goes in the 1st bin, 1 in the 2nd, etc.
```

Technically all algorithm-generated 'random' numbers are called pseudo-random – that is they pass tests for randomness but every time the same code is run you will get the same sequence of pseudorandom numbers, and eventually they will repeat, but not until after a very long time. So in our example, every time we run the code, we will get the same histogram. This property can sometimes be useful if we want to repeat things exactly, but if you want each run to be different you need to change the value of seed manually each time or you can choose the seed in an effectively random way, such as taking it from the system clock. It should be noted that while you may see code online as using **as.double** meaning double precision numbers (i.e. twice as many digits compared to a normal real number), R, unlike some other languages, has no single precision numbers so all are treated as double precision. Thus, in R, **as.double** is identical to **as.numeric**.

Example 1 – Flower Visits by Thai Honey Bee Species

As an example, we will use the data provided by Suwannapong et al. (2011) on the utilization of different flower species by four Thai species of honey bee (*Apis* spp. – viz. *A. cerana*, *A. dorsata*, *A. florea* and *A. mellifera*). To determine whether the flowers visited by the bee species were random (the null hypothesis) or whether the species had evolved to partition floral resources, such that competition was reduced, and conceivably some degree of specialization might evolve, we will use a Monte Carlo approach. The feature we are interested in is how many plants are pollinated by multiple bee species, so we can use a simple measure – we will add up the number of bees utilizing each species of flower, and just square the result, and then add up all these squares. If most plants are visited by only one bee species then most scores will be 1, but if most are visited by all 4 species, most scores will be 16. To make things easy we will enter for each plant/bee combination a 0 if the bee does not utilize the plant and a 1 if it does, so we enter the raw data as follows

```
cerana<-c(1,1,1,1,1,1,1,1,1,0,1,0,1,1,1,0,1,0,1,1,1,0,1,1,0,1,1,0,1,0,0,0,
    1,0,1,0,0,0,0,1, 1,0,1,1,1,1,1,1,1,0,0,0,1,1,1,1,1,1,1,0,1,1,1,1,1,
    1,1,1,1,0,1,0,0,1,1,0,0,0,1,0,1, 1,0,0,1,1,1,0,1,1,1,1,0,1,1,1)
dorsata<-c(0,1,1,1,0,0,1,1,0,0,0,0,0,1,0,1,1,1,0,0,1,0,0,0,0,1,0,1,0,0,
    1,0,0,1,1,1,1,0, 1,0,1,1,1,0,1,0,1,0,0,1,1,0,0,0,0,0,0,0,0,1,1,1,1,1,1,
    1,0,0,0,0,0,0,0,1,0,0,0,0,1,0, 0,0,0,1,0,0,0,1,0,1,0,1,0,0,0,1,0)
florea<-c(0,0,0,1,1,1,0,1,0,1,1,1,0,0,1,0,0,0,0,1,1,0,1,1,0,0,0,1,
    1,1,0,0,0,1,0,0,0,0,1,0,0,1,1,0,1,1,0,0,1,0,0,0,1,1,0,0,0,1,1,1,1,1,1,1,
    0,0,0,1,1,0,0,0,0,1,1,0,1,0,0,1, 0,1,0,0,1,1,0,0,1,1,1,0,0,0,1,0)
mellifera<-c(0,1,0,1,0,0,0,1,1,1,0,1,1,0,0,0,1,1,1,1,1,1,1,1,0,0,
    1,0,0,0,0,1,1,1,1,0,0,1, 1,0,0,1,0,1,1,1,0,1,0,0,1,1,1,0,0,1,0,0,1,0,1,1,1,1,1,1,1,
    0,0,1,1,1,0,0,0,1,1,1,1,1, 1,1,1,1,1,0,0,0,0,1,0,0,0,0,1,1)
data<-cbind(cerana,dorsata,florea,mellifera)
```

These data are available in the online supplementary data file 'Apis_data.txt'. To calculate our observed sum of the squares of the number of bee species per plant we write

```
raw_score<-sum(rowSums(data[,1:4])^2)
raw_score
[1] 525
```

To do the randomizations we will run a loop from 1 to 100,000 (far more than an adequately large number that will only take a few seconds to do in R).

```
n<-nrow(data) # getting the number of plant species
randomizations<-NULL
for(i in 1:100000){
    d1<-sample(data[,1],n,replace=FALSE) # randomizing the 1s and 0s
        for each bee species
    d2<-sample(data[,2],n,replace=FALSE)
    d3<-sample(data[,3],n,replace=FALSE)
    d4<-sample(data[,4],n,replace=FALSE) # technically we only really
        need to shuffle the values for three of the four bee species because
        once three have been randomized, the fourth will be random with
        respect to each of these anyway – but it doesn't matter if we do
    shuffled_score<-sum((d1+d2+d3+d4)^2)
    randomizations<-c(randomizations,shuffled_score)
} # end i loop
```

Now we can look at the results by plotting a histogram of the randomizations and we introduce a couple more R functions: **round**, which allows you to specify how many decimal places are shown, and **paste**, which allows you to merge character vectors or number vectors as characters into a single character string.

> **Information Box 16.1**
>
> *The **paste** function.* **paste** takes two or more items as input and optionally has two arguments ('collapse' and 'sep') specifying how to combine them.
> ```
> a<-1:5
> b<-LETTERS[1:5]
> c<-c("cat","dog")
> paste(c(a,b,c))
> [1] "1" "2" "3" "4" "5" "A" "B" "C" "D" "E" "cat" "dog"
> ```
> The 'collapse' argument combines the contents of all vectors into a single character string
> ```
> paste(c(a,b,c),collapse="") # nothing to separate the elements
> [1] "12345ABCDEcatdog"
> paste(c(a,b,c),collapse=" ") # a space to separate the elements, etc.
> [1] "1 2 3 4 5 A B C D E cat dog"
> ```
> It can include the results of calculations, e.g.
> ```
> paste("The value of p was ",round(1/240,4))
> [1] "The value of p was 0.0042"
> ```
> Paste works on vectors element by element (with recycling) so
> ```
> x<-c("a", "b", "c", "d")
> y<-c("w", "x", "y", "z")
> paste(x,y)
> [1] "a w" "b x" "c y" "d z"
> ```
> The 'sep' argument comes into effect here, e.g.
> ```
> paste(x,y,sep="*")
> [1] "a*w" "b*x" "c*y" "d*z"
> paste(x,y,sep="*",collapse=" ")
> [1] "a*w b*x c*y d*z"
> ```

```
hist(randomizations)
abline(v=raw_score,lwd=2,col=2)
text(528,12000,"observed",srt=270,col="red")
equals<-length(which(randomizations==raw_score))
below<-length(which(randomizations<raw_score))+0.5*equals
above<-length(which(randomizations>raw_score))+0.5*equals
text(480,16000,paste("less than = ",below)) # x positions will have to be
    adjusted because of output from randomization
text(560,16000,paste("greater than = ",above),xpd=NA)
text(560,12000,paste("P = ",round(above/(above+below),4)),col="blue")
```

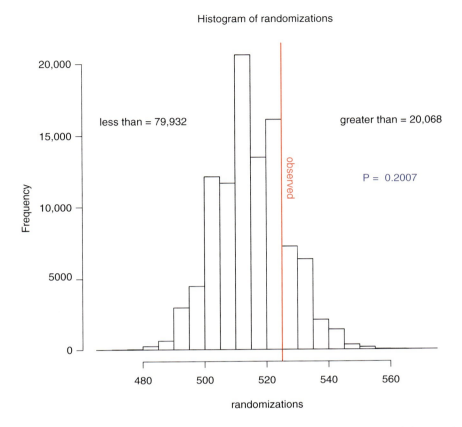

Histogram of randomizations

The observed sum of squares is greater than the median value of the randomizations so some plants are more likely to be visited by multiple bee species. However, there is no significant effect of plant species at the $p = 0.05$ level.

Randomizing Cells in a Matrix

The contents of columns, rows or blocks of a matrix can be randomized easily. For randomizing the elements of a whole matrix there is the function **randomizeMatrix** in the **picante** package. However, here we will write a bit of code in base R to do the same, or just to randomize elements in rows or columns. This may be useful, for example, in randomizing treatments, or setting out randomized experimental plots – though it should be noted that if the number of replicates is small, it may be better to stratify treatments rather than randomizing because 1111100000 can also be a random sequence!

> **Information Box 16.2**
>
> *Stratified design versus random design.* Students are often taught the importance of randomizing treatments so as to avoid inadvertent effects. This is all well and good if we are going to do a large number of replicates, but time, money and/or space often mean that replicates are limited. Even with quite large samples, randomization may still leave some unwelcome structure. For example, if you sow two or more species of plants randomly in a field you may end up with over- or under-representation of particular combinations of species or perhaps more of some species in the wetter part of the field. Stratified design is a way of trying to overcome this by subdividing the total population into a set of similar subsets. For example, instead of randomizing plant species across a whole field, the field is divided into a number of equal-sized areas and the same combination of plant species sown in each smaller area.
>
> *Stratified random sampling.* This follows the same idea. For example, if you randomly select samples from a field you may end up with more samples from the shady part, or from near the edge. So in stratified random sampling, the field would be divided into a number of similar-sized areas and subsamples of the same size picked at random from within each area.

Here we will create a randomized set of treatments to be applied to an 8 × 8 array of experimental plots, and randomly select the plots for each of 4 treatments (A, B, C and D). There will be 16 plots receiving each treatment. We create a vector with the treatments and randomize the order

treatments<-sample(rep(c("A","B","C","D"),16))

then place them in an 8 × 8 array using **noquote** to omit quotation marks
noquote(matrix(treatments,nrow=8))

```
     [,1] [,2] [,3] [,4] [,5] [,6] [,7] [,8]
[1,]  B    A    C    D    C    B    C    D
[2,]  A    B    B    B    C    B    B    A
[3,]  C    B    B    B    B    C    D    D
[4,]  A    A    D    A    C    A    C    D
[5,]  B    A    D    B    D    D    A    D
[6,]  B    D    D    C    B    A    C    A
[7,]  C    A    C    B    C    A    D    D
[8,]  D    D    C    A    C    A    A    C
```

If we start again but without shuffling we get an array with each row comprising a single treatment.
m<-as.data.frame(matrix(rep(c("A","B","C","D"),16),nrow=8))
m

```
  V1 V2 V3 V4 V5 V6 V7 V8
1  A  A  A  A  A  A  A  A
2  B  B  B  B  B  B  B  B
3  C  C  C  C  C  C  C  C
4  D  D  D  D  D  D  D  D
5  A  A  A  A  A  A  A  A
6  B  B  B  B  B  B  B  B
7  C  C  C  C  C  C  C  C
8  D  D  D  D  D  D  D  D
```

Here for example the columns could be treatments and the letters could be species and we might want to randomize the species in each column but keep the total species composition of each column the same. To do this is easy using **sample** and **sapply** (see Chapter 27):

m2<-noquote(sapply(m[,1:8],sample))
m2

```
     V1  V2  V3  V4  V5  V6  V7  V8
[1,] D   D   A   D   A   D   D   A
[2,] C   C   B   B   D   C   B   D
[3,] C   A   D   D   B   A   B   B
[4,] B   A   C   B   B   B   C   C
[5,] A   C   D   C   A   C   A   C
[6,] B   B   C   A   C   A   C   D
[7,] A   B   A   A   C   B   A   A
[8,] D   D   B   C   D   D   D   B
```

Exercise Box 16.1

1) Create a 12 × 12 array and subdivide it into 12 equal 3 × 4 subregions. Select 4 cells at random from each subregion to obtain a list of 48 cells that would constitute a stratified sample. Hint: use **list**.
2) Plot the design in an attractive way.

Tool Box 16.1

- Use collapse= in **paste** to specify how components of vectors are separated.
- Use hist() to plot the frequencies of values.
- Use paste() to combine text or numbers into character strings.
- Use round() to specify the number of decimal places.
- Use sep= in **paste** to specify how components of a pair of pasted vectors are separated.

17 Principal Components Analysis

Summary of R Libraries Introduced
  ade4
  amp
  FactoMineR
  stats

Summary of R Functions Introduced
  autoplot in "ggplot2"
  biplot
  prcomp
  princomp

Despite often being referred to under the heading 'multivariate statistics' principal components analysis is not a statistical technique at all, but rather it is a way of exploring whether a dataset with multiple types of measurements having been made for multiple entities (e.g. physical dimensions of individuals, allele frequencies at different sites), forms a homogeneous group. It weights each of the data types such that it maximizes the variance of the component (the first principal component), and then creates an orthogonal (independent) component by weighting that part of the data not fully explained by the first component, such that it maximizes its variance (the second principal component), and so on. Usually there is no benefit from examining beyond the third or fourth components. The first component is often strongly influenced by absolute size.

To conduct principal components analysis, R has two similar built-in functions **prcomp** and **princomp** in the default **stats** package. Other implementations can be found in various downloadable packages, e.g. the function **PCA** from the package **FactoMineR**, the function **dudi.pca** from the package **ade4** and the function **acp** from the package **amap**. The functions **prcomp** and **princomp** employ different calculation methods but in practice the results they return will be almost identical.

Example 1 – Rock Oyster Allozymes

For this exercise we are using part of the allozyme data for rock oysters presented in Table 4 of Day *et al.* (2000). We have only taken the allele frequencies for the three allozyme loci that they found to be most diagnostic: leucine aminopeptidase (*Lap*), mannose phosphate isomerase (*Mpi*) and phospho-glucose isomerase (*Pgi*). We have created initially a separate vector for each allele,

Principal Components Analysis

eight alleles for *Lap* and six each for *Mpi* and *Pgi*. We have used find-and-replace to insert commas between all the allele frequencies. In order, the allele frequencies are for oyster samples from Trat, Chon Buri, Chumphon (site 5), Chumphon (site 6), Surat Thani, Satun, Trang, Sydney Australia and Magnetic Island Australia, nine sites in total. We could, of course, have pasted the relevant part of the table and edited it in a spreadsheet and loaded it, but it would still have taken quite a bit of editing.

Lap1<-c(0,0,0,0,0,0,0,0.025,0.030); Lap2<-c(0,0,0,0,0,0,0,0,0.030); Lap3<-c (0,0,0,0, 0,0,0,0.350,0); Lap4<-c(0,0.214,0.250,0,0,0.140,0,0.024,0.475,0.156); Lap5<-c(0.167,0.119,0.286,0.278,0.160,0.136,0.119,0,0.345); Lap6<-c(0.833, 0.667, 0.214,0.611,0.580,0.705,0.666,0,0.345); Lap7<-c(0,0,0.250,0.111,0.120, 0.091,0.167,0.150,0.094); Lap8<-c(0,0,0,0,0,0.068,0.024,0,0)

Mpi1<-c(0,0.024,0,0,0.040,0.023,0.050,0,0); Mpi2<-c(0.167,0,0.107,0,0.060, 0.068, 0.025,0,0); Mpi3<-c(0.833,0.262,0.107,0.056,0.360,0.045,0.125, 0.263,0.500); Mpi4<-c(0,0.547,0.607,0.721,0.460,0.546,0.650,0.737,0.406); Mpi5<-c(0,0.143, 0.143,0.167,0.080,0.273,0.150,0,0.094); Mpi6<-c(0,0.024, 0.036,0.056,0,0.045, 0,0,0)

Pgi1<-c(0,0,0.036,0,0,0.040,0,0.190,0.265,0.056);Pgi2<-c(0,0,0,0,0, 0.023,0,0,0); Pgi3<-c(0.667,0.095,0.214,0.111,0.140,0.931,0.525,0.059, 0.167); Pgi4<-c(0.333,0.429, 0.214,0.278,0.240,0.023,0.190,0.529,0.527); Pgi5<-c(0,0.309,0.429,0.444, 0.480,0.023,0.095,0.147,0.250); Pgi6<-c (0,0.167,0.107,0.167,0.100,0,0,0,0)

We need to combine these into a matrix but with sample sites in column 1 and then separate columns for each of the measured variables (allele frequencies). Therefore we need a matrix of 9 rows (one for each site), and 20 columns (20) and we will create this using the function **matrix** and initially fill it with NAs

Allozyme_data<-matrix(NA,nrow=9,ncol=20) # create a blank matrix of appropriate dimensions

Allozyme_data[,1]<-Lap1; Allozyme_data[,2]<-Lap2; Allozyme_data[,3]<-Lap3; Allozyme_data[,4]<-Lap4; Allozyme_data[,5]<-Lap5; Allozyme_data[,6]<-Lap6; Allozyme_data[,7]<-Lap7; Allozyme_data[,8]<-Lap8; Allozyme_data[,9] <-Mpi1;Allozyme_data[,10]<-Mpi2; Allozyme_data[,11]<-Mpi3; Allozyme_ data[,12]<-Mpi4; Allozyme_data[,13]<-Mpi5; Allozyme_data[,14]<-Mpi6; Allozyme_data[,15]<-Pgi1; Allozyme_data[,16]<-Pgi2; Allozyme_data[,17] <-Pgi3;Allozyme_data[,18]<-Pgi4; Allozyme_data[,19]<-Pgi5; Allozyme_ data[,20]<-Pgi6

We now need to assign row and column names

rownames<-c("Trat","Chon_Buri","Chumphon_5","Chumphon_6","Surat_ Thani","Satun", "Trang","Sydney_Aust","Magnetic_Island")

colnames(Allozyme_data)<-c("Lap1","Lap2","Lap3","Lap4","Lap5","Lap6", "Lap7","Lap8", "Mpi1","Mpi2","Mpi3","Mpi4","Mpi5","Mpi6","Pgi1","Pgi2","Pgi3", "Pgi4","Pgi5","Pgi6")

and calculate principal components (our vector pca1) from the transposed matrix (using the function **t** to do the transposing)

pca1<-prcomp(t(Allozyme_data),scale.=TRUE)

head(pca1$rotation,3) # just to check

```
              PC1            PC2            PC3            PC4
              PC5            PC6            PC7            PC8
 [1,]    0.2426138     0.52240926     -0.5406019     0.11964420
        -0.1124592     0.24945448    -0.03318379     0.2779573
 [2,]    0.3899239    -0.07467262     -0.1204928    -0.20442148
         0.5338358    -0.05137909    -0.58582038    -0.3712285
 [3,]    0.3400122    -0.31599560      0.3054013    -0.07686026
        -0.6605697     0.22850332    -0.20995368    -0.1593116
PC9
 [1,]    0.4565369
 [2,]    0.1321369
 [3,]    0.3568391
```

The resulting principal components (the columns in 'pca1$rotation') can now be plotted against one another. We can colour the two Australian samples (blue), differently from the Thai ones (orange), and see that they are generally more divergent.

```
colours1<-c("orange","orange","orange","orange","orange","orange",
    "orange","blue","blue")
```

and we can plot several combinations at the same time using **par** with the argument 'mfrow= ({*number of plot columns*},{*number of plot rows*})'.

```
par(mfrow=c(2,2))
plot(pca1$rotation[,1],pca1$rotation[,2],col=colours1,pch=15)
plot(pca1$rotation[,2],pca1$rotation[,3],col=colours1,pch=15)
plot(pca1$rotation[,1],pca1$rotation[,3],col=colours1,pch=15)
plot(pca1$rotation[,2],pca1$rotation[,4],col=colours1,pch=15)
```

The Australian samples are generally more divergent from the Thai ones especially in the pca2 vs pca3 plot.

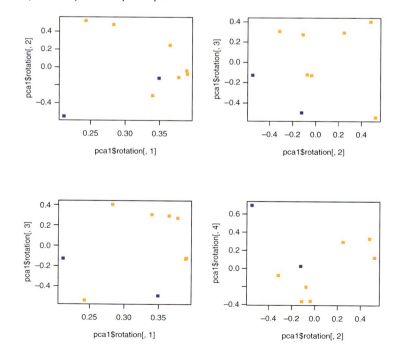

Principal Components Analysis

We can label each point with the locality if we wish (though this can get a little crowded) using **text** and you will inevitably have to experiment with font size and offsets, e.g.

```
localities<-c("Trat","Chon_Buri","Chumphon_5","Chumphon_6",
    "Surat_Thani","Satun","Trang","Sydney_Aust","Magnetic_Island")
plot(pca1$rotation[,1],pca1$rotation[,2],col=colours1,pch=15)
xoffset<-0.01*(max(pca1$rotation[,1])-min(pca1$rotation[,1]))
yoffset<-0.02*(max(pca1$rotation[,2])-min(pca1$rotation[,2]))
text(pca1$rotation[,1]+xoffset,pca1$rotation[,2]+yoffset,localities,cex=0.7,xpd=NA)
```

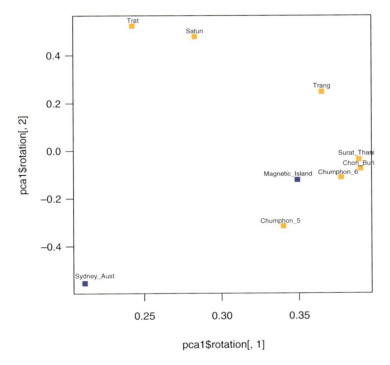

Example 2 – The Iris Dataset

We will demonstrate a couple of other functions on the in-built (and widely used) dataset 'iris'. First we will compare the PCA results of analysing raw continuous variables and logged ones as recommended by Venables and Ripley (2002). We will use the function **biplot** to display all the data points so as to get a better impression of clustering.

```
data(iris)
par(mfrow=c(2,2))
PCA1<-prcomp(iris[,1:4],center=TRUE,scale.=TRUE)
biplot(PCA1)
PCA_log<-prcomp(log(iris[,1:4]),center=TRUE,scale.=TRUE)
biplot(PCA_log)
```

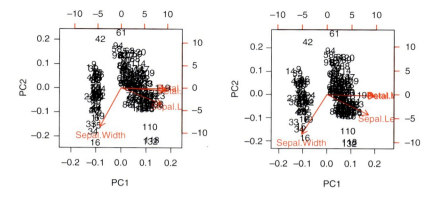

The log transformation has only a small effect; examining the PCA model summaries will show how it has affected the importance of the different flower traits.

Next we will make a rather more informative plot using the function **autoplot** in the **ggplot2** package.

library(ggplot2)
pca2_plot<-autoplot(PCA1,data=iris,colour='Species')
pca2_plot

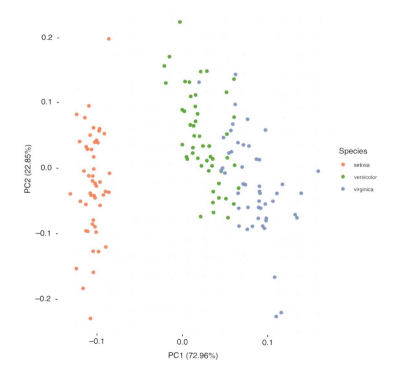

Principal Components Analysis

Exercise Box 17.1

1) Use the following to extract the principal components from the iris analysis and then plot four different PCA combinations against one another in a block of four, colouring points by species. Code is irisPCAs<-predict(PCA1,newdata=iris).
2) Add a legend in the centre. Hint: guestimate the coordinates.
3) There is also a very nice function **scatterplot** in the **car** library. First plot the raw iris data using **scatterplot**. Hint: the syntax is scatterplot(x,y,group={a factor to group and colour by}).
4) Now plot PCA1 versus PCA2 using **scatterplot**. Hint: the values for each PCAvector produced by **prcomp** are stored as columns in the prcomp object, and you can access them using {prcomp object}$x.

Tool Box 17.1

- Use head() to show only the first lines or components of a matrix, dataframe or vector.
- Use matrix() to create a matrix.
- Use ncol= to set the number of columns in a matrix.
- Use nrow= to set the number of rows in a matrix.
- Use prcomp() or princomp() to conduct a principal components analysis.
- Use t() to transpose rows and columns of a matrix.

18 Species Abundance, Accumulation and Diversity Data

Summary of R Libraries Introduced	Summary of R Functions Introduced
coexist	%%
devtools	fisherfit
MBI	grep
SPECIES	rug
vegan	sample
	smooth.spline

Ecologists in particular are often interested in the species richness and diversity of groups of organisms, ranging from studies of small ecosystems to global patterns. In most cases it is not possible to count every individual or to detect every species, and so they use a variety of estimation methods and summary statistics that we will briefly introduce here. This chapter covers estimating species abundance and species richness by looking at accumulation curves. Next we turn to analysing diversity using tests such as the Shannon and Simpson diversity indices. Finally we recreate patterns of niche partitioning using the broken stick model.

Species Accumulation Data

Ecologists, especially in the tropics, are often interested in biodiversity and its estimation. Typically, a group of 'readily' identified organisms may be surveyed on repeated occasions, and the accumulated data used to examine species abundance curves and to predict possible total numbers of species. Here we show an example using transect surveys of butterflies in Papua New Guinea (unpublished data kindly provided by Yves Bassett; see Basset et al., 2013). The total numbers of butterflies of all species observed are:

```
PNG_butterflies<-c(1,1,56,53,1,1,2,8,4,1,15,7,2,22,6,9,1,16,28,49,2,5,562,
    30,63,7,1922,321, 231,59,4,1,1,6,2,2,5,2,3,4,3,163,3,61,2,58,6,5,15,22,
    85,50,8,6,97,3,140,37,4,3,17,2,1, 41,1,1,45,3,4,45,14,5,17,11,1,2,1,17,48,
    24,100,2,4,2,52,17,1,1,1,12,4,9,9,21,11,4,7,37,51,7,9,66,1,4,8,51,39,2,
    19,15,1,3,16,75,10,141,26,21,2,3,689,15,9,9,3,2,6,3,149,1,5,79,1,1,2,
    18,154,10,113,3,3,3,1,1,142,2,9,1,9,16)
par(mfrow=c(2,2)) # setting the plot area to be 2 × 2 graphs
hist(PNG_butterflies,col="grey50")
```
You can see that the distribution is very skewed, partly because a very large number of one species (1922 individuals) were recorded whereas
```
length(which(PNG_butterflies==1))
[1] 24
```
only one individual was seen for 24 of the species. Therefore we can try logging the x axis. First let us try a histogram where we specify the break points at logarithmic intervals using powers of 2 (remember x^0 = 1). We have to make the break points non-integer so that no observed values occur exactly on a break point. We could calculate these in our heads, but we can automate it
```
s<-seq(0:11)-1 # annoyingly seq assumes that the first value is 1 so we have
    to take one off to start at 0
s2<-2^s
s3<-s2-0.5
hist(PNG_butterflies,breaks=s3,col="grey50")
```
Well, the division points are logged but on a linear scale, so we need to log the x axis.
```
log_butterflies<-log(PNG_butterflies,2)
hist(log_butterflies,col="grey50")
```
Very commonly species abundance curves follow a log-normal or log-series distribution (Magurran, 2004). The package **vegan** contains the function **fisherfit**, which will determine which model fits the data best
```
install.packages("vegan")
library(vegan) # use install.packages("vegan") if you have not already
    installed it
fisherfit(PNG_butterflies)
Fisher log series model
No. of species: 150
Fisher alpha: 27.041
```
telling us that the data are best fit by a log-series model with the parameter alpha =27.041, and we can plot it.
```
plot(fisherfit(PNG_butterflies))
```

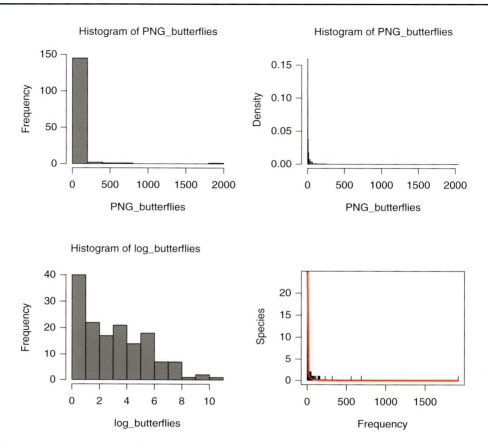

Species Accumulation Curves and Randomization

Basset et al. collected data from many transects but we will present only their first 50 here – the data are saved as a tab delimited text file called 'PNG_transects' so can be read with **read.table**

data<-read.table(file="PNG_transects.txt",header=TRUE)
head(data)

```
                         t1 t2 t3 t4 t5 t6 t7 t8 t9 t10 t11
t12 t13 t14 t15 t16 t17 t18 t19 t20 t21 t22 t23
Badamia_exclamationis    0  0  0  0  0  0  0  0  0   0   0
  0   0   0   0   0   0   0   0   0   0   0   0
Candalides_tringa        0  0  0  0  0  0  0  0  0   0   0
  0   0   0   0   0   0   0   0   0   0   0   0
Cepora_abnormis          0  0  0  0  0  0  0  0  0   0   1
  0   0   0   0   0   0   0   0   0   0   0   0
Chaetocneme_tenuis       0  0  0  0  0  0  0  0  0   0   0
  0   0   0   0   0   0   0   0   0   0   0   0
Deudorix_epirus          0  0  0  0  0  0  0  0  0   0   0
  0   0   0   0   0   0   0   0   0   0   0   0
Deudorix_littoralis      0  0  0  0  0  0  0  0  0   0   0
  0   0   0   0   0   0   0   0   0   0   0   0
```

```
                         t24 t25 t26 t27 t28 t29 t30 t31 t32
t33 t34 t35 t36 t37 t38 t39 t40 t41 t42 t43 t44
Badamia_exclamationis     0   0   0   0   0   0   0   0   0
  0   0   0   0   0   0   0   0   0   0   0   0
Candalides_tringa         0   0   0   0   0   0   0   1   0
  0   0   0   0   0   0   0   0   0   0   0   0
Cepora_abnormis           0   0   0   0   0   0   0   0   0
  0   0   0   0   0   0   0   0   0   0   0   0
Chaetocneme_tenuis        0   0   0   0   0   0   0   0   0
  0   0   0   1   0   0   0   0   0   0   0   0
Deudorix_epirus           0   0   0   0   0   0   0   0   0
  0   0   0   0   0   0   0   0   0   0   0   0
Deudorix_littoralis       0   0   0   0   0   0   0   0   0
  0   0   0   0   0   0   0   1   0   0   0   0
                         t45 t46 t47 t48 t49 t50
Badamia_exclamationis     0   0   0   1   0   0
Candalides_tringa         0   0   0   0   0   0
Cepora_abnormis           0   0   0   0   0   0
Chaetocneme_tenuis        0   0   0   0   0   0
Deudorix_epirus           0   0   0   1   0   0
Deudorix_littoralis       0   0   0   0   0   0
```

Rather than seeing numbers of individuals, to construct a species accumulation curve we need only to know presence/absence data so we will create a new vector replacing values in data >1 with 1, and to do this we can create the presence/absence matrix 'PA' using a trick with the R function **replace**. In this case, our data are passed to a new matrix called PA, and all values in PA that are greater than 1 are replaced with 1.

PA<-replace(PA<-data,PA>1,1)

Now we want a new matrix giving the total number of species recorded up to each column of the matrix (i.e. each transect surveyed). However, this is not simply the sums of each column, so we need to work through the columns one by one, finding which species (rows of PA[,i]) were recorded in that column using **grep**, combining them with the previously recorded species and making sure that there are no duplicates using **unique**.

species_recorded<-NULL
hold_temp<-NULL
for(i in 1:ncol(PA)){ # stepping through the columns
    temp<-grep(1,PA[,i]) # finding which rows in the column have a
        specimen, i.e. a 1
    hold_temp<-c(hold_temp,temp) # adding current column 'temp'
    species_recorded<-c(species_recorded,length(unique(hold_temp)))
        # removing duplicates to get total number of species recorded up to
        and including current column
}
plot(species_recorded,xlab="Transect number",ylab="Total number of
    species")

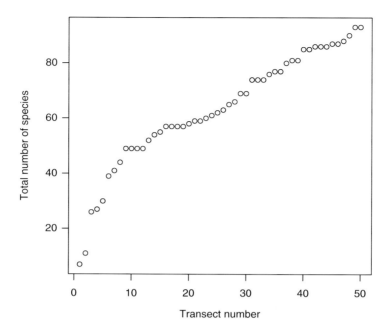

We can see that the number of species rises steeply for the first few transects, but then seems to become fairly linear after about the first 20. However, the randomness with which species occur in transects means that we do not have a smooth curve, but a rather stepped one, and it is harder to appreciate what exactly the trend might be. Had the transects been performed in a different order the plot could have looked different. We would like to generate a smooth curve in a meaningful way. This is not the same as fitting a smoothed line through the points using something like the function **smooth.spline**, rather it is to try to see what would have been obtained as an average species accumulation curve if the same transects and results had been repeated many times, but in different random orders. For this we are going to use **sample**, which picks *n* values from a set in random order, either with or without replacement. If 'replace=FALSE', no number will be sampled more than once. Here what we want to do is to pick the order randomly in which transects (columns) are added. We have data for 50 transects in the matrix 'PA', corresponding to columns 1:50. We will select those 50 columns (transects) in a random order numerous times, calculating the values of the species accumulation curve for each, remembering them, and then plotting the mean and some estimate of variation, e.g. the range.

To program this we are going to need to put the loop used above, within another loop for the number of random replicates – doing 100 replicates is probably sufficient, call it 'N'. And we need a matrix to store the accumulation curves for each randomized addition sequence, call it 'M'. The rest we did above.

```
N<-100
M<-matrix(NA,nrow=N,ncol=ncol(PA)) # to hold the results of N (100) random
    sequences
for(i in 1:N){
    to_add_seq<-sample(1:50,50,replace=FALSE)
    species_recorded<-NULL
    hold_temp<-NULL
    for(j in to_add_seq){ # automatically steps through all values in the list
        to_add_seq
        temp<-grep(1,PA[,j])
        hold_temp<-c(hold_temp,temp)
        species_recorded<-c(species_recorded,
            length(unique(hold_temp))) }
    M[i,]<-species_recorded} # end i loop
head(M)
```

```
     [,1]  [,2]  [,3]  [,4]  [,5]  [,6]  [,7]  [,8]  [,9]  [,10]
     [,11] [,12] [,13] [,14] [,15] [,16] [,17] [,18]
[1,]  14    18    22    34    34    34    48    50    50    51
      52    55    59    65    66    66    68    68
[2,]  13    23    27    31    37    42    47    51    51    51
      53    53    53    53    54    55    55    56
[3,]   7    17    22    24    26    34    34    45    47    52
      56    61    61    64    66    66    68    72
[4,]  11    14    15    18    26    32    37    40    45    52
      54    54    54    54    60    60    62    65
[5,]  13    21    29    33    37    38    48    50    56    58
      63    63    64    67    68    74    75    77
[6,]   7    14    21    25    26    27    34    34    34    38
      41    43    45    45    47    53    55    56
     [,19] [,20] [,21] [,22] [,23] [,24] [,25] [,26] [,27]
     [,28] [,29] [,30] [,31] [,32] [,33] [,34] [,35]
[1,]  70    70    72    73    73    74    74    74    74
      74    76    79    79    80    82    82    84
[2,]  57    57    58    58    59    62    62    62    65
      67    67    72    73    73    74    79    79
[3,]  72    73    73    73    73    74    74    77    78
      79    80    80    81    81    81    85    87
[4,]  65    67    67    68    68    74    76    76    77
      79    79    79    81    81    82    83    84
[5,]  77    77    77    78    81    82    83    84    84
      84    84    84    84    85    86    86    87
[6,]  57    57    57    58    59    59    60    61    61
      68    69    70    70    72    73    75    75
```

```
       [,36] [,37] [,38] [,39] [,40] [,41] [,42] [,43] [,44]
 [,45] [,46] [,47] [,48] [,49] [,50]
[1,]    84    85    86    87    88    88    88    88    89
   92    92    92    92    93    93
[2,]    79    79    80    81    82    82    87    87    88
   88    88    88    90    90    93
[3,]    87    87    88    90    91    92    92    93    93
   93    93    93    93    93    93
[4,]    84    85    85    85    87    87    87    87    87
   87    88    88    93    93    93
[5,]    87    87    87    87    88    89    90    90    92
   92    92    93    93    93    93
[6,]    77    78    80    80    80    82    82    82    85
   88    88    88    88    88    93
```

This shows that the code worked as expected, with each row increasing monotonically to the total of 93 species recorded up to the end 50 transect surveys. First, let us plot all these simulations on the same graph as small points – we can do this by using a loop to plot each row thus:

plot(NULL,xlim=c(1,50),ylim=c(1,93),xlab="Transect number",ylab="Total number of species")

for(i in 1:100) points(c(1:50),M[i,],pch=15,cex=0.2)

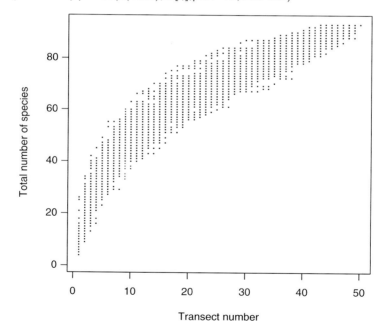

Now we can use **colMeans** to show the mean values
points(c(1:50),colMeans(M),col="red",pch=16)

Species Abundance, Accumulation and Diversity Data

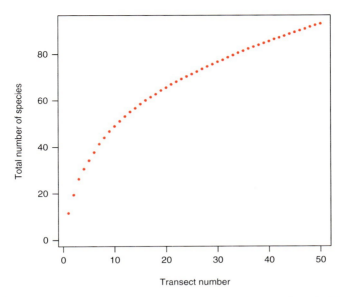

It is clear that the curve is nowhere near plateauing with only 50 samples. We want to fit a smooth curve to these means, which we can justifiably do since we know that it is a simple monotonic relationship but don't know the precise underlying mathematical equation, we can use **smooth.spline** to do this. This function has two relevant arguments: the logical argument 'all.knots=' determines whether the fitted spline must pass through all the data points, and 'spar=' determines the degree of smoothing and is found by trial and error to give you the curve you like most.

```
lines(smooth.spline(c(1:50),colMeans(M),spar=0.4,all.knots=F),lwd=2,
    lty=1,col="blue")
```

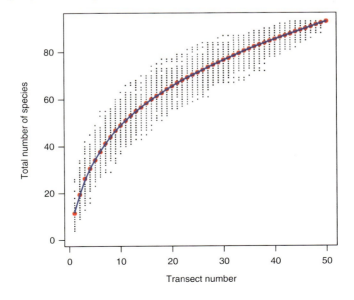

Species Richness Estimation

Numerous methods have been proposed to take species abundance and accumulation data and try to estimate from them the likely total number of species in the sampled pool. A widely used program *EstimateS* (pronounced estimate S) combines many of these, but there is also an R package called **SPECIES** which does much the same, obviating the need to export data and analyse them with another package. On a related matter, there is a package **coexist** for analysing species coexistence data, and **MBI** for multiple site biodiversity data.

Species Diversity Indices

Apart from total number of species, ecologists often employ a number of indices that also take into account the numbers of individuals (recorded) of those species. Two of the commonest are Simpson's D and Shannon–Weaver (or Shannon–Wiener) H indices. These are of course broadly related and consider evenness of numbers of individuals of each species within a community. Here we will write short piece of code to calculate both but they are also both calculated by the function **diversity** available in the library **vegan** (e.g. **diversity**(x,index="shannon",MARGIN=1, base=exp(1)).

Some academics feel strongly about these, but essentially Shannon and Simpson indices are strongly correlated and only differ really in their sensitivities to rare species (Simpson's index is more strongly influenced by a few dominant species).

Simpson's index is given by

$$D = 1 - \Sigma(p_i^2)$$

(or sometimes D/ Σp_i^2 sometimes called inverse Simpson)
and Shannon's H by

$$H = -\Sigma(p_i \ln(p_i))$$

where p_i is the proportion of the *i*th species. Shannon's equation is based on information theory. To get a list of proportions divide each species representation by the total. Here we use the vector 'PNG_butterflies' that was created earlier in this chapter (see p. 201).
propPNGbutts<-PNG_butterflies/sum(PNG_butterflies)
Thence
Simpson<-1-sum(propPNGbutts^2)
Simpson
[1] 0.8976594
and
ShannonH<--sum(propPNGbutts*log(propPNGbutts,2))
ShannonH
[1] 4.76282

We can now easily compare the two indices across all of the transects. First we will plot both indices against the total number of butterfly species,

```
species_observed<-NULL; Simpson_transect<-NULL; Shannon_transect<-NULL
for(i in 1:length(data)){
    temp<-data[which(data[,i]>0),i]
    species_observed<-c(species_observed,length(temp))
    Simpson_transect<-c(Simpson_transect,1-sum((temp/sum(temp))^2))
    Shannon_transect<-c(Shannon_transect,-sum(sum((temp/sum(temp))
        * log(temp/sum(temp),2))))}
par(mfrow=c(2,2))
plot(species_observed,Simpson_transect,xlab="Number of butterflies spp
    in transect",ylab="Simpson's Index",pch=18,col="cyan3")
plot(species_observed,Shannon_transect,xlab="Number of butterflies spp in
    transect",ylab="Shannon's Index",pch=18,col="chocolate")
```

and now we can plot the two diversity indices against one another

```
plot(Shannon_transect,Simpson_transect,xlab="Shannon",ylab="Simpson",
    col="red")
```

and, not surprisingly, there is a pretty strong correlation especially for high values of both indices.

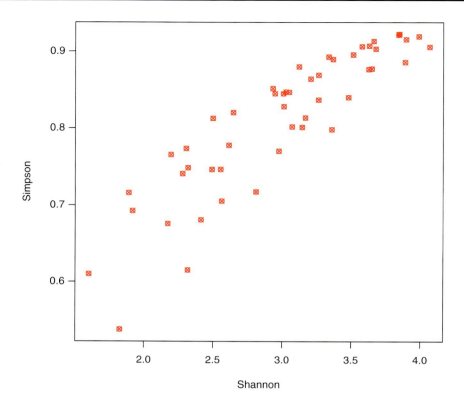

A Note to Be Cautious about Logarithms in Functions

In our calculation of the Shannon index we used logs base 2, which are assumed by the logic of information theory, but some workers use logarithms base e (i.e. base=2.718281828459) or even base 10. Which you use is only important if you are comparing between different studies, but of course, we may wish to do so.

Now we will calculate the Shannon index using the version in the package **vegan**.

install.packages("vegan")# if not already installed
library(vegan)
SW<-diversity(propPNGbutts,index="shannon",base=exp(1))
SW
[1] 3.301336

Hmmm! Not the answer that we got above. So now we can see that the function **diversity** in **vegan** uses logs to base e and not logs base 2 as we had used.
ShannonH<--sum(propPNGbutts*log(propPNGbutts))   # log(propPNGbutts, 2.718281828459)); also note the minus sign before **sum**
ShannonH
[1] 3.301336

Of course, the documentation in R (type '?diversity' at the prompt) does tell you that it uses the base of natural logarithms by default, and this can be changed, e.g. using the argument 'base=2' or the special function **log2**. The point here is that it is important to check.

Exercise Box 18.1

1) Using the number of species in the dataset 'SpeciesArea' from the **Stat2Data** package, determine whether any of the islands listed is anomalous for species number.

Broken-stick Models

Tool Box 18.1 ✕

- Use base= to specify the base that the log function uses.
- Use diversity() in vegan to calculate a species diversity index.
- Use fisherfit() to determine which model fits the data best.
- Use index= in **diversity** to set which diversity index to use.
- Use lend= to specify the shape of the end of the line(s).
- Use replace=TRUE/FALSE to replace a value/only sample a value once.
- Use replace() to replace values in a vector with a new value.
- Use side= to specify which side of the plot box the rug will be plotted: 1 = bottom, 3 = top.

Originally proposed by MacArthur (1957, 1961) as a null model for species abundance distributions the broken-stick model assumes that a resource can be thought of as randomly partitioned by species. The broken stick corresponds to a community colonized simultaneously by all species which partition a single resource randomly (J.B. Wilson, 1993). Some related hypotheses predict that relative abundances of such things such as the numbers of species in each of a set of genera/subfamilies/families etc. might also follow a broken stick model.

To generate a random stick distribution, let us say, for species in a number of genera within a group such as a subfamily, we take the total number of species and randomly partition them into the number of genera. This gives an opportunity to introduce vectorization for doing calculations.

For data we will use the checklist of British ladybird beetles (Coleoptera: Coccinellidae) from coleoptera.org.uk/family/coccinellidae (accessed 23 April 2020). The data are also available in print (Booth, 2012).

```
cocc<-read.table("British_ladybirds.txt",header=TRUE)
head(cocc)
        Genus              Species
1       Adalia           bipunctata
2       Adalia         decempunctata
3       Anatis              ocellata
4   Anisosticta    novemdecimpunctata
5     Aphidecta            obliterata
6       Calvia  quattuordecimguttata
```

```
Number_of_species<-nrow(cocc)
Number_of_species
[1] 53
Number_of_genera<-length(levels(cocc$Genus))
Number_of_genera
[1] 30
```
To summarize this we can use the function **table** to convert our data to a row for each genus and a column for each species (we suggest you have a look at 'table(cocc)' by itself, it is a bit too big to print here). The total of each row will be the number of species in each genus, but not presented on the screen in a very easily to summarize format (try 'rowSums(table(cocc)))'. But we can coerce this into a neat table using as.data.frame.

```
species_per_genus<-as.data.frame(rowSums(table(cocc)))
head(species_per_genus)
            rowSums(table(cocc))
Adalia             2
Anatis             1
Anisosticta        1
Aphidecta          1
Calvia             1
Chilocorus         2
```

'hist(rowSums(table(cocc)),breaks=30)' shows us that most genera have just a single British representative. To create a random broken stick model we consider all 53 ladybird species in a line and then select 29 break points between them randomly without repetition. This will give us 30 clusters. It is a good idea to check code by making just one set of breaks before setting everything in a loop.

a<-sort(sample(c(2:Number_of_species),Number_of_genera-1)) # the default function **sample** has replace=FALSE, which is what we want; the function **sort** arranges elements into ascending order

```
c<-a-0.5
c
[1]  1.5  3.5  4.5  7.5 10.5 11.5 14.5 17.5 21.5 23.5 25.5
    26.5 27.5 28.5 29.5 30.5 32.5 33.5 35.5 38.5 39.5 41.5
    42.5 45.5 46.5 48.5 49.5 51.5 52.5
```

We subtract the 0.5 because the breaks have to go 'between' species, so we set breaks to be at intervals of 0.5 from 1.5 to 52.5. There are other ways that do not involve fractional break points but using them is most intuitive. We will now use a **for** loop to create a list of numbers of species in each segment of the broken stick.

hold<-floor(c[1]) # setting the number of species in the first genus to the smallest random number generated

for(i in 1:length(c)-1) hold<-c(hold,c[i+1]-c[i])

hold<-c(hold,Number_of_species-floor(c[i+1])) # here we are adding the last species to the end of the last genus

Let us check that we have the correct number of species and the correct number of genera in our randomly partitioned pieces.
length(hold)==Number_of_genera
[1] TRUE
sum(hold)==Number_of_species
[1] TRUE

Now we want to generate a number of replicated random splits, for example 100 randomizations, which will allow us to test whether our coccinellid beetle data fit a random stick model. Open a new blank plot window and we will use the function **rug** to mark tick points delimiting each species along the bottom of the plot area. The function **cumsum** returns the cumulative sum of the values in a vector.

```
par(mfrow=c(2,2)) # we will be plotting the results of using two different approaches
plot(NULL,ylim=c(0,100),xlim=c(-0,Number_of_species+1),axes= FALSE,ylab="",xlab="")
rug(0:53,ticksize=0.03,side=1,lwd=0.5) # the first 53 small tick marks
rug(c(5,10,15,20,25,30,35,40,45,50),ticksize=0.05,side=1,
    lwd=2,line=0.5,lend="square") # second bigger and fatter tick marks every five species
twocolours<-c(rep(c("red","yellow"),1+Number_of_genera/2)) # set up a list of alternating colours
remember<-NULL # 'remember' will store all our results for analysis at the end
for(j in 1:100){ # start 100 random replicates loop
    a<-sort(sample(c(2:Number_of_species),Number_of_genera-1))
    c<-a-0.5
    hold<-floor(c[1])
    for(i in 1:length(c)-1){ # a loop nested within the j loop
        hold<-c(hold,c[i+1]-c[i])
        } # end i loop
    hold<-c(hold,Number_of_species-floor(c[i+1]))# Now to plot each replicate
    m<-0
    for(i in 1:Number_of_genera){
        lines(c(m,cumsum(hold)[i]),c(j,j),lwd=3.5,col=twocolours[i],
            lend="square")
        m<-cumsum(hold)[i]
        } # end i loop
    remember<-rbind(remember,hold)
    } # end of the 100 randizations j loop
```

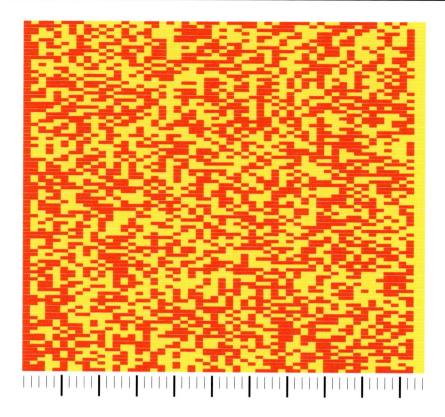

However, such loops, and especially loops within loops, are slow in R. Further, the speed with which such incrementally built plots appear may also be adversely affected on devices with very high-resolution graphics.

A Much Faster Approach Using Vectorization

Let us use a different example, the numbers of species in each genus of the plant family Styracaceae (http://www.theplantlist.org/browse/A/Styracaceae/; accessed 21 July 2019). The website uses a classification with 12 recognized genera (*Alniphyllum, Bruinsmia, Changiostyrax, Halesia, Huodendron, Melliodendron, Pamphilia, Parastyrax, Pterostyrax, Rehderodendron, Sinojackia, Styrax*) though in some classifications *Parastyrax* is synonymized into *Styrax*. We will use a vectorized approach creating two lists based on a random set of break points, which has a length of one less than the number of genera

Styracaceae<-c(3,2,1,2,3,1,1,2,2,5,8,87) # numbers of spp. in same order as genus list
Styracaceae<-sort(Styracaceae)
num_genera<-length(Styracaceae)
nsp<-sum(Styracaceae)
list3<-c(1:(nsp-1)) # note the list stops one short of the total number of species
mem<-NULL

Species Abundance, Accumulation and Diversity Data

```
for(i in 1:100){
    breaks<-sample(list3,(num_genera-1)) # creating break points that are
        integers between 1 and num_species minus 1
    breaks<-sort(breaks)
    # if a random break happens to be number 1, the vector 'd' has a zero
        added to its left-hand end so that one minus zero will give the
        number of species in the first random genus
    d<-c(min(list3)-1,breaks)
    # if a random break point happens to be the number of species minus
        one, the last length of the broken stick must just include the last
        species. This is allowed for in 'e' by adding max(list3)+1 (i.e. nsp).
    e<-c(breaks,max(list3)+1)
    r<-sort(e-d)
    mem<-rbind(mem,r)} # end of *i* loop
```

When writing any code that is at all complicated, as this is, you should check that the code has worked as expected. For example, look at the contents of our vector 'mem' to make sure that there are no impossible values, such as zeros, and look at the row totals of 'mem' to make sure that they are all the same as the total numbers of species. Having convinced ourselves that it is working OK, we can now plot it

```
plot(1:num_genera,Styracaceae,xlab="Rank genus based on no.
    spp.",ylab="Number of species",ylim=c(0,nsp),col="blue",pch=7)
    for(i in 1:100) points(1:num_genera,mem[i,],col="red",pch=4)
```

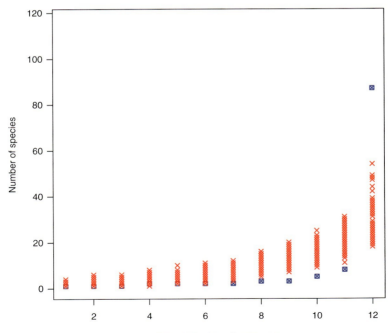

Two methods can be used to determine what proportion out of the 100 randomized genus sizes are larger than the observed values for Styracaceae genera. The first is probably the most intuitive, looping through the columns of our memory vector 'mem' and counting.

```
random_is_bigger<-NULL
for(i in 1:num_genera){
    random_is_bigger<-c(random_is_bigger,length(which
        (mem[,i]> Styracaceae[i]))/100)} # end i loop
random_is_bigger
[1]  0.32 0.71 0.92 0.87 0.94 0.98 1.00 1.00 1.00 1.00
    1.00 0.00
```

The second, much faster, using the function **apply** (note we used the function **t** to transpose the output of **apply** because the straight output is columns by rows, which is not intuitively what we want).

```
temp<-t(apply(mem,1,function(x){ifelse(x>Styracaceae,1,0)}))
prop_out_of_100<-colSums(temp)/100
prop_out_of_100
[1] 0.32 0.71 0.92 0.87 0.94 0.98 1.00 1.00 1.00 1.00 0.00
```

Of course, the results of both methods are the same.

In books and online tutorials you often see the use of in-built functions with **apply**, such as mean, max, etc., but you can define your own function as was done here. The **ifelse** used here gives us a zero or one answer, which we can add to summarize the results of our 100 random stick-breaking exercises using the **colSums** on the following line. If we had just used 'function(x) x > Styracaceae' we would have got a set of TRUE or FALSES which we could not add up quite so straightforwardly.

There is now a prewritten broken stick function in the **devtools** package from github.

```
install.packages("devtools")
devtools::install_github("hafen/hbgd")
devtools::install_github("stefvanbuuren/brokenstick")
```

Exercise Box 18.2

1) Which genus of Styracaceae has the largest number of recognized species?
2) Try to estimate using randomizations, just how improbably large the largest genus is if indeed the numbers of species were to follow a broken stick model.
3) What reasons might there be for the big difference between it and the next largest?
4) Can you think why there might be different dominant explanations for the departure from the broken stick model in the Styracaceae and the ladybird beetles examined earlier?
5) Using **proc.time** (see Chapter 27), and a large number of replicates, compare the speed of using a loop versus using **apply**, for summarizing these.

Tool Box 18.2

- Use all.knots= in **smooth.spline** to send the line through all points.
- Use breaks= in **hist** to set the number categories.
- Use cumsum() to calculate the cumulative sum of a list of numbers.
- Use diversity() to calculate a species diversity index.
- Use exp() to raise e to the power x.
- Use floor() to obtain the largest integer less than a given real number.
- Use grep() to find which rows have a particular value in them.
- Use rug() to mark tick points along an axis.
- Use sample() to pick random components from a vector.
- Use smooth.spline() to fit a smooth curve to a set of points.
- Use sort() to arrange elements of a vector in ascending or descending order.
- Use spar= in **smooth.spline** to set the degree of smoothing.
- Use ticksize= in **rug** to set the size of the lines.
- Use unique() to remove duplicates.

19 Survivorship

Summary of R Packages Introduced	Summary of R Functions Introduced
dplyr	coxph in "survival"
survival	Filter
survminer	Negate
	Surv in "survival"
	survfit in "survival"

Biologists are often interested in survival or differences in survival based on census data. The data usually consist of records of individuals such as tree saplings or bird nests. Analysis of survivorship data is particularly prone to difficulties and we advise reading up on the topic, certainly before thinking of publishing on it. There is perhaps a bigger issue with the relationship between observations and causal interpretation. For example, several studies have shown that the probability of alcohol-related death is greater in younger people, but why? One possibility is that there are genuine physiological reasons, but another is that this is a result of censoring, i.e. the older cohorts have probably already survived years of alcohol consumption during which time the more vulnerable individuals have been removed from the population.

Example 1 – Survival of Killdeer Nests

Here we will use data on the survival of nests of a bird called the killdeer, *Charadrius vociferus*: those nesting at gravelled oil pads and those that nested on native grass cover in western Oklahoma, USA (Atuo *et al.* 2018).
Input=("
day grass_cover oil_pad

```
1   16   28
2   16   28
3   16   28
4   16   28
5   16   28
6   16   28
7   16   28
8   16   27
9   16   26
10  16   25
11  16   24
12  16   24
13  16   24
14  16   24
15  16   24
16  16   23
17  16   21
18  16   20
19  15   19
20  15   17
21  15   17
22  14   14
23  14   12
24  14   11
25  14    8
26  14    8
27  12    8
28   9   NA
")
killdeer<-read.table(textConnection(Input),header=TRUE)
```

We will check this visually in a simple plot

```
par(mfrow=c(2,2))
plot(killdeer$day,killdeer$grass_cover,ylim=c(0,28),pch=16,col=
    "darkgreen", xlab="Survival (days)",ylab="Number of nests")
points(killdeer$day,killdeer$oil_pad,pch=16,col="brown")
```
now plotting as proportion surviving
```
plot(killdeer$day,killdeer$grass_cover/max(killdeer$grass_cover),ylim =c(0,1),
    pch=16,col="darkgreen",xlab="Survival (days)",ylab="Number of nests")
```

There is a slight problem with calculating the proportion surviving with the oil_pad numbers because there is a missing value for the last day, represented by NA. So asking R to calculate the maximum value of 'killdeer$oil_pads' returns all NAs. Therefore, knowing we have an NA in the data we have to tell R to exclude it for the calculation of max by inserting 'na.rm=TRUE' into the function call.

```
points(killdeer$day,killdeer$oil_pad/max(killdeer$oil_pad,
    na.rm=TRUE),pch=18, col="brown")
par(mfrow=c(1,1))
```

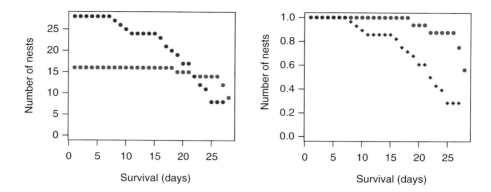

The standard way of presenting survival data graphically is called a Kaplan Meier plot, which shows the proportion of the population still surviving after successive times as a stepped line. The R package **survival** includes most of the functions you are likely to need. So we will use it to plot survival curves for us
install.packages("survival")
You will now be asked a question:
There is a binary version available but the source version is later:
binary source needs_compilation
survival 3.1-11 3.2-3   TRUE
Do you want to install from sources the package which needs compilation?
(Yes/no/cancel)
We suggest you answer no.
library(survival)
The format of our data in the table 'killdeer' is not in the format required by the **survival** package and therefore we need to create a new object with the appropriate structure. Specifically a table in which each row represents an individual, in this case a killdeer nest, then a column representing how long it survived (here we call it 'time') and another column specifying whether at that time the nest had survived (by convention a zero) or not (by convention a one), and optional additional columns, which in our case we use to specify habitat type (here the options are 1="grass_cover" and 2="oil_pad"). Looking at the 'killdeer' table above we see that nine nests in the grassy habitat survived for all 28 days of the study so the first nine rows are filled with the number of surviving days (28) and zeros in column 2 meaning survived. Three nests did not survive between the 27th and 28th days, so the next three rows have the time 27 and a one in column 2 meaning lost/died/predated, and so on.

Convert our 'killdeer' table to the 'killdeerSurv' format

First we will deal with the grass data and then with the 'oil_pad' data, this being slightly complicated because of the NA. To follow the code below we suggest you remove the 'for *i*' loop and manually increment it checking the values of d and diff as you go along.
grass_survive<-killdeer[nrow(killdeer),2] # the number of nests remaining at the end

```
data<-c(rep(killdeer[nrow(killdeer),1],grass_survive),rep(0,grass_survive),
    rep(1,grass_survive))
grass_survival<-matrix(data,ncol=3,byrow=FALSE)  # creating matrix with
    just surviving nests
z<-length(levels(as.factor(killdeer[,2])))-1 # minus 1 because we have already
    filled the matrix with the first level
temp<-grass_survive
for(i in 1:z){
    d<-max(which(killdeer[,2]>temp))
    diff<-killdeer[d,2]-temp
    for(j in 1:diff) grass_survival<-rbind(grass_survival,c(d,1,1))
    temp<-killdeer[d,2]}
colnames(grass_survival)<-c("time","status","habitat")
```

We need to know the final value of surviving nests for the 'oil_pad' with the NA. Of course, we could just enter the number by hand but it is better to have a more general function. Here we introduce the useful function **Filter** to do this. **Filter** can be passed a function called **Negate**, which removes whatever value is passed to it from a vector including a specific argument 'anyNA', which means all NAs. Combining these we can now extract the non-NA values from killdeer$oil_pad as follows

```
oil<-Filter(Negate(anyNA),killdeer$oil_pad)
oil
[1] 28 28 28 28 28 28 28 27 26 25 24 24 24 24 24 23 21 20
    19 17 17 14 12 11  8  8  8
```

There are many other ways of achieving the same result, e.g.

```
# unlist(lapply(killdeer$oil_pad,function(x) x[!is.na(x)]))
# killdeer$oil_pad[sapply(killdeer$oil_pad,Negate(anyNA))]
```

Now we run the same code modified for the 'oil_pad' data

```
oil_survive<-oil[length(oil)]
data<-c(rep(killdeer[length(oil),1],oil_survive),rep(0,oil_survive),rep(2,
    oil_survive))
oil_survival<-matrix(data,ncol=3,byrow=FALSE)
temp<-oil_survive
for(i in 1:13){
    d<-max(which(killdeer[,3]>temp))
    diff<-killdeer[d,3]-temp
    for(j in 1:diff) oil_survival<-rbind(oil_survival,c(d,1,2))
    temp<-killdeer[d,3]} # end i loop
killdeerSurv<-as.data.frame(rbind(grass_survival,oil_survival))
killdeerSurv # these are the data in the format required by the function Surv
    time   status   habitat
1    28      0         1
2    28      0         1
3    28      0         1
4    28      0         1
5    28      0         1
6    28      0         1
7    28      0         1
```

```
 8  28  0  1
 9  28  0  1
10  27  1  1
11  27  1  1
12  27  1  1
13  26  1  1
14  26  1  1
15  21  1  1
16  18  1  1
17  27  0  2
18  27  0  2
19  27  0  2
20  27  0  2
21  27  0  2
22  27  0  2
23  27  0  2
24  27  0  2
25  24  1  2
26  24  1  2
27  24  1  2
28  23  1  2
29  22  1  2
30  22  1  2
31  21  1  2
32  21  1  2
33  21  1  2
34  19  1  2
35  19  1  2
36  18  1  2
37  17  1  2
38  16  1  2
39  16  1  2
40  15  1  2
41  10  1  2
42   9  1  2
43   8  1  2
44   7  1  2
```

The function **Surv** is used first to convert your input table into the necessary format, called a 'survival object'. Then the function **survfit** generates the Kaplan-Meier estimate of the survival function. The argument passed to it must be a function (i.e. it requires a tilde).

par(mfrow=c(1,1))
kdfit<-survfit(Surv(killdeerSurv$time,killdeerSurv$status) ~ killdeerSurv$habitat)
plot(kdfit,xlab="Days",ylab="Nest survivorship",lwd=2,lty=c(1,2),col=
    c("darkgreen", "brown"))

Survivorship

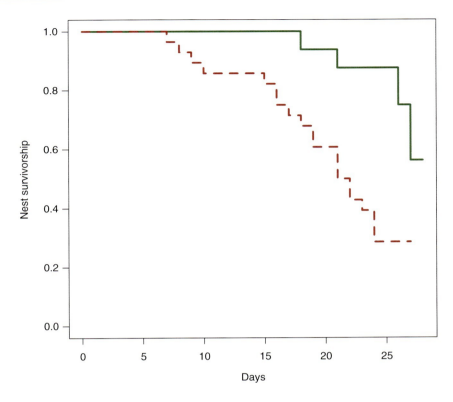

```
summary(kdfit)
Call: survfit(formula = Surv(killdeerSurv$time, killdeer
    Surv$status) ~
killdeerSurv$habitat)
                    killdeerSurv$habitat=1
time n.risk n.event survival std.err lower 95% CI
upper 95% CI
    18     16      1    0.938  0.0605           0.826
         1.000
    21     15      1    0.875  0.0827           0.727
         1.000
    26     14      2    0.750  0.1083           0.565
         0.995
    27     12      3    0.562  0.1240           0.365
         0.867
                    killdeerSurv$habitat=2
time n.risk n.event survival std.err lower 95% CI
upper 95% CI
     7     28      1    0.964  0.0351           0.898
         1.000
     8     27      1    0.929  0.0487           0.838
         1.000
```

9	26	1	0.893	0.0585	0.785
					1.000
10	25	1	0.857	0.0661	0.737
					0.997
15	24	1	0.821	0.0724	0.691
					0.976
16	23	2	0.750	0.0818	0.606
					0.929
17	21	1	0.714	0.0854	0.565
					0.903
18	20	1	0.679	0.0883	0.526
					0.876
19	19	2	0.607	0.0923	0.451
					0.818
21	17	3	0.500	0.0945	0.345
					0.724
22	14	2	0.429	0.0935	0.279
					0.657
23	12	1	0.393	0.0923	0.248
					0.623
24	11	3	0.286	0.0854	0.159
					0.513

To see the 95% confidence limits plotted we calculate the Kaplan-Meier estimates (Kaplan and Meier, 1958) separately for each habitat (i.e. first rows 1 to 16, and second rows 17 to 46).

kdgrass<-Surv(killdeerSurv$time[1:16],killdeerSurv$status[1:16])~killdeerSurv$habitat[1:16]
kdoil<-Surv(killdeerSurv$time[17:44], killdeerSurv$status[17:44])~killdeerSurv$habitat[17:44]
g95<-survfit(kdgrass)
o95<-survfit(kdoil)
plot(g95,col="darkgreen",xlab="Days",ylab="Nest survivorship",main="Kaplan-Meier estimate with 95% confidence bounds",lwd=2)
lines(o95,col="brown",lwd=2)

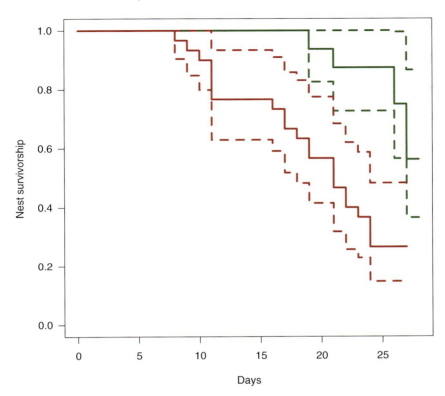

Kaplan-Meier estimate with 95% confidence bounds

There is no overlap between the 95% confidence limits so survivorship is probably significantly different between habitats. We will use Cox's Proportional Hazards to test this formally. This is very much the same as an ordinary ANOVA because we have a categorical explanatory variable ('habitat') but the method also would do an ANCOVA if we had a mix of categorical and continuous explanatory variables and allows specifying interactions and model simplification (see Chapter 14). We use the Cox's proportional hazards function **coxph**
kdc<-coxph(Surv(killdeerSurv$time,killdeerSurv$status) ~ killdeerSurv$habitat)
summary(kdc)

```
Call:
coxph(formula = Surv(killdeerSurv$time, killdeerSurv$status) ~
killdeerSurv$habitat)
n= 44, number of events= 27
                        coef exp(coef) se(coef)      z
Pr(>|z|)
killdeerSurv$habitat  1.0262    2.7904   0.4438  2.312
  0.0208 *
---
Signif. codes:  0 '***' 0.001 '**' 0.01 '*' 0.05 '.' 0.1 ' ' 1
```

```
                        exp(coef)    exp(-coef)     lower .95
upper .95
killdeerSurv$habitat     2.79         0.3584         1.169
    6.659
Concordance= 0.642 (se = 0.04 )
Likelihood ratio test= 6.06 on 1 df, p=0.01
Wald test            = 5.35 on 1 df, p=0.02
Score (logrank) test = 5.8 on 1 df, p=0.02
```

So we can see that there is a significant difference in survival between habitats at the 0.05 level, our *p*-value being 0.0208. This may not seem staggering but the sample sizes are rather small and there were rather few mortality events in the smaller grassland sample.

Information Box 19.1

Grammar of Graphics survival plots. If you install the packages **survminer** and **dplyr** you can create a wider range of survival plots with numerous extra options. With our killdeer data you can get a simple plot using with probability and pretty confidence intervals added, e.g.
ggsurvplot(kdfit, data=killdeerSurv, pval=TRUE, pval.method=TRUE, conf.int=TRUE, palette=c("springgreen","orangered1"))
or you can plot events on the *y* axis
ggsurvplot(kdfit, data=killdeerSurv, fun="event", linetype="strata", pval=TRUE, conf.int=TRUE, pval.method=TRUE, palette="Brewer")

Exercise Box 19.1

1) Using the **leuk** dataset in the **MASS** package provides survival times for patients diagnosed with acute myelogenous leukaemia (data originally from Feigl and Zelen, 1965). Determine whether the survival time after diagnosis in months is correlated with the presence of Auer rods (the column labelled 'ag' in leuk). Hint: pass the time of survival to **Surv** and plot the survfit. How does the presence of Auer rods correlate with survival expectancy?
2) The **leuk** dataset also includes the patients' white blood cell count ('wbc'). Use the Cox's proportional hazards to determine the combined correlation Auer rod status and white blood cell count. Hint: use an additive model.

Tool Box 19.1

- Use coxph() to carry out a Cox's Proportional Hazards test (ANOVA).
- Use Filter() with Negate() to remove types of elements from a list, e.g. remove NAs.
- Use surv() to convert an input table to be used with **survfit**.
- Use survfit() to generate the Kaplan-Meier estimate of survival.
- Use survival package and library to carry out survivorship analyses.

20 Dates and Julian Dates

Summary of R Libraries Introduced
  erer
  ggpubr
Summary of R Functions Introduced
  as.Date
  as.POSIXlt
  date
  density
	format
	julian
	loess
	lowess
	scan
	Sys.Date
	Sys.time
	Sys.timezone

We have left this topic until rather late in the book because dates and times in R can be a bit of a nightmare.

Today's date (technically your computer system's date) can be found using either of two functions: **date**() or **Sys.Date**(). However, they give it in slightly different formats.

date()
[1] "Sat Aug 31 15:04:41 2019"
Sys.Date()
[1] "2019-08-31"

The formats can be modified quite easily, e.g.
format(Sys.Date(),"%d %b %Y")
[1] "31 Aug 2019"
or
format(Sys.Date(),"%d %m %Y")
[1] "31 08 2019"
or
format(Sys.Date(),"%d %b %y")
[1] "31 Aug 19"

The format is specified using %d for day, month as a word not number using %b and year %y; if month was given as a number 1 to 12 we would specify %d%m%y. If we want the year to four digits then we use %Y or if we only want two digits then lower case y. A capital letter D gives us an abbreviated date in US order, i.e. month, day, year.
format(Sys.Date(),"%D")

```
[1] "08/31/19"
```
The function **Sys.Date**() just gives a pre-specified subset of the output of **Sys.time**()
Sys.time()
```
[1] "2019-08-31 15:09:42 +07"
```
and hence,
format(Sys.time(),"%d %b %Y")
```
[1] "31 Aug 2019"
```
Except for the POSIXlt class, dates in R are stored internally as the number of days or seconds from some reference date.
date<-as.POSIXlt(Sys.time(),format="%Y/%m/%d")
date
```
[1] "2019-07-24 17:53:20 +07"
```
The '+07' means hours different from Greenwich mean time (GMT).
Sys.timezone(location=TRUE)
```
[1] "Asia/Bangkok"
```
Sys.time always has 23 characters so substring can be used to extract elements, e.g. year as two digits
substring(Sys.time(),3,4)
```
[1] "19"
```
date<-as.POSIXlt(Sys.Date(),format="%Y/%m/%d")
date
```
[1] "2019-08-31 UTC"
``` # UTC stands for Coordinated Universal Time
julian(date)
```
Time difference of 18101.45 days
```
The default Julian start day in R (but this might be version dependent) is 01 Jan 1970. The value returned by 'julian' is the number of days since then. You can set the start date using the argument 'origin=' (e.g. 'origin=as.Date("1970-01-01")').
tmp<-as.Date("31Aug19",format="%d%b%y")
format(tmp,"%j")
```
[1] "243"
```
i.e. today is the 243rd day of the year.

To illustrate some date handling we will look at nest building and laying dates for breeding pairs of the blue tit (*Cyanistes caeruleus*) in Europe, on the mainland and on the island of Corsica (Blondel *et al.*, 1990). The data have been transcribed and are in a comma separated format in file 'Corsican blue tit.csv'.
CorsBT<-read.csv("Corsican_blue_tit.csv")
CorsBT
```
       Building      Laying        Location
1  27 March 1986  26 April 1986   mainland
2  31 March 1987  22 April 1987   mainland
3  17 March 1987  27 April 1987   mainland
4  15 March 1988  23 April 1988   mainland
5  28 March 1988  22 April 1988   mainland
```

```
6   4 April 1988  29 April 1988  mainland
7  24 April 1986  10 May   1986  Corsica
8  29 April 1987   8 May   1987  Corsica
9   5 May   1987  11 May   1987  Corsica
10 25 April 1988  14 May   1988  Corsica
11  5 May   1988  10 May   1988  Corsica
12  8 May   1988  20 May   1988  Corsica
```

First let us see what happens when we try to convert one date in the new format to a date that R will recognize as a date. Our input data have the full month name so we must specify that using %B rather than %b, the latter being only for three letter abbreviated month names. Similarly, we are inputting year with century so we must use %Y not %y.

dat<-as.Date("27March1986",format="%d%B%Y")
dat
[1] "1986-03-27"

If our dates have day, month, year separated by slashes or hyphens we must specify it in the format argument otherwise we will get unexpected and wrong output

dat<-as.Date("27-March-1986",format="%d-%B-%Y")
dat
[1] "1986-03-27"

Similarly, you can use format="%d|%B|%Y", format="%d/%B/%Y" or format=%m_%d_%y or format="%d %B %Y", etc.

as.Date(CorsBT[2,2],format="%d %B %Y")
"1987-04-22"

but

CorsBT[,1:2]<-apply(CorsBT[,1:2],2,function(x){x<-as.Date(x,format="%d %B %Y")})
CorsBT

```
   Building Laying Location
1      5929   5959 mainland
2      6298   6320 mainland
3      6284   6325 mainland
4      6648   6687 mainland
5      6661   6686 mainland
6      6668   6693 mainland
7      5957   5973 Corsica
8      6327   6336 Corsica
9      6333   6339 Corsica
10     6689   6708 Corsica
11     6699   6704 Corsica
12     6702   6714 Corsica
```

This is probably not what you were expecting. The numbers are dates of a sort, they are the number of days since 1 Jan 1970. We therefore have to introduce a second conversion step. The class is numeric, not 'Date'.

CorsBT$Building=as.Date(CorsBT$Building,origin='1970-01-01')
   CorsBT$Laying=as.Date(CorsBT$Laying,origin='1970-01-01')

```
CorsBT
     Building      Laying Location
1  1986-03-27 1986-04-26 mainland
2  1987-03-31 1987-04-22 mainland
3  1987-03-17 1987-04-27 mainland
4  1988-03-15 1988-04-23 mainland
5  1988-03-28 1988-04-22 mainland
6  1988-04-04 1988-04-29 mainland
7  1986-04-24 1986-05-10  Corsica
8  1987-04-29 1987-05-08  Corsica
9  1987-05-05 1987-05-11  Corsica
10 1988-04-25 1988-05-14  Corsica
11 1988-05-05 1988-05-10  Corsica
12 1988-05-08 1988-05-20  Corsica
```
class(CorsBT$Building)
```
[1] "Date"
```
The Julian date within a year is the number of days into that year, and is a useful way of examining phenological data such as flight periods, breeding dates, emergence dates, etc., especially when it comes to comparing between years and seeing if there are any trends.

Because our blue tit data are for different years, we should be precise and specify the starting date for each year separately, 1988 is after all, a leap year with 29 days in February.

start_each_year<-c("1986-01-01", "1987-01-01", "1987-01-01", "1988-01-01", "1988-01-01", "1988-01-01", "1986-01-01", "1987-01-01", "1987-01-01", "1988-01-01", "1988-01-01", "1988-01-01")

building<-NULL # a new variable with lower case "b"
laying<-NULL # ditto
for(i in 1:nrow(CorsBT)){
    building<-c(building,CorsBT$Building[i]-as.Date(start_each_year[i]))
    laying<-c(laying,CorsBT$Laying[i]-as.Date(start_each_year[i]))
    }
building
```
[1]  85  89  75  74  87  94 113 118 124 115 125 128
```
laying
```
[1] 115 111 116 113 112 119 129 127 130 134 130 140
```
plot(NULL,xlab="Pair number",ylab="Nest building to laying period", xlim=c(1,length(building)),ylim=c(min(c(building,laying)),c(max(c(building,laying)))))
linecolours<-c(rep("darkgreen",6),rep("orange",6))
for(i in 1:length(building)) lines (c(i,i),c(building[i],laying[i]),lwd=4, col=linecolours[i])
text(3,136,"Mainland",col="darkgreen",cex=2)
text(10,86,"Corsica",col="orange",cex=2)

If we want to automate writing the year above each line, we need to extract the year part of the dates in 'start_each_year'. As they are all in exactly the same format we could extract the year using

```
substring(start_each_year,1,4)
[1] "1986" "1987" "1987" "1988" "1988" "1988" "1986"
    "1987" "1987" "1988" "1988" "1988"
```
and this is probably the simplest, most pragmatic method. Though we want to use them as class Date objects we can.
```
years<-format(as.Date(start_each_year,format="%Y-%m-%d"),"%Y")
years
[1] "1986" "1987" "1987" "1988" "1988" "1988" "1986"
    "1987" "1987" "1988" "1988" "1988"
```
Replacing the final %Y with %y will generate a list of two-digit dates.

Adding them to our graph is easy either way.
```
for(i in 1:length(building)) text(i,laying[i]+3,years[i],cex=0.5)
```

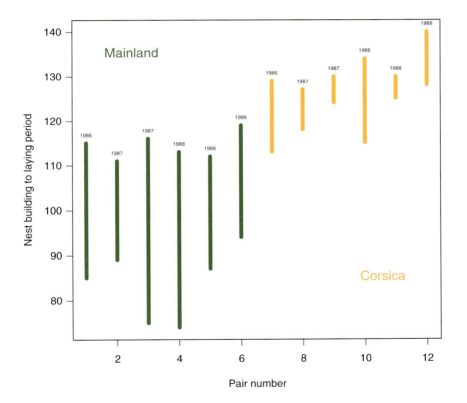

The durations in days between starting nesting and laying may be of interest and can be extracted by simple subtraction if the vectors in the dataframe are of class Date
```
CorsBT$Laying-CorsBT$Building
Time differences in days
[1] 30 22 41 39 25 25 16 9 6 19 5 12
```

> **Exercise Box 20.1**
>
> 1) Test whether there is a significant difference between locations and the interval between initiation of nest building and start of laying.
> 2) Use ANCOVA to investigate whether there is an interaction between location and nest building start date on the interval between starting nest building and laying.

Problem with Two-digit Dates and POSIX: A Date of Burial Example

A common student exercise in many biology departments is to take students to a large often old cemetery and have them record date of birth/death, and gender and then analyse their data to determine whether people are living longer nowadays than they did 100 or more years ago. The results seldom show a marked difference but there are potentially many biases about who got buried in a cemetery in the past. Here we use data available online for burials at the Hope Cemetery, Derbyshire, UK. The data files can be found in the online resources.

Although sex was not given it was possible to add this to the file based on first name with very high confidence.

```
hope<-read.csv("HOPE_Date_Order_Table_1.csv",header=TRUE)
head(hope,12)
      Reg..No. Death.date Burial.date
                      Names Sex Age      X Address Grave.No.
1           1             7/3/34
         John Edwin Hunt    m 59 Years              D32
2           2             1/19/38
         John Arthur Allott m 84 Years              D01
3           3             5/10/39
            Harriet Allott  f 71 Years              D01
4           4             3/19/55
            Lydia Bramwell  f 78 Years              D02
5           5             8/21/61
       Amy Dorothy Zealley  f 71 Years              D31
6           6             7/2/62
       George Carl Williams m 60 Years              D03
7           7             11/6/64
   Frances Ann Green Roberts f 78 Years             D29
8           8             4/1/65
         John Arnold Roberts m 88 Years             D29
9           9             11/4/66
              Fanny Hewitt  f 86 Years              D27
10         10             5/20/67
       David Thomas Bramwell m 89 Years             D02
```

```
11         11                 12/12/68
           Sally Ann Roberts  f 21 Years           D23
12         12                 1/30/69
           John Thomas Eyre   m 80 Years           D04
```
We are only interested in three of the columns, 'Burial.date' (as a surrogate for the incomplete Death.date column), 'Sex' and 'Age'. However, before we can combine just these into our dataframe, we need to deal with the date that is given in a particular format by telling R that these have the attribute that they belong to a class of R object called 'Date' and what the date structure is.

burial<-as.Date(hope$Burial.date,"%m/%d/%y")
head(burial,14)
```
[1] "2034-07-03" "2038-01-19" "2039-05-10" "2055-03-19"
    "2061-08-21" "2062-07-02" "2064-11-06" "2065-04-01"
    "2066-11-04" "2067-05-20" "2068-12-12" "1969-01-30"
    "1969-04-03" "1969-04-12"
```
Oh dear! There seems to be a bit of a problem with the first few dates. Because the input date only gives a two-digit year, years from 00 to 68 are prefixed by default with 20. This is part of the relevant (2008 POSIX) standard and, of course, very annoying. Based on a Stack Overflow discussion the best we can probably do is manually edit the 11 offending dates. The function **sub** used here differs from **gsub** in that it only substitutes the first occurrence of a pattern. It is not necessary in this case but if one of the dates had been 2020, **gsub** would have changed it to 1919.

burial[1:11]<-sub("20","19",burial[1:11])
head(burial,12)
```
[1] "1934-07-03" "1938-01-19" "1939-05-10" "1955-03-19"
    "1961-08-21" "1962-07-02" "1964-11-06" "1965-04-01"
    "1966-11-04" "1967-05-19" "1968-12-12" "1969-01-30"
```
Use **data.frame** to create a dataframe of the three columns we need

data<-data.frame(burial,hope$Sex,hope$Age)
head(data)
```
      burial hope.Sex hope.Age
1 1934-07-03        m       59
2 1938-01-19        m       84
3 1939-05-10        f       71
4 1955-03-19        f       78
5 1961-08-21        f       71
6 1962-07-02        m       60
```
We are probably only interested in the year that each person died so we can extract that from the full date object as shown above

year<-format(as.Date(data$burial,format="%d/%m/%Y"),"%Y")
head(year)
```
[1] "1934" "1938" "1939" "1955" "1961" "1962"
```
We can add these on to our dataframe as a new numeric column simply by giving the column a name, data$Year and assigning it the values

data$Year<-as.numeric(year)

> **Exercise Box 20.2**
>
> 1) Create separate histograms of age at death classified according to sex and those born before 1923. Note, there are some NAs among the Age data. Hint: use the set operator function **intersect** (see Chapter 8) OR the **%in%** operator.

> **Tool Box 20.1**
>
> - Use as.Date() to convert character strings to objects of class Date.
> - Use date() to show the date as per your system clock.
> - Use density() to smooth a histogram into a curve.
> - Use format() to specify the layout.
> - Use julian({*a date*}) to show the number of days since 1/1/1970.
> - Use substring() to extract elements of a character string.
> - Use Sys.Date() to show the system clock date.
> - Use Sys.time to() show the system clock time.

Phenology and the *density* Function

Studies of phenology often have records for observations of individuals for given dates. The file gives the days into the year 2003 when each individual adult moth of the European corn borer was collected at a light trap (data extracted from Keszthelyi et al., 2008). The numbers were saved using write({*the list of numbers as text character*},{*filename*}). this might seem logical but is a bit of a problem in R.

```
obs1<-readLines("European_corn_borer_2003.txt")
head(obs1)
[1] "143 143 143 143 143" "143 143 143 143 143" "143 148
    148 148 148" "148 148 148 148 148" "148 148 148 148
    148" "148 148 148 148 148"
```

By default, the function **write** in R saves vectors of numbers in groups of five, so each line of the data file contains character representations of five numbers. If you need to write a list with each value on a separate line, there is the function **write.list** in the **erer** package. Of course, we could write code to split each line at every space, but there is no need because R has another read function called **scan**.

```
obs2<-scan("European_corn_borer_2003.txt")
Read 469 items
head(obs2,40)
 [1] 143 143 143 143 143 143 143 143 143 143 143 148 148
     148 148 148 148 148 148 148 148 148 148 148 148 148
     148 148 148 148 148 148 148 148 148 148 148 148
[39] 148 148
```

We can now see that by using **scan** each value in the vector 'obs2' is treated separately.

```
par(mfrow=c(1,2))
hist(obs2,breaks=max(obs2)-min(obs2)+1) # breaks is set to the total number
    of days covered
plot(density(obs2))
par(mfrow=c(1,1))
```

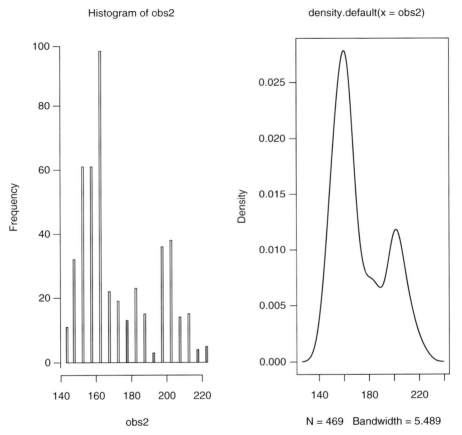

The function **density** uses a rule of thumb for smoothing, which is the default shown here. The default equation for smoothing factor, called the bandwidth 'bw', is

$$bw = \frac{range\ (x)}{1 + 2 * (log_2(number.of.data.points))}$$

The rule of thumb 'bw' works pretty well, and we can see that we get two nice smooth peaks. There are similar functions, e.g. **ggdensity** in the **ggpubr** package, and **geom_density** in **ggplot2**, that have some additional features (see Chapter 7).

Extracting Day and Month from Julian Days

The numbers in our data file are Julian days but we probably want to have some indication of dates and months. The year 2003 is not a leap year so the numbers of days in each month can be assigned to a vector using the familiar adage 'thirty days hath September, April, June and November, all the rest have thirty-one, save poor February alone, which has twenty-eight days clear, but twenty-nine each leap year'.

MD<-c(31,28,31,30,31,30,31,31,30,31,30,31)
sum(MD) # just to check
[1] 365

So the first days of each month in a non-leap year are:
FD<-c(1,1+cumsum(MD)[-length(MD)])
FD
[1]   1 32 60 91 121 152 182 213 244 274 305 335

To work out what the above line does we must consider a number of things: the first Julian day of the year is 1, the first of January. The cumulative sum of MD (cumsum(MD)) has 12 values but we are not interested in the last day. Therefore we can remove the cumulative sum for the end of December using minus indexing 'cumsum(MD)[-length(MD)]', which removes the 12th value because length of 'MD' is 12. The last day of each month following January in the cumulative sum is the last day of the month, therefore we add 1 to get the Julian day of the first day of the following month. To find out which firsts of months occur within our range of Julian dates for corn borer captures

months_in_range<-which(FD>=min(obs2) & FD<=max(obs2))
months_in_range
[1] 6 7 8 9

Now we will use a neat function called **density** to create a smoothed curve representing average density of observation estimates; we just use the default arguments, which work quite well.

plot(density(obs2))

R has month names stored in the predefined vector 'month'
MN<-month.name[1:12]
MN
[1]   "January"   "February"   "March"   "April"   "May"   "June"
      "July"   "August"   "September"   "October"   "November"
      "December"

MN[months_in_range] # or more directly month.name[months_in_range]
[1] "June" "July" "August" "September"

and they commence on
FD[months_in_range]
[1] 152 182 213 244

We can use the function **rug** to place ticks on the x axis indicating the first days of each month

rug(FD[months_in_range],lwd=2,lend="square")

and now we can write text above each tick giving the date

Dates and Julian Dates

```
FirstOfMonth<-paste("01",MN[months_in_range]) # paste automatically puts
    a space between the items being pasted unless we use collapse=""
FirstOfMonth
[1] "01 June" "01 July" "01 August" "01 September"
text(FD[months_in_range],rep(0.0005,length(months_in_range)),FirstOfMonth,
    cex=0.9, col="blue")
```

and finally we can add 'Julian date in 2003' below the x axis position by trial and error

```
text(mean(range(obs2)),-0.0032,"Julian date in 2003",xpd=NA,cex=1.2)
    # mean(range(x)) gives the midpoint of the x axis
```

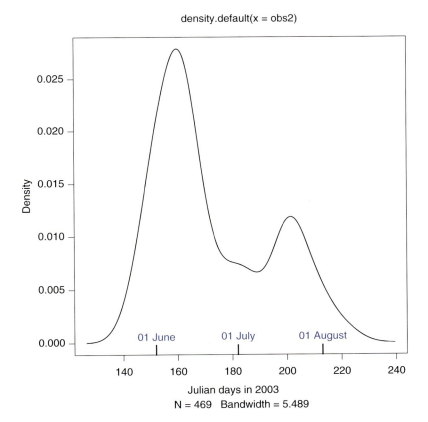

Exercise Box 20.3

1) The dataset CornBorer_1992.txt provides data for an earlier year. Plot density functions for the 1992 moths and the 2003 moths on the same plot. Hints: (i) consider the extents of the axes; (ii) use density({*object*})$y to see y values of density curve; (iii) use **lines**; and (iv) you may want to adjust the bandwidth of the superimposed plot.

Seasonal Patterns and Other Smoothing Curves

Jonkers and Kučera (2015) present data on the abundance (shell influx) of the foraminiferan *Turborotalita quinqueloba* amassed over a nearly 3-year sampling period at a given site.

day_in_year<-c(0.9023162,1.9475052,1.9111508,17.947441,17.61407,
    19.799334,33.986637, 33.615097,39.078434,49.624847,50.02184,
    58.024895,65.629875,65.61569,76.569275,81.67198,82.0228,
    95.510284,97.67119,98.09835,114.81702,113.7471,113.76201,
    133.79655,129.08539,145.89676,145.54774,152.80917,161.92287,
    171.76254,177.9617,191.43936,194.37025,209.6631,209.6791,
    228.60011,225.7012,247.55856,251.5441,254.8538,270.53964,
    270.5327,286.1364,286.54538,289.83145,302.56824,302.5624,308.39548,
    318.93423,318.5678,327.66803,334.57318,334.956,346.61084,
    350.22012,350.6073)
log_number<-c(1.434,0.454,-0.294,-0.123,0.513,0.506,0.109,-0.041,-0.056,
    -0.415, 0.258,0.034,-0.887,-1.179,-0.655,-0.596,-0.872,-0.678,-1.187,
    0.108,-0.671,-0.199,0.107,0.099,0.601,1.73,2.044,1.55,1.69,1.782,
    1.916,1.909,2.253,2.118,2.447,2.012,2.327,2.35,1.907,2.55,3.01,2.86,1.63,2.55,
    2.707,2.445,2.325,2.422,1.906,1.86,1.726,1.397,1.779,1.741,1.0525,1.524)

The function **lowess** does a polynomial fit to a set of data points and the user needs to decide what parameter values to use in order to achieve what is essentially a plot that is pleasing to the eye by adjusting the values of 'f', 'iter' and 'delta'. The argument 'f' is called the smoother span

plot(day_in_year,log_number)
lines(lowess(day_in_year,log_number,f=.2,iter=5,delta=1),col="red",lwd=2)

The default **lowess** is more smoothed

m<-lowess(day_in_year,log_number)
lines(m,col="green",lwd=2)

The similar function **loess** is created by statistical modelling of the data. Using the above existing plot with the red **lowess** line, we can calculate a fit model with **loess**

model<-loess(log_number ~ day_in_year)
xv<-seq(0,366,0.1) # create a sequence 0:366 in increments of 0.1
yv<-predict(model,xv) # use the loess model to predict y values for the xv
    date values
lines(xv,yv,col="blue",lwd=2)

All three fit curves are shown here in one plot.

Dates and Julian Dates

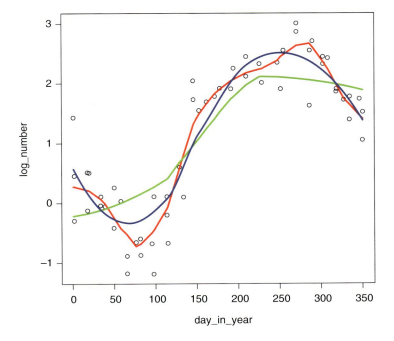

Tool Box 20.2 🛠

- Use lapply() to apply a function to a table and return a list.
- Use loess() to fit a polynomial curve by local fitting.
- Use lowess() to plot a locally weighted polynomial curve.
- Use month.name[] to get a vector of months.
- Use scan() to read each value of a vector separately.

21 Mapping and Parsing Text Input for Data

Summary of R Packages Introduced
 dismo
 maptools
 raster
 rworldmap

Summary of R Functions Introduced
 col.map in "maptools"
 complete.cases
 country2Region in "rworldmap"
 data in "maptools"
 gbif() in "dismo"
 joinCountryData2Map in "rworldmap"
 mapBubbles in "rworldmap"
 mapCountryData in "rworldmap"
 numeric
 plot in "maptools"
 polygon
 subset
 vignette()

Increasingly, data are becoming available about the distributions of organisms around the world and are being collated as freely available online resources in various formats. In this chapter we introduce the **maptools** library and plot distributions of taxa at country level on maps. Here we combine mapping and a good deal more on text parsing, because often the coordinates you need are in lists in papers and may have a load of different formats.

install.packages("maptools")
library(maptools)
data(wrld_simpl)
plot(wrld_simpl)
countries<-wrld_simpl$NAME # contains a list of all the country names in the file wrld_simpl

In sequence we have told R to load the library called **maptools**, to access the data and call it 'wrld_simpl', plot the simple country outline of the world map in a plot window, and list the country names that **maptools** recognizes. Note that Myanmar is still called Burma.

countries[18]
[1] Burma
246 Levels: Aaland Islands Afghanistan Albania Algeria
    American Samoa Andorra Angola Anguilla Antarctica
    Antigua and Barbuda Argentina Armenia Aruba Australia
    ... Zimbabwe

It will often be the case that if you access country lists of taxa by country, the names will differ so if you are plotting them with **maptools** map, you will need to change the name of the country in your data file to match that in **maptools**. The **gsub** function we met in Chapter 4 is often appropriate for this.

Here we will plot a map shaded according to the relative number of butterfly species for a small selection of countries. Peru, with some 3700 species has by far the largest number of any country in the world

Australia<-400/3700 # proportion of Peru's total in Australia
France<-260/3700
Japan<-300/3700
Nigeria<-1319/3700
Peru<-3700/3700
Thailand<-1287/3700
USA<-750/3700

To shade the map relative to the numbers of species, calculate a depth of colour for each country from white to fully red (or green or blue or cyan…) we will use **plot(wrld_simpl)** and specify as a parameter the colour we want using the **rgb** system (red, green, blue), where the intensity of each colour is specified on the range 0:1. Which country is plotted that colour is specified by its position in the list of country names **wrld_simpl$NAME**.

nc<-length(countries)
col.map<-numeric(nc) # this just creates a vector of as many zeros as there are countries (nc), i.e. it's the same as "col.map<-c(rep(0,nc))"

The slight catch is that the default amount of colour in an argument 'rgb' is none, i.e. black, because 'rgb(0,0,0)' means no red, blue or green, whereas 'rgb(1,1,1)' means white. There are a lot of possible ways to overcome this: we will use rgb({*amount of colour*},1-{*amount of colour*},0), which will range colours from green(least), through brown, to red (most).

col.map[which(countries=="Australia")]<-rgb(Australia,1-Australia,0)
col.map[which(countries=="France")]<-rgb(France,1-France,0)
col.map[which(countries=="Japan")]<-rgb(Japan,1-Japan,0)
col.map[which(countries=="Nigeria")]<-rgb(Nigeria,1-Nigeria,0)
col.map[which(countries=="Peru")]<-rgb(Peru,1-Peru,0)
col.map[which(countries=="Thailand")]<-rgb(Thailand,1-Thailand,0)
col.map[which(countries=="United States")]<-rgb(USA,(1-USA),0)
plot(wrld_simpl,col=col.map)

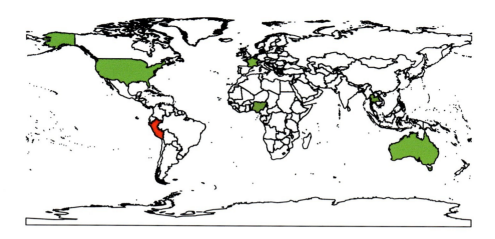

Exercise Box 21.1

1) On a map of Africa, colour the distribution of the Egyptian cobra, *Naja haje*. These are: Tanzania, Kenya, Somalia, Ethiopia, Uganda, South Sudan, Sudan, Cameroon, Nigeria, Niger, Burkina Faso, Mali, Senegal, Mauritania, Morocco, Algeria, Tunisia, Libya and Egypt.

For more detailed regional maps use package installer to download the package **rworldmap** from CRAN written by South (2011). This package includes a lot of country level data such as biodiversity, which are specified in 'namecolumntoplot'.
install.packages("rworldmap")
library(rworldmap)
Welcome to rworldmap
For a short introduction type :   vignette('rworldmap')
vignette is an R function which allows you to see a brief overview of some R packages. If you want to use **rworldmap**, you are strongly recommended to look at the vignette. The included country level data includes 80 different variables which you can see by entering **countryExData**; here we will use ENVHEALTH and BIODIVERSITY.
data(countryExData) # using the data function in base R to make data available by name
country2Region(regionType="Stern") # to report which countries make up regions, not shown
sternEnvHealth<-country2Region(inFile=countryExData,nameDataColumn= "ENVHEALTH",joinCode="ISO3", nameJoinColumn="ISO3V10",regionType= "Stern",FUN="mean")
Then create the 'SpatialPolygonsDataFrame' (**sPDF**) as follows:
sPDF<-joinCountryData2Map(countryExData,joinCode="ISO3", nameJoinColumn= "ISO3V10") # ISOs are international standardized country and region codes, see https://en.wikipedia.org/wiki/ISO_3166-1
149 codes from your data successfully matched countries in the map
0 codes from your data failed to match with a country code in the map
94 codes from the map weren't represented in your data

```
mapCountryData(sPDF,nameColumnToPlot="BIODIVERSITY",
    mapTitle="World Biodiversity",oceanCol="lightblue",missingCountryCol=
    "white",borderCol="black")
```

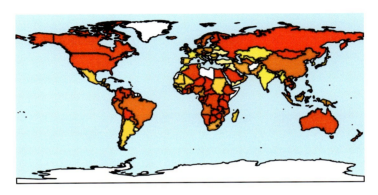

or for the more precisely pre-defined Asian region
```
mapCountryData(sPDF,nameColumnToPlot="BIODIVERSITY",mapTitle="Asia",
    oceanCol="lightblue",missingCountryCol="white",mapRegion="Asia",
    borderCol="black")
```

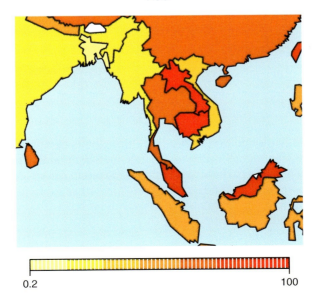

Other data of interest to biologists include ENVHEALTH, ECOSYSTEM, CLIMATE.1...

You can zoom in or out more by replacing the argument 'mapregion' by longitude and latitude (in degrees) using 'xlim' and 'ylim' so set longitude and latitude ranges, respectively. Thus to get Australia use:

mapCountryData(sPDF,nameColumnToPlot="BIODIVERSITY",mapTitle= "Australia", oceanCol="lightblue",missingCountryCol="white",xlim=c(110, 160),ylim=c(-48,-5), borderCol="black") # for southern hemisphere ylim goes from south to north

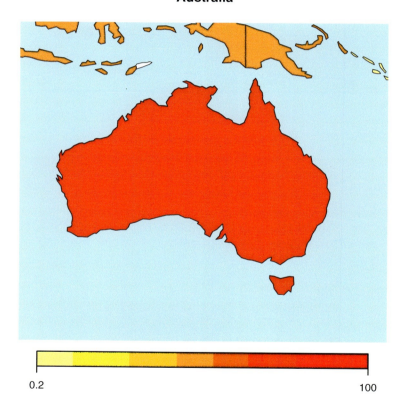

You can easily plot points on this map using points or text, with degree latitude and degree longitude coordinates, in fact anything that you can do on a normal plot. In addition, the package also includes its own functions such as **mapBubbles**, **mapBars** and **mapPies**. See South (2011) for further explanation.

To get a world country outline map with no colours on which you can plot things, you can cheat by

mapBubbles(dF=getMap())

and ignore

```
Error in `[.data.frame`(dF, , nameZSize) : undefined
    columns selected
```

Mapping and Parsing Text Input for Data

Which might be OK for some purposes, but we can make it a bit prettier
mapBubbles(dF=getMap(),oceanCol="lightblue",landCol="wheat")

For a list of rworldmap country names use getMap()$NAME
head(getMap()$NAME)
[1] Afghanistan   Angola   Albania   United   Arab   Emirates
    Argentina Armenia
243 Levels: Afghanistan Aland Albania Algeria American
    Samoa Andorra Angola Anguilla Antarctica Antigua and
    Barb. ... Zimbabwe

To plot some bubbles create a dataframe, here we use the number of mammal species in a few selected counties just for purposes of illustration. The data are from https://rainforests.mongabay.com/03mammals.htm).

```
mammals<-read.table(text="
Country 'Mammal species' 'threatened spp' Lat Lon
Canada 426 40 60 -95
Brazil 648 80 -34 -64
Mexico 923 96 24 -102
Peru 647 63 -9 -75
Colombia 442 58 3.9 -73
'United States' 440 40 39 -99
Argentina 374 38 -35 -65
Ecuador 372 47 -1 -78
Bolivia 362 21 -17 -64
Chile 143 19 -37 -72
",header=TRUE) # read.table treats sets of spaces like a tab, and does not need textConnection
mapBubbles(dF=mammals,nameZSize="Mammal.species",nameZColour="Country", oceanCol="lightblue",landCol="wheat",nameX="Lon", nameY="Lat",main="Number of Mammal species",symbolSize=0.5)
```

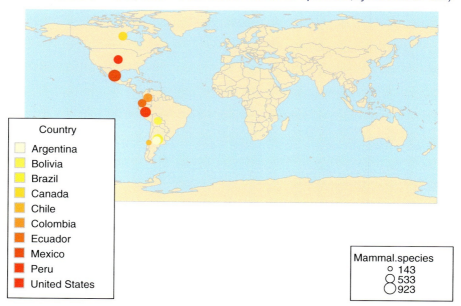

Exercise Box 21.2

1) Kays *et al.* (2020) present results of a massive study of domestic cat home ranges. Here we will look at just the data for 442 radio-tracked cats in the vicinity of Adelaide, Australia. Load the file 'Pet-Cats-Australia.csv' and write code to plot the tracks of each individual cat in a different colour using **rworldmap**. Hints: (i) determine the maximum and minimum latitudes and longitudes; (ii) extract the row numbers for each individual cat; and (iii) use **lines** to join each cat's home range.
2) The total sample is scattered over too wide an area to see much detail so zoom in by selecting a smaller area and replot.

Mapping and Parsing Text Input for Data

> **Tool Box 21.1** ✗
>
> - Use borderCol= in **mapCountryData** to set the colour of the borders.
> - Use Country2Region() to list the regions of the map.
> - Use countryExData to see the 80 data types available in **rworldmap**.
> - Use date() to get current time and date.
> - Use getMap()$NAME to get a list of country names.
> - Use joinCode= in **rworldmap** to specify how countries are referenced (options are: "ISO2","ISO3","FIPS","NAME", "UN").
> - Use joinCountryData2Map() in "ISO2","ISO3","FIPS","NAME", "UN" to join user data referenced by country codes or names to an internal map, ready for plotting using **mapCountryData**, can be used to colour countries.
> - Use mapBubbles() to create a map outline and to plot 'bubbles' on it.
> - Use mapCountryData() to draw a map of country level data.
> - Use missingCountryCol= in **mapCountryData** to set the colour of countries with no data.
> - Use nameColumnToPlot= in **mapCountryData** to specify the column with data you want to plot.
> - Use nameJoinColumn= in **joinCountryData2Map** to link data to a map ready for plotting.
> - Use oceanCol= in **mapBubbles** to set the colour of the oceans.
> - Use RegionType= to specify how regions are defined.
> - Use rgb() to set the colour scale.
> - Use vignette() to see an overview of a package.

Creating Our Own Map from Digitized Coordinates

We could not find one ready-made for R so you will need to do one yourself – it's not difficult (depending on the complexity of the border you need). We did this ourselves in a way that anyone can easily do. We found a map of Thailand on the Internet, screen-grabbed it, saved it as a jpeg file and used a freely available application (on a MacOSX computer you can use **PlotDigitizer**, for example) to digitize its border and saved coordinates as a .csv file. Digitizing the map outline took one of us about 10 minutes.

map<-read.csv("Thailand_Border.csv",header=TRUE)

Check that everything seems OK

head(map)

```
     longitude    latitude
1    0.2570303    0.9961125
2    0.2714608    0.9947726
3    0.2810845    0.9947596
4    0.2834904    0.9947566
5    0.2907081    0.9947476
6    0.3003377    0.9960544
```
....

Adjust so that the digitized maximum and minimum Thai border coordinates have exactly 0 and 1 values respectively. If you have set the digitizer to give precise degrees you don't need to do this.

```
map[,1]<-(map[,1]-min(map[,1])*(1/(max(map[,1])-min(map[,1]))))
map[,2]<-(map[,2]-min(map[,2])*(1/(max(map[,2])-min(map[,2]))))
```
We can plot it quickly to make sure there are no mistakes:
```
plot(NULL,xlim=c(0,1),ylim=c(0,1),axes=F,xlab="",ylab="",main="")
polygon(map)
```

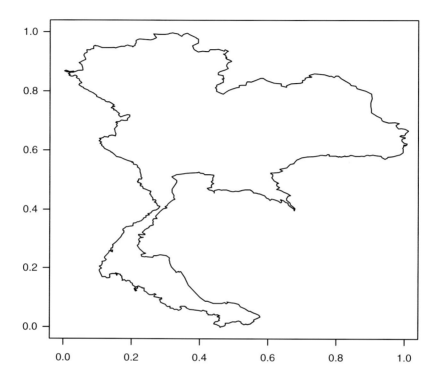

Oops, so here is Thailand's border and immediately we see that we need to adjust the axes to get the relative north–south and east–west proportions correct because at the moment both are scaled 0:1 whereas Thailand is much taller than it is wide. We'll then convert the x and y coordinates to longitude and latitude. Since both are scaled 0:1 we can simply multiply the longitude (x) and latitude (y) values by the number of degrees longitude and latitude that Thailand occupies. A quick search of Wikipedia (e.g. http://www.wikiwand.com/en/List_of_countries_by_southernmost_point) will give us the latitudinal and longitudinal extremities of Thailand in degrees and minutes (60ths of a degree) – we need first to convert these to decimal degrees – you can write a script to do this in R but with so few values, it's easier to use your head or a calculator (minutes*100/60); then we need to add the longitude (x) and latitude (y) of the western-most and southern-most parts of mainland Thailand (Table 21.1).

Table 21.1. Latitudinal and longitudinal extremities and ranges of Thailand.

Extremities	degrees and minutes	decimal degrees	latitude/longitude extent
northernmost	20° 27'	20.45	**= 20.45–5.612**
southernmost	5° 37'	5.612	**= 14.838**
westernmost	97° 20'	97.33	**= 105.67–97.33**
easternmost	105° 40'	105.67	**= 8.34**

Therefore, in order to transform our latitude and longitude coordinates
map[,1]<-map[,1]*8.34+97.33 # remember multiplications are always calculated before additions
map[,2]<-map[,2]*14.838+5.612
Now we need to create a new plot with an appropriate range of latitude and longitude and redraw our map of Thailand:
plot(NULL,xlim=c(94,108.83),ylim=c(5.5,20.5),axes=F,xlab="",ylab="", main="Thailand") # to keep proportions correct we use the same overall range for longitude
polygon(map,col="palegreen1")

Thailand

In biology we might use this to plot species distributions, or anything that can be represented by a point or shape (polygon) within the plotting area. We recently described many new Thai species of a parasitic wasp genus called *Aleiodes* (Hymenoptera: Braconidae) (Butcher *et al.*, 2012) and from this we can extract from this latitude and longitude of all the mentioned specimens.

There are no hard and fast rules for dealing with such unstructured text in R but here is what we would do. First assign the text to a variable

placidus<-readLines("A_placidus.txt")
head(placidus)
```
[1] "Aleiodes placidus sp. nov."
[2] "(Fig. 133)"
[3] "Holotype ♀, Thailand, Phetchabun Province, Khao Kho NP,
    mixed deciduous, 26.ii.2007, 16° 39.550 N, 101° 08.123
    E, 230m, Chacumnan & Singtong (voucher BCLDQ00228 ,
    Genbank JQ388468) (QSBG)."
[4] "Paratypes: 1 ♂, Thailand, Phetchabun Province, Khao Kho
    National Park, office, 26.vii.2006, 16° 39.550 N, 101°
    98.134 E, Chacumnan & Singtong, (voucher BCLDQ00158,
    Genbank JF962554) (QSBG) ; 1 ♀, Thailand, Chaiyaphum, Tat
    Tone NP, beside Sapsomboom substation, 2.vi.2007, 16°
    00.792' N, 101° 58.472' E, 641m, Tawit Jaruphan & Orowan
    Budsawong (voucher BCLDQ01552........
```

We are now interested in the latitude and longitude values, which, in this case, are given as degrees followed by decimal minutes – that means we have two things to deal with. We need to automatically recognize lat-longs AND then do a conversion to decimal degrees; annoying, but not difficult. The consistency in the text presentation is very helpful – if there are inconsistencies you may be better off transcribing data by hand, but that should be avoided because you can introduce errors and it would be immensely tedious if there were hundreds of data points.

We now need to 'pull out' the data that we need. Here we only want co-ordinates, so we look to see what formatting they all have in common, e.g. '20° 3.649' N, 99° 8.852' E, '. There are several things we can potentially use here – there are capital letters N and E each followed by a comma and space, there is the degrees symbol preceded by one or two integers and followed by a space, etc. To process such a text file we need to be a bit inventive. Our solution, which we explain step-by-step, is as follows.

1. Split the text into 'words' using **strsplit** (= string split) with a space as the split character:

words<-strsplit(placidus," ")
words
```
[[1]]
[1] "Aleiodes" "placidus" "sp." "nov."
[[2]]
[1] "(Fig." "133)"
[[3]]
[1] "Holotype" "♀," "Thailand," "Phetchabun" "Province," "Khao"
    "Kho" "NP," "mixed" "deciduous," "26.ii.2007," "16°"
    "39.550"
[14] "N," "101°" "08.123" "E," "230m," "Chacumnan" "&" "Singtong"
    "(voucher" "BCLDQ00228" "," "Genbank" "JQ388468)"
[27] "(QSBG)."
```

Mapping and Parsing Text Input for Data

```
[[4]]
[1] "Paratypes:""1""♂,""Thailand,""Phetchabun"
    "Province,""Khao""Kho"
[9] "National""Park,""office,""26.vii.2006,"
    "16°""39.550""N,""101°"….
```

The double square brackets ('[[1]]') tell us we have created a nested list, which is normally inconvenient to work with, so we will 'unlist' it.

words<-unlist(words)

head(words, 10)

```
[1] "Aleiodes""placidus""sp.""nov.""(Fig.""133)"
    "Holotype""♀,""Thailand,""Phetchabun"
```

Now we can access each 'word' using **grep**

   2. Find the positions of the degree signs in the list:

degrees.All<-grep("°",words)

To see what we have found:

words[degrees.All]

```
 [1] "16°""101°""16°""101°""16°""101°""16°""101°"
     "16°""101°""16°""101°""16°""101°"
[15] "16°""101°""16°""101°""16°""101°"
```

looks good, so now to obtain the decimal minute. These are always the next piece of text after the degrees so to extract them we can add 1 to the indices of the degrees:

dec.mins<-words[degrees.All+1]

dec.mins

```
 [1] "39.550" "08.123" "39.550" "98.134" "00.792'"
     "58.472'" "39.550" "08.123" "39.550" "08.123"
[11] "39.479" "08.105" "39.257'" "7.945'" "39.587"
     "08.134" "44.896" "27.874" "50.540" "43.663"
```

   3. Make a table pairing the degrees and decimal minutes:

placidus.coords<-cbind(words[degrees.All],dec.mins)

placidus.coords

```
 [1,]  "16°"   "39.550"
 [2,]  "101°"  "08.123"
 [3,]  "16°"   "39.550"
 [4,]  "101°"  "98.134"
 [5,]  "16°"   "00.792'"
 [6,]  "101°"  "58.472'"
 [7,]  "16°"   "39.550"
 [8,]  "101°"  "08.123"
 [9,]  "16°"   "39.550"
[10,]  "101°"  "08.123"
[11,]  "16°"   "39.479"
[12,]  "101°"  "08.105"
[13,]  "16°"   "39.257'"
[14,]  "101°"  "7.945'"
[15,]  "16°"   "39.587"
[16,]  "101°"  "08.134"
```

```
[17,]  "16°"   "44.896"
[18,]  "101°"  "27.874"
[19,]  "16°"   "50.540"
[20,]  "101°"  "43.663"
```
Now we have to convert the degrees and decimal minutes to decimal degrees. First strip the apostrophes from the second column:
placidus.coords[,2]<-gsub("'","",placidus.coords[,2])
similarly the degrees symbol in placidus.coords[,1]
placidus.coords[,1]<-gsub("°","",placidus.coords[,1])
The input is not completely consistent but that does not matter in this case. Note that values in our table are given within " ", which means that R is currently treating them as characters rather than numbers so we have to coerce them to be numbers using **as.numeric**.
placidus.coords[,2]<-as.numeric(placidus.coords[,2])*(100/60) # 60 minutes in a degree, so now they are all decimals of a degree
We need to remove the decimal point from the minutes column because these are now the numbers that will follow the degree value BUT unfortunately we cannot do this using 'placidus.coords[,2]<- gsub(".","", placidus.coords[,2])' because the decimal point character is a wildcard in R so we have to precede such special characters by '\\'.
placidus.coords[,2]<-gsub("\\.","",placidus.coords[,2])
Now we can combine the degrees and decimal degrees that were minutes using 'paste' (which is like **c** but you can select to omit spaces between the elements by using 'sep=""' and place a decimal point after the degrees):
new.degrees<-paste(placidus.coords[,1],".",placidus.coords[,2],sep="")
We have bulk processed the degree and minute coordinates without paying attention to whether they are latitude or longitude. We therefore have 10 numbers in 'new.degrees' the 1st, 3rd, 5th… of which are latitudes and the 2nd, 4th, 6th… longitudes and we now want to plot the locality data for *Aleiodes placidus* on our map of Thailand so we need to separate these. There are many ways to do this; here we create vectors for the positions of the x and y values within 'new.degrees'
lat<-(c(1:10)*2)-1 # there are 10 rows in the table
long<-(c(1:10)*2)
points(as.numeric(new.degrees[long]),as.numeric(new.degrees[lat]), pch=10)
mtext ("Aleiodes placidus",font=3) # this is an easy way to add a subtitle to a graph

Thailand

Aleiodes placidus

The **dismo** package has many functions built-in for spatial modelling. Its function **gbif** downloads data from the GBIF (Global Biodiversity Information Facility) data portal. Here we will analyse the available data for the Asian great Bhutan butterfly (*Bhutanitis lidderdalii*: Papilionidae). We chose this butterfly because there are only a few records, for many common species there are many thousands

install.packages("dismo")
library(dismo)
bhut<-gbif("Bhutanitis","lidderdalii")
201 records found
0-201 records downloaded
colnames(bhut) # there are a lot of columns, some with long names, which is why the output is ugly

```
 [1] "ISO2"                    "acceptedNameUsage"
     "acceptedScientificName"
 [4] "acceptedTaxonKey"        "accessRights"      "adm1"
 [7] "adm2"                    "associatedSequences"
     "basisOfRecord"
[10] "bibliographicCitation"   "catalogNumber"
     "class"
[13] "classKey"                "cloc"
     "collectionCode"
[16] "collectionID"            "continent"
     "coordinateUncertaintyInMeters"
```

```
 [19] "country"                 "crawlId"
      "datasetID"
 [22] "datasetKey"              "datasetName"
      "dateIdentified"
 [25] "day"                     "disposition"
      "dynamicProperties"
 [28] "endDayOfYear"            "eventDate"
      "eventID"
 [31] "eventTime"               "family"
      "familyKey"
 [34] "fieldNumber"             "fullCountry"
      "gbifID"
 [37] "genericName"             "genus"
      "genusKey"
 [40] "geodeticDatum"           "georeferenceProtocol"
      "georeferenceRemarks"
 [43] "georeferenceSources"     "georeferenceVerification
      Status" "georeferencedBy"
 [46] "georeferencedDate"       "higherClassification"
      "higherGeography"
 [49] "http://unknown.org/occurrenceDetails"
      "http://unknown.org/recordId" "identificationID"
 [52] "identificationQualifier" "identificationRemarks"
      "identifiedBy"
 [55] "identifier"              "individualCount"
      "informationWithheld"
 [58] "installationKey"         "institutionCode"
      "institutionID"
 [61] "key"                     "kingdom"
      "kingdomKey"
 [64] "language"                "lastCrawled"
      "lastInterpreted"
 [67] "lastParsed"              "lat"
      "license"
 [70] "lifeStage"               "locality"
      "locationAccordingTo"
 [73] "locationID"              "lon"
      "modified"
 [76] "month"                   "nomenclaturalCode"
      "occurrenceID"
 [79] "occurrenceRemarks"       "order"
      "orderKey"
 [82] "organismID"              "otherCatalogNumbers"
      "ownerInstitutionCode"
 [85] "parentNameUsage"         "phylum"
      "phylumKey"
 [88] "preparations"            "previousIdentifications"
      "protocol"
```

```
 [91] "publishingCountry"       "publishingOrgKey"
      "recordedBy"
 [94] "references"              "rights"
      "rightsHolder"
 [97] "scientificName"          "sex"
      "species"
[100] "speciesKey"              "specificEpithet"
      "startDayOfYear"
[103] "taxonConceptID"          "taxonID"
      "taxonKey"
[106] "taxonRank"               "taxonRemarks"
      "taxonomicStatus"
[109] "type"                    "verbatimCoordinateSystem"
      "verbatimEventDate"
[112] "verbatimLocality"        "verbatimTaxonRank"
      "vernacularName"
[115] "year"                    "downloadDate"
```

Different species have different subsets of data depending on the specimens from which data have been harvested. We can see that at least some records have latitude ('lat') and longitude ('lon') data because those are present in the column names. First create a subset of the 'bhut' dataframe with only latitude and longitude data using the built-in **subset** function to which we pass the columns to include using the argument 'select='.

localities<-subset(bhut,select=c("lat","lon"))
head(localities)

```
        lat        lon
1  25.46008   94.27281
2  27.53908   93.89472
3  27.54765   93.89844
4  27.54851   93.89780
5  27.54753   93.89781
6       NA         NA
```

Here we see that not all of the records on GBIF have any lat long data and so we filter out the ones with missing data using another built-in R function, **complete.cases**, which outputs only cases with all values

b<-complete.cases(subset(bhut,select=c("lat","lon")))
localities[b,]

```
          lat        lon
1    25.46008   94.27281
2    27.53908   93.89472
3    27.54765   93.89844
4    27.54851   93.89780
5    27.54753   93.89781
130  26.10000   94.26000
131  26.10000   94.26000
134  20.00000   77.00000
135  27.50000   90.50000
```

```
152  26.36071  94.95435
174  22.50000  93.50000
178  26.36071  94.95435
198  26.36071  94.95435
```
newmap<-getMap(resolution="low")
plot(newmap)
points(localities$lon,localities$lat,col="red",cex=2) # use pch to change point type

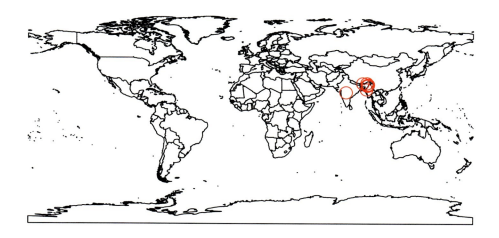

Not too surprising that *Bhutanitis* is mostly known from Bhutan and adjacent areas (sadly it is now extinct in Thailand).

Tool Box 21.2 ✖

- Use complete.cases() to remove incomplete data.
- Use gbif() to download data from the Global Biodiversity Information Facility.
- Use grep() to find which rows have a particular character.
- Use mapRegion= to only show the map of a particular region.
- Use mapTitle= to set the main title.
- Use mtext() to add a subtitle below a graph title.
- Use polygon() to draw a shape based on x y coordinates.
- Use select=in subset to choose what goes into each subset.
- Use strsplit() to split a character string.
- Use subset() to create a subset of a dataframe based on columns or values in a vector.
- Use unlist() to remove nesting of a list.

22 More on Manipulating Text

Summary of R Functions Introduced
 .+? - a wildcard
 map
 regexpr
 switch
 unname

You will often need to edit a few items in a text file in order to extract information in a desired format. If for a given project you will only need to do it once or twice, it may well be simpler just to use Microsoft Word and/or Excel (or similar programs). However, if there are many such things to do, or you are likely to want to do similar things in the future, then doing it by hand will be both time-consuming and liable to introduce errors. R has many very useful tools for carrying out text manipulations. We have already met **grep** and **gsub**; we will use them a lot here.

Example 1 – Standardizing Names in a Phylogenetic Tree Description

Here we will use R to reformat taxon names, create lists, sort data and use wildcards for when some things we are interested in don't have exactly the same length. The example tree description here concerns parasitoids of caterpillars at a study site that have been DNA barcoded and their possible taxonomic identities added automatically. As a consequence, each taxon name is rather long and unwieldy. The tree description also contains rather long numbers following a colon, and these are branch lengths for the taxon or clade. We will use only a small part of the tree description to illustrate processes. First, we assign this piece of text to a variable, 'tr1':

tr1<-"(((((BBTH004_16_CCDB_24026_A4_91Tachinidae[90Drosophila]:
    1.14675336480776657311,(BBTH006_16_CCDB_24026_A6_91
    Belvosia:0.59297714057027572920,BBTH055_16_CCDB_24026_

E7_92Tachinidae[91Belvosia]:0.52736765819178677006):
0.32998852276617751667):0.31714205003432383023,((BBTH084_16_
CCDB_24026_G12_97Tachinidae:0.00006954085997979414,BBTH085_1
6_CCDB_24026_H1_97Tachinidae:0.00006954085997979414):
0.00006954085997979414,(BBTH080_16_CCDB_24026_G8_97
Tachinidae:0.00006954085997979414,BBTH091_16_CCDB_24026_
H7_97Tachinidae:0.00006954085997979414)"

The round brackets and commas show the inferred relationships between the taxa and so their precise positioning relative to the taxa is vital. Each taxon entry contains a series of code numbers relating to the specimen (e.g. BBTH004), the year of the sequencing (e.g. 16), the DNA sequencing plate number (e.g. 24026), the well number in that plate (e.g. A4), a percentage similarity to the nearest sequence on the DNA sequence library GenBank (e.g. 91), the name of the most similar taxon (e.g. Tachinidae) and the length of the branch on the tree (e.g. :1.14675336480776657311). In some cases, a second close taxon match is given in square brackets. This is all important information in terms of keeping track of your data, but unwieldy for some presentations of the tree we might want to show in a paper. The important thing is that these bits of information are presented in a consistent format, which means that we can change it all in one go. In real life you may be getting DNA sequences from multiple sources, and the formats of names will differ between sources, but they will usually be consistent within a source.

The first thing we might want to do is remove superfluous information from the names to make viewing and reading a tree easier. What is superfluous will depend on the situation, but here we can see that all names contain the character string '16_CCDB_24026_' and we can remove these from all names in the tree using **gsub**.

tr2<-gsub("16_CCDB_24026_","",tr1)
tr2
[1]
"(((((BBTH004_A4_91Tachinidae[90Drosophila]:
    1.14675336480776657311,(BBTH006_A6_91Belvosia:
    0.59297714057027572920,BBTH055_E7_92Tachinidae
    [91Belvosia]:0.52736765819178677006):0.3299885227
    6617751667):0.31714205003432383023,((BBTH084
    _G12_97Tachinidae:0.00006954085997979414,
    BBTH085_H1_97Tachinidae:0.00006954085997979414):
    0.00006954085997979414,(BBTH080_G8_97Tachinidae:
    0.00006954085997979414,BBTH091_H7_97Tachinidae:
    0.00006954085997979414)"

Alternatively, the BBTH part and the plate number (24026) may be important, but the text in-between might not be, but this text is not the same for each taxon. This is where we will use a wildcard (.+?) to represent the variable part in each name starting with BBTH and ending with 24026, and we replace with the shorter character string 'BBTH24026'.

tr3<-gsub("BBTH.+?24026","BBTH24026",tr1)
tr3

More on Manipulating Text

```
[1]
"(((((BBTH24026_A4_91Tachinidae[90Drosophila]:
    1.1467533648077665311,(BBTH24026_A6_91Belvosia:
    0.59297714057027572920,BBTH24026_E7_92Tachinidae
    [91Belvosia]:0.52736765819178677006):0.3299885227661
    7751667):0.31714205003432383023,((BBTH24026_G12_97
    Tachinidae:0.00006954085997979414,BBTH24026_H1_
    97Tachinidae:0.00006954085997979414):0.00006954085997979414,
    (BBTH24026_G8_97Tachinidae:0.00006954085997979414,
    BBTH24026_H7_97Tachinidae:0.00006954085997979414)"
```

Similarly we can use a wildcard to remove the second best taxon matches that are given in square brackets, however, the square bracket characters, [and], belong to a small set of characters in R called metacharacters – because in the R language they represent various functions, in this case subsetting a vector. In text manipulation these metacharacters have to be dealt with in a special way, by specifying their nature using two forward slashes before them, e.g. \\[. The other metacharacters in R are \ (a single forward slash), |, (,), {, }, ^, $, *, + and ?

Method 1 with Wildcards

Thus to remove the square bracket parts of names including the use of a wildcard for the contents of the square brackets we use – the wildcard used here is '.+?', but more generally the dot character acts as a wildcard (see below)
tr4<-gsub("\\[.+?\\]","",tr3) # this means remove everything between the square brackets, [and]
tr4
```
[1]
"(((((BBTH24026_A4_91Tachinidae:1.14675336480776657311,
    (BBTH24026_A6_91Belvosia:0.59297714057027572920,
    BBTH24026_E7_92Tachinidae:0.52736765819178677006):
    0.32998522766177751667):0.31714205003432383023,
    ((BBTH24026_G12_97Tachinidae:0.00006954085997979414,
    BBTH24026_H1_97Tachinidae:0.00006954085997979414):
    0.00006954085997979414,(BBTH24026_G8_97Tachinidae:
    0.00006954085997979414,BBTH24026_H7_97Tachinidae:
    0.00006954085997979414)"
```
To remove the branch lengths is slightly more difficult, because although each follows a colon, they can be followed by either a comma or a ')'. There are several different ways that we can achieve this and we will show you three methods.

Method 1 – split the character string to each individual character and find ':' and ',' and ')'.
tr5<-strsplit(tr4,"")
Because this is now a long vector we'll only show the beginning
tr5
[[1]]

```
[1]  "("  "("  "("  "("  "("  "B"  "B"  "T"  "H"  "2"  "4"  "0"  "2"
     "6"  "_"  "A"  "4"  "_"  "9"  "1"  "T"  "a"  "c"  "h"  "i"  "n"
     "i"  "d"  "a"  "e"  ":"  "1"  "."  "1"
[35] "4"  "6"  "7"  "5"  "3"  "3"  "6"  "4"  "8"  "0"  "7"  "7"  "6"
     "6"  "5"  "7"  "3"  "1"  "1"  ","  "("  "B"  "B"  "T"  "H"  "2"
     "4"  "0"  "2"  "6"  "_"  "A"  "6"  "_"
[69] "9"  "1"  "B"  "e"  "l"  "v"  "o"  "s"  "i"  "a"  ":"  "0"  "."
     "5"  "9"  "2"  "9"  "7"  "7"  "1"  "4"  "0"  "5"  "7"  "0"  "2"
     "7"  "5"  "7"  "2"  "9"  "2"  "0"  ","  ......
```

Note that because the vector 'tr4' had one level and we have split it, all the elements are now a second level – if you ask the length of 'tr5' R will tell you it has length 1, as indicated by [[1]]. To access the individual elements, we need to unlist it

tr5<-unlist(tr5)

head(tr5,40) # the 40 tells R how many items in the list tr5 to show

```
[1]  "("  "("  "("  "("  "("  "B"  "B"  "T"  "H"  "2"  "4"  "0"  "2"
     "6"  "_"  "A"  "4"  "_"  "9"  "1"  "T"  "a"  "c"  "h"  "i"  "n"
     "i"  "d"  "a"  "e"  ":"  "1"  "."  "1"
[35] "4"  "6"  "7"  "5"  "3"  "3"
```

> **Information Box 22.1**
>
> *Use of **regexpr** (regular expression) to extract positions of patterns in strings.* An alternative to splitting a string into individual characters in order to find the positions in the string where a particular character (or pattern) occurs is to use **regexpr** or **gregexpr**. The first gives the position of the first occurrence of the character (or pattern) and its number of occurrences, the second gives the positions of all occurrences. For example:
>
> x<-"attgacccggaggaggaatta"
> first_G<-regexpr("g",x)
> first_G
> [1] 4
> attr(,"match.length")
> [1] 1
> attr(,"index.type")
> [1] "chars"
> attr(,"useBytes")
> [1] TRUE
>
> So the first occurrence of "G" in the string is x[1], i.e. 4. In this case the output is listed so to extract and use the positions we can either
> all_GGA<-gregexpr("gga",x)
> unlist(all_GGA)
> [1] 9 12 15
> or, rather more cumbersome, explicitly select the element of the list, i.e. 1, that we want
> all_GGA[[1]][1:length(all_GGA[[1]])]
> [1] 9 12 15

In practise we would probably have written tr5<-unlist(strsplit(tr4,"")) in the first place. Note that we cannot strsplit using ')' or ':' because the character string being used as the splitter in **strsplit** is deleted, and we need to keep them all so that we can reassemble the tree when we have removed the branch lengths.

Finding the beginning of the branch lengths is easy because they are all preceded by a colon (which we will also want to remove). Use **grep** to find the positions of the colons:

colon<-which(tr5==":")
colon
```
[1]  31  79 128 152 176 228 277 301 351 400
```
To find the end of the number now we need to know which is the first comma or) that follows each colon.

close<-which(tr5==")")
comma<-which(tr5==",")
close
```
[1] 151 175 300 423
```
comma
```
[1]  54 102 199 251 324 374
```
Here we can see that the first comma or) following the first colon is position 54. We can check that our code is working by looking at the elements we have identified

tr5[colon[1]:(comma[1]-1)] # we have to use () to take the 1 only off the value of comma[1]
```
[1] ":" "1" "." "1" "4" "6" "7" "5" "3" "3" "6" "4"
    "8" "0" "7" "7" "6" "6" "5" "7" "3" "1" "1"
```
So we want to step through each of the colons and find the position in tr5 of the first) or , that follows it. We can then define a substring starting with the colon, and ending one position before either the next , or) and delete it. EASY, but BEWARE, there is a catch. If you start at the beginning, after the first removal you will change the positions of all subsequent colons, commas and brackets in the vector. We could do an inelegant cheat here, by replacing all these characters by some other unique character that does not occur in the tree, for example an exclamation mark, and then in one go, remove all the exclamation marks. OR, we can write a loop so that we recreate the vectors of colon, comma and close each time, which for heuristic purposes is what we will do here.

We can use a '**for**' loop which needs to know the number of occurrences of branch lengths (or colons as here), or a repeat-until loop, which is what we will show here. Repeat loops carry on until a specified condition is met and the loop is told to break. Each time R cycles through the loop, it finds the positions (indices) of the colons, commas and close brackets (lines 3–5). As we will have deleted the previous branch length and its associated colon each time, the next branch length will always start at the first colon in the list. The comments in the code below show the values the first time through the repeat loop

```
tr4<-gsub("\\[.+?\\]","",tr3)
repeat{
    ss<-unlist(strsplit(tr4,""))
    colon<-which(ss==":") # [1] 31 79 128 152 176 228 277 301 351 400
    L<-length(unlist(colon)) # 10
    if(L==0) break # condition to break out of repeat loop
    close<-which(ss==")") # [1] 151 175 300 423
    comma<-which(ss==",") # [1] 54 102 199 251 324 374
    a<-which(comma>colon[1]) # [1] 1 2 3 4 5 6
    b<-which(close>colon[1]) # [1] 1 2 3 4
    c<-min(comma[a[1]],close[b[1]]) # 54
    n<-substr(tr4,colon[1],c-1) # [1] ":1.14675336480776657311"
    tr4<-gsub(n,"",tr4) # removing substring n from tr4
    } # end repeat loop
tr4
[1]
"(((((BBTH24026_A4_91Tachinidae,(BBTH24026_A6_91Belvosia,
    BBTH24026_E7_92Tachinidae)),((BBTH24026_G12_
    97Tachinidae,BBTH24026_H1_97Tachinidae),(BBTH24026_
    G8_97Tachinidae,BBTH24026_H7_97Tachinidae)"
```

Method 2 Based on Fixed Character String Length

As all the branch length numbers in this case have precisely 21 characters including the decimal point, we can use a mask:
```
tr4<-gsub("\\[.+?\\]","",tr3)
tr5<-unlist(strsplit(tr4,""))
colons<-grep(":",tr5)
ends<-colons+22 # 22 including the colon
for(i in length(colons):1) tr4<-gsub(substr(tr4,colons[10],ends[10]),"",tr4)
tr4
[1]
"(((((BBTH24026_A4_91Tachinidae:1.14675336480776657311,
    (BBTH24026_A6_91Belvosia:0.59297714057027572920,
    BBTH24026_E7_92Tachinidae:0.52736765819178677006):
    0.32998852276617751667):0.31714205003432383023,
    ((BBTH24026_G12_97Tachinidae,BBTH24026_H1_
    97Tachinidae),(BBTH24026_G8_97Tachinidae,BBTH24026_
    H7_97Tachinidae)"
```

Method 3 Using a Vector of Positions

Alternatively, we can create a vector of all the positions representing the branch length numbers in tr5,
```
tr4<-gsub("\\[.+?\\]","",tr3)
tr5<-unlist(strsplit(tr4,""))
```

```
colons<-grep(":",tr5)
ends<-colons+22 # 22 including the colon
positions<-NULL
for(i in 1:length(colons)){
    temp<-colons[i]:ends[i]
    positions<-c(positions,temp)}
tr5<-tr5[-positions]
```
and now we can recreate tr4 by collapsing (collapse) the remaining elements of tr5 into a single level vector using **paste**
```
tr4<-paste(tr5,collapse="")
tr4
[1]
"(((((BBTH24026_A4_91Tachinidae,(BBTH24026_A6_9
   1Belvosia,BBTH24026_E7_92Tachinidae)),((BBTH24026_G12_
   97Tachinidae,BBTH24026_H1_97Tachinidae),(BBTH24026_G8_
   97Tachinidae,BBTH24026_H7_97Tachinidae)"
```
Now suppose we want to extract the taxon names and to sort them by well number of the DNA plate, i.e. A1, A2, A3 … H11, H12. First we need to extract the names of the taxa, and you will note that they are always preceded by a '(' or a , '[' or they are always followed by a ')' or a ']' therefore we can string split the vector at these and remove any other punctuation that is not part of a name.
```
names<-gsub("\\(","",tr4) # removing (
names<-unlist(strsplit(names,")")) # splitting string at )
names<-unlist(strsplit(names,",")) # splitting string at commas
names
[1] "BBTH24026_A4_91Tachinidae" "BBTH24026_
    A6_91Belvosia" "BBTH24026_E7_92Tachinidae" ""
[5] "BBTH24026_G12_97Tachinidae" "BBTH24026_
    H1_97Tachinidae""" "BBTH24026_G8_97Tachinidae"
[9] "BBTH24026_H7_97Tachinidae"
```
We have successfully isolated each of the taxon names from the tree BUT we have some empty items in the list (""), and to remove these we can use!="" to keep in the list only the non-empty values
```
names<-names[which(names!="")]
names
[1] "BBTH24026_A4_91Tachinidae" "BBTH24026_A6_91Belvosia"
    "BBTH24026_E7_92Tachinidae" "BBTH24026_G12_97Tachinidae"
[5] "BBTH24026_H1_97Tachinidae" "BBTH24026_
    G8_97Tachinidae" "BBTH24026_H7_97Tachinidae"
```
To rearrange each name so that it starts with its well number, we will process each name in turn using a for loop, split each name at the underscore character, swap the order of the first part 'BBTH24026' with the second part (the well number) and recombine. We have added:
```
rearrange<-NULL
for(i in 1:length(names)){
    temp<-unlist(strsplit(names[i],"_"))
    # temp = "BBTH24026" "A4" "91Tachinidae"
    rearrange<-c(rearrange,paste(temp[2],temp[1],temp[3],sep="_"))
```

} # end *i* loop
```
rearrange
[1] "A4_BBTH24026_91Tachinidae" "A6_BBTH24026_91Belvosia"
    "E7_BBTH24026_92Tachinidae" "G12_BBTH24026_97Tachinidae"
[5] "H1_BBTH24026_97Tachinidae" "G8_BBTH24026_97Tachinidae"
    "H7_BBTH24026_97Tachinidae"
```
We can use the R function **sort**() to order the elements of a vector in increasing numerical or alphabetical order (the default). If we want to have the elements in decreasing order we would set the sort parameter decreasing=TRUE. The final hurdle here is caused by the fact that the sort function will consider A10, A11 and A12 to come alphabetically after A1 but before A2. In our example G12 appears in the sorted list before G8.
```
sort(rearrange)
[1] "A4_BBTH24026_91Tachinidae" "A6_BBTH24026_91Belvosia"
    "E7_BBTH24026_92Tachinidae" "G12_BBTH24026_97Tachinidae"
[5] "G8_BBTH24026_97Tachinidae" "H1_BBTH24026_97Tachinidae"
    "H7_BBTH24026_97Tachinidae"
```
We therefore have one final bit of parsing to do to convert A1 through A9 to A01 through A09 and so on. We will use **sub** NOT **gsub** because the former only substitutes the first instance of a character string and not all instances, and it is conceivable that some taxon names could contain the same string as a well name
```
for(i in 1:length(rearrange)){
    if(substr(rearrange[i],3,3)=="_") {
        m<-substr(rearrange[i],1,1)
        n<-substr(rearrange[i],2,3)
        new<-paste(m,"0",n,sep="")
        rearrange[i]<-sub(substr(rearrange[i],1,3),new,rearrange[i])} # end if
    } # end i loop
rearrange
[1] "A04_BBTH24026_91Tachinidae" "A06_BBTH24026_91Belvosia"
    "E07_BBTH24026_92Tachinidae" "G12_BBTH24026_97Tachinidae"
    "H01_BBTH24026_97Tachinidae" "G08_BBTH24026_97Tachinidae"
[7] "H07_BBTH24026_97Tachinidae"
```
Now applying the function **sort** we achieve our goal.
```
sort(rearrange)
[1] "A04_BBTH24026_91Tachinidae" "A06_BBTH24026_91Belvosia"
    "E07_BBTH24026_92Tachinidae" "G08_BBTH24026_97Tachinidae"
    "G12_BBTH24026_97Tachinidae" "H01_BBTH24026_97Tachinidae"
[7] "H07_BBTH24026_97Tachinidae"
```

Example 2 – Substrings of Unknown Length

In this example we will search for a numeric substring of unknown length but with a standard prefix. It will also let us show how to avoid errors when a test is not applicable in some cases and yields a character0 output. Our data are some

DNA sequences from a set of *Aleiodes* wasps, and we want to move, when needed, the specimen code 'MRSXXX' to the beginning of each line and then to sort the dataset accordingly. However, there are some sequences that lack MRS numbers and not all the MRS numbers are followed by the same number of digits.

First enter the data using **textConnection** and **readLines**. We can do this in a single command. 'X' is often used instead of 'N' to mean an unknown number of missing bases.

data<-readLines(textConnection("Aleiodes_bicolor_MRS1008 TATTTTATA
    TTTTTTATTT
Aleiodes_praetor_UK_MRS67_   XXXXXXXXXGTTTTATAT
Aleiodes_rugulosus_CollHH1599_Norway   xxxxxxxxxxATTTTGTATTTTTT
Aleiodes_seriatus_MRS252_France   xxxxxxxxxxxxxxxxxxxxxxxxxx
Aleiodes_seriatus_MRS254_France   xxxxxxxxxxxxxxxxxxxxxxxxxxxx
Aleiodes_seriatus_MRS263_France   xxxxxxxxxxATTTTATACTTTTTATTTGG
Aleiodes_seriatus_MRS264_France   xxxxxxxxxxATTTTATACTTTTTA
    TTTGG
Aleiodes_seriatus_MRS136_France   GATATTGGAATTTTATATTT
MRS239_Aleiodes_seriatus_Russia   xxxxxxxxxxGTTTTATACTTCTTATTT
Aleiodes_seriatus_MRS222_Germany   xxxxxxxxxxATTTtaTaCTTTTTATT
Aleiodes_sibiricus_MRS313_Sweden   xxxxxxxxxxxTTTTGTATTTTTTATT
Aleiodes_signatus_MRS378_Sweden   xxxxxxxxxxxTTTATATTTTTTATT
Aleiodes_signatus_MRS712_Sweden   GATATTGGTATTTTATATTTTTA
Aleiodes_unipunctator_CollHH1603_Norway   xxxxxxxATTTTATATTTT
    TTATG"))

For these sorts of operations there are a number of 'regular expression' functions that are supported by R. We already know **grep**,

grep("MRS",data)
[1] 1 2 4 5 6 7 8 9 10 11 12 13

this tells us which elements of our data contain the character string 'MRS'. The **regexpr** function gives more information. The –1s (minus 1s) show that 'MRS' was not found in some elements, and the other numbers in the first part are the position along each row where the first character of 'MRS' occurs. The second row gives us how many characters there are in the found search string

regexpr("MRS",data)
 [1] 18 21 -1 19 19 19 19 19 19 19 20 19 19 -1
attr(,"match.length")
 [1] 3 3 -1 3 3 3 3 3 3 3 3 3 3 -1
attr(,"index.type")
[1] "chars"
attr(,"useBytes")
[1] TRUE

To find which occurrences of 'MRS' are followed directly by some numbers use **gregexpr**, which returns a list (hence the need to **unlist** the result)
unlist(gregexpr(pattern ="MRS[0-9]+",data))
[1] 18 21 -1 19 19 19 19 19 19 19 20 19 19 -1
This is the same as the **regexpr** result because in our data every MRS is followed by a number. Then to find what that number is (in fact the whole sample code) we combine **gregexpr** with **regmatches**
unlist(regmatches(x=data,gregexpr("MRS[0-9]+",text=data)))
[1] "MRS1008" "MRS067" "MRS252" "MRS254" "MRS263" "MRS264" "MRS136" "MRS239" "MRS222" "MRS313" "MRS378" "MRS712"
Though this does not show the lines that did not contain 'MRS' followed by a number, this is a lot easier than using a loop and incrementing substring after the 'MRS' until we stop finding numbers.

Our aim was to find the MRS numbers and move them to the beginning of the line, deleting where they originally were. We could loop through the lines and in each case test if there was an MRS number combination, extract it, paste it at the beginning of the line and delete it. We will do this but using **lapply** for simplicity (see Chapter 27 for more on **apply** family functions). However, it is often a good idea while you are learning to actually start off with the loop method, because if some unexpected error occurs it makes it easy to identify where, so that you can correct it.

Since we wish to do a number of things with each line we need to create our own function to call from within the **lapply** statement. First we will try something that does not work:
startwithMRS<-function(b){
    mrs<-unlist(regmatches(x=b,gregexpr("MRS[0-9]+",text=b)))
    z<-gsub(mrs,"",b)
    z<-paste(mrs,"_",z,sep="")
    return(z)} # end startwithMRS function
MRSatstart<-unlist(lapply(data, function(x) startwithMRS(x)))
Error in gsub(mrs, "", b) : invalid 'pattern' argument
What has gone wrong? Well, if we had used a loop we would have found that the problem occurred on line 3, and a glance at the data would prompt you that the problem might be because there is no 'MRS' in line 3. Following this up, we would run through the lines of our function with just the third row of data to see if we can replicate the error
mrs<-unlist(regmatches(x=data[3],gregexpr("MRS[0-9]+",text=data[3])))
mrs
character(0)
Aha! We have identified the problem and now need a solution. It would be nice if we could simply test for this by say
mrs==character(0)
logical(0)
but that returns logical(0) rather than the TRUE or FALSE we would have liked. This result is perfectly logical because character(0) is a character of zero length and thus the == test gives an answer that it is not logically possible to compare

it with another vector. There are two obvious work-arounds. Which to use depends a bit on the situation. Both logical(0) and character(0) have zero length, so we can test our code by doing different things depending on whether the length of 'mrs' is 0 or >0.

```
startwithMRS<-function(b){
     mrs<-unlist(regmatches(x=b,gregexpr("MRS[0-9]+",text=b)))
     if(length(mrs)>0) { # the MRS string occurs on the line
          z<-gsub(mrs,"",b) # remove the original MRS number occurrence
          z<-paste(mrs,"_",z,sep="") # paste the MRS number at the
               beginning
          } # end if
     else z<-b # the MRS string does not occur on the line
     z # because the function has produced just the one value, 'z', we can
          omit 'return(z)'
     } # end startwithMRS function
MRSatstart<-unlist(lapply(data,function(x) startwithMRS(x)))
MRSatstart
 [1] "MRS1008_Aleiodes_bicolor_  TATTTTATATTTTTATTT "
 [2] "MRS67_Aleiodes_praetor_UK__  XXXXXXXXXGTTTTATAT"
 [3] "Aleiodes_rugulosus_CollHH1599_Norway xxxxxxxxxATTT
     TGTATTTTTT"
 [4] "MRS252_Aleiodes_seriatus__France xxxxxxxxxxxxxxxxx
     xxxxxxxxx"
 [5] "MRS254_Aleiodes_seriatus__France xxxxxxxxxxxxxxxxxx
     xxxxxxxxxx"
 [6] "MRS263_Aleiodes_seriatus__France xxxxxxxxxATTTTAT
     ACTTTTTATTTGG "
 [7] "MRS264_Aleiodes_seriatus__France xxxxxxxxxATTTTAT
     ACTTTTTATTTGG "
 [8] "MRS136_Aleiodes_seriatus__France GATATTGGAATTTT
     ATATTT"
 [9] "MRS239__Aleiodes_seriatus_Russia xxxxxxxxxGTTTTAT
     ACTTCTTATTT"
[10] "MRS222_Aleiodes_seriatus__Germany xxxxxxxxxATTTta
     TaCTTTTTATT"
[11] "MRS313_Aleiodes_sibiricus__Sweden xxxxxxxxxxTTTTG
     TATTTTTTATT"
[12] "MRS378_Aleiodes_signatus__Sweden xxxxxxxxxxxTTT
     ATATTTTTATT "
[13] "MRS712_Aleiodes_signatus__Sweden GATATTGGTATTTTA
     TATTTTTA "
[14] "Aleiodes_unipunctator_CollHH1603_Norway xxxxxxxATTT
     TATATTTTTATG"
```

To put the lines in alphabetical order simply apply **sort** to the vector. An alternative to 'length(mrs)>0' is to use the function **identical**, e.g. 'if(!identical(mrs,character(0)))'.

Trimming White Spaces and/or Tabs

In our wasp DNA data example above (in the vector 'MRSatstart' that we created) you will notice that there are a variable number of white spaces after the chunk of DNA sequence data. These are not only ugly but can upset formatting or make extracting the DNA sequences alone rather more complicated. R has a very useful function **trimws** (=trim white spaces) that will remove these from our strings. If we do not specify 'r' or 'l' as an argument in the function, white spaces (also tabs, line feeds or carriage returns) will be removed from both the beginning and end of the string.

Using Wildcards to Locate Internal Letter Strings

In addition to being able to find, substitute and sort by the first letter and first letter strings, there may be occasions when we need to find character strings at certain positions within a word or list.

The '^' symbol in character searches specifies the empty string at the beginning of a character string. Therefore '^.' is the first character of the string, no matter what it is, '^..' is the second, etc.

> **Information Box 22.2**
>
> *Regular expression code using wildcards to replace multiple spaces in a string with a single space.* Often when dealing with DNA sequence matrices we may wish to replace a run of spaces with a single space rather than removing all spaces. The following by David Arenburg on Stack Overflow (2014) shows one way of doing this
>
>   with_single_space<-gsub("^ *|(?<=) | *$", "", data, perl = TRUE).
>
> How this works was explained by Rick Measham if you want to try and follow it.
>
> | Node | Explanation |
> |---|---|
> | ^ | the beginning of the string |
> | * | ' ' (0 or more times (matching the most amount possible)) |
> | \| | logical OR |
> | (?<= | look behind to see if there is a: |
> | | ' ' |
> |) | end of look-behind |
> | | ' ' |
> | \| | logical OR |
> | * | ' ' (0 or more times (matching the most amount possible)) |
> | $ | before an optional \n, and the end of the string |

Exercise Box 22.1

1) There is more tidying of the file required to make our data presentable. As an exercise replace double underscores with single ones and delete any underscores at the ends of the names. After that, write a function to count the spaces after each name and the number of characters in the name, and then standardize the starting position of each of the sequences. The code in the Information box above might be helpful.

Finding Suffixes, Prefixes and Specifying Letters, Numbers and Punctuation

The last character of a character string is denoted in text searches by $. This can be useful if we want to find elements with particular character or character string endings that might also occur inside or at the beginning of words. To illustrate, 'ini' in the following example occurs also in 'initially', and 'ina' inside names ending in 'inae', and that means you cannot just use **grep**.

text1<-"We want to write some R code recognizes animal or plant family, subfamily, tribe or subtribe names in a piece of text. For animals these always end in -idae, -inae, -ini and -ina respectively. The Noctuidae, Erebidae, Sphingidae and Saturniidae are families of moths, the Lymantriinae is a subfamily of Erebidae, the Lymantriini, Leucomini, Orgyiini and Nygmiini are tribes of the Lymantriinae, etc."

First, we will isolate the 'words' using **strsplit** with a space character as the split value, but there are some complications, because commas, full stops, semicolons, etc., that adjoin words will be left as the final character of a word and will need to be dealt with separately.

text2<-unlist(strsplit(text1," "))

Prefixes can generally be found easily, for example to find words beginning with 'Ereb' we can use 'grep' specifying that this character string is at the beginning of a word:

lymantrids<-grep("^Lyman",text2)
lymantrids
[1] 46 53 62

For suffixes such as family names that end always (in zoology) in -idae, we use

families<-text2[grep("idae$",text2)]
families
[1] "Sphingidae" "Saturniidae"

but here our search only finds two and misses Noctuidae and Erebidae. The reason is that the **strsplit** based on the space character left punctuation marks, such as commas, attached to the ends of some of the words. We can remove all punctuation characters in one fell swoop by using '[[:punct:]]$'. The dollar sign makes sure that we are only deleting punctuation at the end of a word and not, for example, the hyphens at the beginning.

temp<-gsub("[[:punct:]]$","",as.character(text2))
temp

```
[1] "We" "want" "to" "write" "some" "R" "code"
[8] "recognizes" "animal" "or" "plant" "family" "subfamily"
    "tribe"
[15] "or" "subtribe" "names" "in" "a" "piece" "of"
[22] "text" "For" "animals" "these" "always" "end" "in"
[29] "-idae" "-inae" "-ini" "and" "-ina" "respectively" "The"
[36] "Noctuidae" "Erebidae" "Sphingidae" "and" "Saturniidae"
    "are" "families"
[43] "of" "moths" "the" "Lymantriinae" "is" "a" "subfamily"
[50] "of" "Erebidae" "the" "Lymantriini" "Leucomini"
    "Orgyiini" "and"
[57] "Nygmiini" "are" "tribes" "of" "the" "Lymantriinae"
    "etc"
```

Now we can find all families ending in -idae,
families<-temp[grep("idae$",temp)]
families
```
[1] "Noctuidae" "Erebidae" "Sphingidae" "Saturniidae"
    "Erebidae"
```
However, doing it this way means that we cannot put the punctuation marks back, so, if our intention was perhaps to flag the family names in the text, for example by adding editorial marks, we would have to do so by hand. So instead of deleting them we need to preserve them in place. An easy way to do this is to separate them from the word by a space character, using the 'nchar' function to obtain the number of characters in the word and then split into two substrings, one up to 'nchar(word)–1' and the other, the last character, i.e. the punctuation mark.

punctuation<-grep("[[:punct:]]$",as.character(text2))
for(i in 1:length(punctuation)){
    word<-text2[punctuation [i]]
    word<-paste(substr(word,1,nchar(word)-1),substr(word,nchar(word),
       nchar(word)),sep=" ")
text2[punctuation [i]]<-word} # end *i* loop
text2<-unlist(strsplit(text2," "))
text2
```
[1] "We" "want" "to" "write" "some" "R" "code"
[8] "recognizes" "animal" "or" "plant" "family" ","
    "subfamily"
[15] "," "tribe" "or" "subtribe" "names" "in" "a"
[22] "piece" "of" "text" "." "For" "animals" "these"
[29] "always" "end" "in" "-idae" "," "-inae" ","
[36] "-ini" "and" "-ina" "respectively" "." "The" "Noctuidae"
[43] "," "Erebidae" "," "Sphingidae" "and" "Saturniidae"
    "are"
[50] "families" "of" "moths" "," "the" "Lymantriinae" "is"
[57] "a" "subfamily" "of" "Erebidae" "," "the" "Lymantriini"
[64] "," "Leucomini" "," "Orgyiini" "and" "Nygmiini" "are"
[71] "tribes" "of" "the" "Lymantriinae" "," "etc" "." "."
```

More on Manipulating Text

All the words and punctuation are now isolated, we can find all the words with specified endings, modify them if we wish, and we can reassemble the original text relatively easily.

newpunct<-grep("^[[:punct:]]$",as.character(text2)) # the ^ and $ make sure it is both the first and last character
newtext<-text2[1] # we assume the first "word" is not a punctuation mark
for(i in 2: length(text2)){
    ifelse(i %in% newpunct,newtext<-paste(newtext,text2[i],sep=""),newtext<-paste(newtext," ",text2[i],sep=""))} # end if
newtext
[1] "We want to write some R code recognizes animal or plant family, subfamily, tribe or subtribe names in a piece of text. For animals these always end in -idae, -inae, -ini and -ina respectively. The Noctuidae, Erebidae, Sphingidae and Saturniidae are families of moths, the Lymantriinae is a subfamily of Erebidae, the Lymantriini, Leucomini, Orgyiini and Nygmiini are tribes of the Lymantriinae, etc."

Manipulating Character Case

The R language is case sensitive, so a vector called x is not the same as a vector called X. When handling text we may wish to change the case (upper or lower) of a character or string, or we may wish to search for text independent of its case. To change all text to upper case there is the function **toupper**, and its converse **tolower**. See also **ignore.case=T** below.
Fly<-"Tachinid"
Fly<-toupper(Fly)
Fly
[1] "TACHINID"
Fly<-tolower(Fly)
Fly
[1] "tachinid"
alternatively there is a function 'casefold' that changes the case, e.g.
casefold(Fly,upper=TRUE)
[1] "TACHINID"
To test whether a string contains at least one upper case character we can use a logical version of **grep** called **grepl** which returns TRUE/FALSE
grepl("[[:upper:]]","Fly")
[1] TRUE
grepl("[[:upper:]]",tolower(Fly))
[1] FALSE
Using wildcards, we can change the case of given characters. There are several ways of doing this – one uses PERL, a computing language available from within R in the library **PCRE** (which does not usually need to be loaded separately), and the argument 'perl=TRUE' automatically loads this.

```
parasitoids<-c("tachinid","braconid","ichneumonid","eulophid")
parasitoids<-gsub("(\\w)(\\w*)","\\U\\1\\L\\2",parasitoids,perl=TRUE)
parasitoids
[1] "Tachinid" "Braconid" "Ichneumonid" "Eulophid"
```
In the second argument the U means make uppercase, the 1 means first character of string and the \L\\2 means the rest lower case. If you are not familiar with PERL this will probably be somewhat confusing, so we will achieve the same using **substr** and **toupper** where the logic is hopefully clearer:
```
parasitoids<-c("tachinid","braconid","ichneumonid","eulophid")
    for(i in 1:length(parasitoids)){
parasitoids[i]<-gsub(substr(parasitoids[i],1,1),toupper(substr(parasitoids
    [i],1,1)), parasitoids[i])} # end i loop
parasitoids
[1] "Tachinid" "Braconid" "Ichneumonid" "Eulophid"
```
If we had not put these in a loop, only the first element would have been modified.

In biological nomenclature, adjectival forms of taxon names are not normally capitalized, but formal hierarchical levels from genus upwards are. So, we might want to change them to Tachinidae, Braconidae, etc. Here we have changed the list so that some elements of the list are adjectival and some are family names.
```
parasitoids<-c("tachinid","braconidae","ichneumonid","eulophidae")
```
Suppose we wish to capitalize only the family names as is standard practice, and not the adjectival forms that do not end in 'ae'. We can check the last letters of the name using $ as a specific wildcard identifier of the last letter of a string:
```
family_names<-grep(".ae$",parasitoids)
family_names
[1] 2 4
substr(parasitoids[family_names],1,1)<-toupper(substr(parasitoids
    [family_names],1,1))
parasitoids
[1] "tachinid" "Braconidae" "ichneumonid" "Eulophidae"
```
Normally in this book we have used **grep** to obtain the indices of the search term in the vector being searched, for example, family names are elements 2 and 4 in the vector 'parasitoids'. However, if we wish to see them without writing another line of code, we can add 'value=T' to the **grep** statement thus:
```
family_names<-grep(".ae$",parasitoids,value=TRUE)
family_names
[1] "braconidae" "eulophidae"
```

Ignoring Character Case

If you are handling plain text, words you may wish to find or change might not always be consistently in one case or another. Words that would not normally start with a capital letter, will do so, for example, at the beginning of a sentence. We can find character strings irrespective of case by incorporating the argument 'ignore.case=TRUE' as in **grep** and similar functions.

Exercise Box 22.2

1) Write code to capitalize the first letter of the genus names and put the rest of the genus and species names in lower case for each species in the following vector: sp_list<-"common birdwing, troides helena; golden birdwing, TROIDES AECUS; Malayan birdwing, Troides Amphrysus; mountain birdwing, troides cuneifera". Hint: you will probably need **substring**, **strsplit**, **paste**, **toupper**, **tolower**, **trimws**, **grep**.

Specifying Particular and Modifiable Character Classes

If we want to remove or locate all of a certain class of characters in a string we can create or modify the class. Above we used '[[:punct:]]' to find all punctuation marks in the text. There are several other built-in categories in R:

[:alnum:] – all upper and lower case letters plus digits 0:9
[:alpha:] – comprises lower case [:lower:] and upper case [:upper:] characters
[:digit:] – 0:9
[:space:] – space, tab

The set **[:punct:]** is defined as
punct<-'[]\\?!\"\'#$%&(){}+*/:;,._`|~\\[<=>@\\^-]'
and we can modify it (a version of it) to exclude certain characters from it, for example:
punct2<-sub("%","",punct) # removing the % symbol
punct2
```
[1] "[]\\?!\"'#$&(){}+*/:;,._`|~\\[<=>@\\^-]"
```
We can then use this vector to remove all punctuation from a piece of text except for % signs:
tx<-"Numbers, and percentages, of the fish were: 9 (14%), 12 (19%), and 2 (2%), respectively"
gsub(punct2,"",tx)
```
[1] "Numbers and percentages of the fish were 9 14% 12 19%
    and 2 2% respectively"
```

Tool Box 22.1

- Use .+? as a wildcard.
- Use [[:punct:]]$ to locate punctuation following character strings.
- Use \\ to treat metacharacters as ordinary characters (also called escaping).
- Use ^ followed by a specific number of to specify how many wildcard characters follow the initial value.
- Use attr() to find the named attribute of an R object.
- Use casefold() to reverse the case.
- Use gregexpr() to find all occurrences of a character.
- Use grepl() to return matches and non-matches as TRUE or FALSE.
- Use gsub() to substitute all occurrences in a string.
- Use ignore.case to locate characters irrespective of case.
- Use nchar() to obtain the number of characters in a word.
- Use perl= to use the computing language PERL.
- Use regexpr() to find the first occurrence of a character.
- Use sort() to order elements of a vector.
- Use sub() to substitute the first occurrence in a string.
- Use substr() to extract or replace substrings in a character vector.
- Use tolower() to change text to lower case.
- Use toupper() to change text to capitals.
- Use unname() to remove names of rows from a dataframe.
- Use value=TRUE with **grep** to return the elements found rather than their indices.

23 Phylogenies and Trees

Summary of R Packages Introduced
　ape
　phytools

Summary of R Functions Introduced
　attributes
　drop.tip in "ape"
　nodelabels in "phytools"
　packageVersion
　plot in "ape"
　plotTree in "phytools"
　read.tree in "ape"
　root in "ape"
　rtree in "ape"
　str
　tiplabels in "phytools"

Several packages have been developed to allow R-users to work with phylogenetic trees, something that most biologists will need to do at some point in their careers. The most basic is the **ape** package, which stands for Analysis of Phylogenetics and Evolution (Paradis and Schliep, 2018). Here we give some of the basics of handling 'trees' in R and show things that we can calculate with them – this is a huge field and there is vastly more than we can cover here, so we just want to get readers started. Alongside **ape**, another package with extra capabilities we will introduce is **phytools**.

　install.packages("ape")
　library(ape)

Warning messages about masked objects. Depending upon what packages/libraries you may have previously have installed, you may sometimes see messages such as the one below:
Attaching package: 'ape'
The following object is masked from 'package:Biostrings':
complement
The following objects are masked from 'package:raster':
rotate, zoom
This is telling us that some functions with identical names are already active and so will not be available to use in the newly installed package/library. Here, for example, the function **complement** is already in use in the **Biostrings** package for complementing DNA sequences. If you need to use those functions in the new package, you need to uninstall the previous package (see Chapter 7).

Here we will create a simple phylogenetic tree description for some familiar groups of insects. A basic bracketed (parenthetical) notation is used to describe relationships between taxa and clades; this is often referred to as Newick notation. This format is recognized by most specialist phylogenetic software programs (e.g. PAUP, phylip, MrBayes, RaxML, Figtree.). The inner-most pairs of bracketed taxa are sister groups and the relationships proceed hierarchically until all taxa are included. The basal-most taxon is often the outgroup, though it might be more than one entity. As with many R packages, we may find changes, additions, corrections from time to time so you may want to make note of which particular version of the package you are using

packageVersion("ape")

```
[1] '5.3'
```

Here we create a character string called 'insecta', which is a bracketed description of a phylogenetic tree:

insecta<-"(dragonflies,(cockroaches,(bugs,(wasps&bees,((beetles,lacewings), (flies, butterflies&moths))))));" # note the semicolon at the end of the line and that the taxon names do not need to be in inverted commas

insectTree<-read.tree(text=insecta) # or we can read from a file, read.tree ({file name})

Here we have created an object of class phylo called 'insectTree'. This class name is stored in the object and tells packages that recognize that class how to deal with it. We started with a simple set of taxa arranged in a nested set of round brackets. However, the object that the ape function **read.tree** creates does not look like that at all. We can use some base R functions to explore its structure.

What does an R object of class phylo comprise? We can use a few familiar and some new R functions to examine it and to tease it apart. Let's start with **summary**

summary(insectTree)

```
Phylogenetic tree: insectTree
    Number of tips: 8
    Number of nodes: 7
    No branch lengths.
    No root edge.
    Tip labels: dragonflies
                cockroaches
                bugs
                wasps&bees
                beetles
                lacewings
                flies
                butterflies&moths
    No node labels.
```

summary tells us the number of tips (terminal taxa), the number of nodes, etc. You can also see from this example that there are missing items that can be in an object of class phylo, such as branch lengths or a specified root. The function **str** (meaning structure) gives a different summary

```
str(insectTree)
List of 3
 $ edge  : int [1:14, 1:2] 9 9 10 10 11 11 12
12 13 14 ...
 $ Nnode : int 7
 $ tip.label: chr [1:8] "dragonflies" "cockroaches"
"bugs" "wasps&bees" ...
 - attr(*, "class")= chr "phylo"
 - attr(*, "order")= chr "cladewise"
```
The function **attributes** tells us some more of the structure
attributes(insectTree)
```
$names
[1] "edge"  "Nnode"  "tip.label"

$class
[1] "phylo"

$order
[1] "cladewise"
```
An obvious question that you might ask is, where is the tree description that we entered? Well, it is there but in a different format. Now to look at the internal elements in a bit more detail. The attribute 'edges' has one row for each terminal and one for each of the internal nodes of the tree.
insectTree$edge
```
        [,1] [,2]
 [1,]    9    1
 [2,]    9   10
 [3,]   10    2
 [4,]   10   11
 [5,]   11    3
 [6,]   11   12
 [7,]   12    4
 [8,]   12   13
 [9,]   13   14
[10,]   14    5
[11,]   14    6
[12,]   13   15
[13,]   15    7
[14,]   15    8
```
To understand how the elements in an object of class phylo work better, we will use a related package **phytools**.
install.packages("phytools")
library(phytools)
plotTree(insectTree,offset=1) # this phytools function allows us to add some other information
tiplabels() # rather like adding points or lines to an open plot in base R
nodelabels()

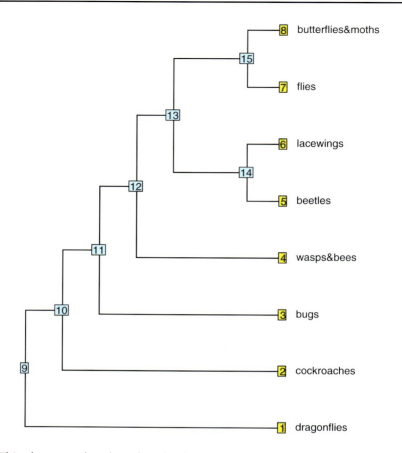

This shows us that the edges leading to the original terminals dragonflies to butterflies&moths are numbered 1 to 8 in order from the base. Then the nodes of the tree are numbered starting with 9 from the root of the tree. So now look at 'insectTree$edge' The first row says that node 9 (the base) leads to node 1 (dragonflies, the most basal of the taxa included), the second row says that the node 9 leads to node 10 (which here includes all the non-dragonflies, and so on.

The programs that you will use to work with objects of class phylo obviously don't require you to understand all the structure, but it may help to give a little demonstration
plot(insectTree)

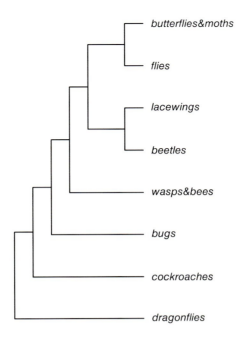

Here you see that the **ape** library has a function called **plot**, but if you search for help with ?plot you will only see information about the **plot** function in base R. The **plot** function in **ape** has many specialist arguments. To find the help page for these use:
??plot
which shows all the different **plot** function versions in various packages, and then select one of the 'ape::' help pages.

Branch Lengths

Our trees above are cladograms, they only show relationships. We can also assign lengths to each branch. These may be based on estimated distances from phylogenetic analyses or from fossil calibrations or both. This is done by creating a vector 'edge.length' within the 'phylo' object, or by adding the branch lengths to each node of the tree description following a colon, e.g. adding ages of nodes in millions of years.
insectTree$edge.length
 <-c(370,20,350,40,310,50,260,80,50,130,130,80,100,100)
write.tree(insectTree,"Insect_tree.txt")
readLines("Insect.tree")
[1] "(dragonflies:370,(cockroaches:350,
    (bugs:310,(wasps&bees:260,((beetles:130,
    lacewings:130):50,(flies:100,butterflies&mo
    ths:100):80):80):50):40):20);"

Random Trees

We can generate a random tree topology with random branch lengths using the function **rtree** (=random tree), which takes the number of taxa you want as its argument, e.g.
random_tree25<-rtree(n=25)
random_tree25 # you probably won't get the same answer because the tree is random
```
Phylogenetic tree with 25 tips and 24 internal nodes.

Tip labels:
t13, t8, t25, t21, t6, t12, ...

Rooted; includes branch lengths.
```
plot(random_tree25)

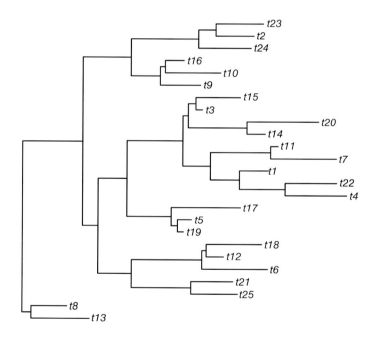

To view its separate components of 'random_tree25' we look at the indexed parts, e.g. part 2
random_tree25 [2]
$tip.label
```
[1] "t13" "t8"  "t25" "t21" "t6"  "t12" "t18" "t19" "t5"
    "t17" "t4"  "t22" "t1"  "t7"  "t11" "t14" "t20" "t3"
    "t15" "t9"  "t10" "t16" "t24" "t2"
[25] "t23"
```
or by name
random_tree25$tip.label

```
[1] "t13" "t8"  "t25" "t21" "t6"  "t12" "t18" "t19" "t5"
    "t17" "t4"  "t22" "t1"  "t7"  "t11" "t14" "t20" "t3"
    "t15" "t9"  "t10" "t16" "t24" "t2"
[25] "t23"
```

Different Types of Plots in *ape*

par(mfrow=c(2,2))
plot(random_tree25,edge.width=2,label.offset=0.1,type ="cladogram")
plot(random_tree25,edge.width=2,label.offset=0.1,type ="phylogram")
plot(random_tree25,edge.width=2,label.offset=0.1,type ="fan")
plot(random_tree25,edge.width=2,label.offset=0.1,type ="radial")
A fifth option is 'unrooted', and the default is 'phylogram'.

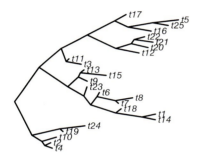
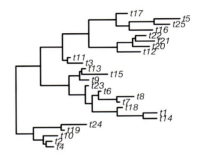
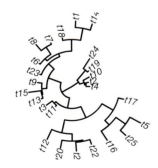
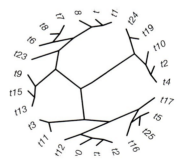

We can easily rename the tips
random_tree25$tip.label[3]
[1] "t25"
random_tree25$tip.label[3]<-"apple"
or remove tips,
Nodonata<-drop.tip(insectTree,"dragonflies")
or reroot the tree.
wrong<-root(Nodonata,7,resolve.root=TRUE)
plot(wrong)

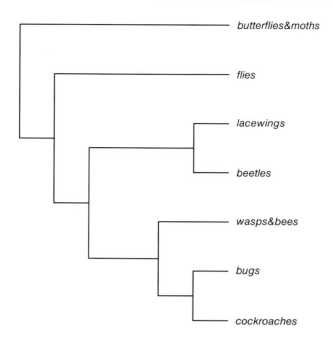

Exercise Box 23.1

1) The following tetrapod phylogenetic string is based on Huggall *et al.* (2007): ((tuatara, (lizards, snakes)), (turtle, (crocodiles, (birds, non.avian_dinosaurs)))). Present it as an unrooted tree, then root it on the squamates (tuatara+lizards+snakes). Hint: add extra round brackets to force a basal group to be monophyletic.
2) The Dinosauria are paraphyletic with respect to birds so add another taxon, saurischian dinosaurs as the sister group to birds.
3) Create a dated phylogeny give the approximate ages (mya) for the clades as follows: tuatara 240; lizards 220; snakes 220; dinosaurs 230; birds 160, crocodiles+Dinosauria 245, turtle+Crocodiles+Dinosauria 270, and the root of the tree 283. Hints: (i) you need to specify branch lengths for all terminals and clades (i.e. internal nodes); and (ii) the sums of lengths for each path from an extant terminal to the root should all add to the root age.
4) Optionally, shorten the extinct dinosaur branches so that they terminate 65 million years ago.

Tool Box 23.1 ✕

- Use attributes() to show the structure of an R object.
- Use drop.tip() in ape to delete a tip from a tree.
- Use node.labels() in **phytools** to number the nodes on a tree.
- Use packageVersion() to determine which version of a package is running.
- Use plot() in ape to draw a phylogenetic tree.
- Use plotTree() in **phytools** to plot a phylogenetic tree.
- Use read.tree() in ape to read a phylogenetic tree(s) from a parenthetic format text string.
- Use root() in ape to reroot a tree.
- Use rtree() in ape to randomize the topology.
- Use str() to show the structure of an R object.
- Use summary() to show the structure of an R object.
- Use tiplabels() in ape to number the tips.
- Use type= to set the style of tree, e.g. cladogram/phylogram/radial/fan/unrooted.

24 Working with DNA Sequences and Other Character Data

Summary of R Packages Introduced	DNAStringSet in "Biostrings"
BiocManager	intToUtf8
Biostrings	map[]
Seqinr	mode
	rev
Summary of R Functions Introduced	s2c in "seqinr"
all.equal	sapply
c2s in "seqinr"	switch
comp in "seqinr"	utf8ToInt
complement in "Biostrings"	

There are of course numerous free web-based resources for manipulating nucleotide sequences such as reverse-complementing and complementing, translating DNA to RNA, converting base sequences to amino acids, but it can be educational to run through a few examples, and in doing so, introduce some more useful R functions. For those unfamiliar with DNA data, each of the four bases, A, T, G and C pairs with (binds to) its complementary base on the opposite strand along Crick and Watson's famous double helix (see the table below for details).

To complement a DNA sequence, you need to replace each base by its complement. Starting with a DNA sequence vector called 'p' we first split it into its components

```
p<-unlist(strsplit("ATCTCGGCGCGCATCGCGTACGCTACTAGC",""))
p
[1] "A" "T" "C" "T" "C" "G" "G" "C" "G" "C" "G" "C" "A"
    "T" "C" "G" "C" "G" "T" "A" "C" "G" "C" "T" "A" "C"
    "T" "A" "G" "C"
```

Then we define a mapping function using 'map'. This function has a different syntax to most. It was written initially to map factors but it is generic and so can also be applied to characters as here.

```
map=c("A"="T","T"="A","G"="C","C"="G")
map[p]
```

```
 A   T   C   T   C   G   G   C   G   C   G   C   A   T   C   G   C   G   T   A   C   G   C   T   A   C   T   A   G   C
"T" "A" "G" "A" "G" "C" "C" "G" "C" "G" "C" "G" "T" "A"
"G" "C" "G" "C" "A" "T" "G" "C" "G" "A" "T" "G" "A"
"T" "C" "G"
```

In the same way we could use 'map' to translate DNA to RNA, which has uracil (U) instead of thymine (T). To do this we change the mapping thus, map=c("T"="U", "A"="T", "G"="C", "C"="G"). All bases (letters) have to be included in the map statement otherwise R will return NAs. In our output each position in the remapped version is given the name (top row), which is probably not desirable. We can remove names (or dimnames) using the function **unname**, hence

unname(map[p])
```
[1] "T" "A" "G" "A" "G" "C" "C" "G" "C" "G" "C" "G" "T"
    "A" "G" "C" "G" "C" "A" "T" "G" "C" "G" "A" "T" "G"
    "A" "T" "C" "G"
```

Another way of doing this would be to use **sapply** with the function **switch**
sapply(p,switch, "A"="T","T"="A","G"="C","C"="G")
```
 A   T   C   T   C   G   G   C   G   C   G   C   A   T   C   G   C   G   T   A   C   G   C   T   A   C   T   A   G   C
"T" "A" "G" "A" "G" "C" "C" "G" "C" "G" "C" "G" "T" "A"
"G" "C" "G" "C" "A" "T" "G" "C" "G" "A" "T" "G" "A"
"T" "C" "G"
```

Writing your own code to do things such as reverse complementing sequences is quite easy. Here is a function, combining **map**[] to change the bases and a line for reversing the sequence

RCOMP<-function(x){
   r<-sapply(lapply(strsplit(x,NULL),rev),paste,collapse="") # this line splits the sequence to individual bases, reverses the order and recombines
   y<-paste(map[unlist(strsplit(r,NULL))],collapse="") # this line replaces the bases with their complements
   return(y)} # end RCOMP function

However, there are packages that have simple functions for doing this sort of thing, for example **seqinr**. Install the **seqinr** library, which enables easy exploratory data analysis and visualization for DNA and protein sequence data.

install.packages("seqinr")
library(seqinr)

The **seqinr** function **comp** gives complement, e.g.

comp(p)
```
[1] "t" "a" "g" "a" "g" "c" "c" "g" "c" "g" "c" "g" "t"
    "a" "g" "c" "g" "c" "a" "t" "g" "c" "g" "a" "t" "g"
    "a" "t" "c" "g"
```

But it can also work with strings of bases, so

rev(comp(p)) # gives the reverse complement, i.e. the complement read in the reverse direction
```
[1] "g" "c" "t" "a" "g" "t" "a" "g" "c" "g" "t" "a" "c"
    "g" "c" "g" "a" "t" "g" "c" "g" "c" "g" "c" "c" "g"
    "a" "g" "a" "t"
```

> **Information Box 24.1**
>
> *Reversing string using utf8ToInt.* Here we have written a generic bit of code that splits a string into its individual elements, reverses the order of the elements and then pastes them into a new string. However, for DNA which has four bases you can use an in-built pair of R functions: **utf8ToInt** and **intToUtf8**, which convert character strings to a vector of integers representing each separate character, e.g.
> utf8ToInt("abcdefghijklmnopqrstuvwxyz")
> ```
> [1]  97  98  99 100 101 102 103 104 105 106 107 108 109 110 111 112 113
>     114 115 116 117 118 119 120 121 122
> ```
> and convert them back again
> intToUtf8(utf8ToInt("abcdefghijklmnopqrstuvwxyz"))
> ```
> [1] "abcdefghijklmnopqrstuvwxyz"
> ```
> Because **utf8ToInt** yields a list, we can apply functions such as **rev** to the list and then convert back to a string with **intToUtf8**.

The package **seqinr** has two useful shortcut functions: **c2s** (meaning 'character to string') converts a vector of characters into a string, in other words what we did with **paste**(X,sep="",collapse=""), and its reverse **s2c** (meaning 'string to character'), which we achieved in base R using **strsplit**(X,""). Combining these with the function comp we can now complement all the bases in a character string:

c2s(comp(s2c("AAAATTTTGGGGCCCC"))) # comp accepts upper or lower case but returns lower case
```
[1] "ttttaaaaccccgggg"
```
c2s(rev(comp(s2c("aaaattttggggcccc"))))
```
[1] "ggggccccaaaatttt"
```
However, non-basic DNA characters including ns and xs are replaced by NA.
allbases<-s2c("atgcmynnsk")
comp(allbases) # NAs are produced for non-ATGC bases
```
[1] "t" "a" "c" "g" NA  NA  "n" "n" NA  NA
```
To overcome the NA issue, comp can be told 'ambiguous=TRUE', which then allows ambiguity codes to be employed. The full set of ambiguity codes is shown in Table 24.1.
comp(allbases,ambiguous=TRUE) # no more NAs
```
[1] "t" "a" "c" "g" "k" "r" "n" "n" "s" "m"
```
There are also several alternative ways of achieving the same goal in R, for example, there is another package called **Biostrings**, with a function called complement.
install.packages("BiocManager")
library(Biostrings)
dna=DNAStringSet(c("ATCTCGGCGCGCATCGCGTACGCTACTAGC", "ACCGCTA"))
complement(dna)

```
A DNAStringSet instance of length 2
    width seq
[1]    30 TAGAGCCGCGCGTAGCGCATGCGATGATCG
[2]     7 TGGCGAT
```

Table 24.1. International Union of Pure and Applied Chemistry (IUPAC) DNA base codes.

IUPAC code	Base meaning	Complement
A	A	T
C	C	G
G	G	C
T	T	A
M	A or C	K
R	A or G	Y
W	A or T	W
S	C or G	S
Y	C or T	R
K	T or G	M
V	A or C or G	B
H	A or C or T	D
D	A or G or T	H
B	C or G or T	V
N	A or T or C or G	N

Exercise Box 24.1

1) Write a function to reverse, complement and reverse complement a DNA sequence using the **seqinr** library detecting whether the sequence passed to the function is lower case or upper case and returning the reverse and complements in the same case as was supplied to it.
2) Write a function to reverse complement a DNA sequence that also allows for ambiguity codes and 'x' or 'X', which is often used to designate not just an unknown base but also an unknown number of them, and is accepted by a number of sequence analysis and phylogeny reconstruction programs. Hint: Use **map[]**.

Tool Box 24.1

- Use ambiguous=T in **comp** to allow ambiguity codes for bases.
- Use c2s() to convert a vector of characters into a string.
- Use comp() in **seqinr** to give the complementary bases.
- Use intToUtf8() to convert integer values to their character equivalents.
- Use map= to map one character to another.
- Use map[ ] to specify character substitutions to be made in a vector.
- Use rev() to reverse the order of a vector.
- Use s2c() to convert a string into a vector of single characters.
- Use sapply() to apply a function to multiple items and return a vector.
- Use switch() to swap (which) pairs of values in a vector.
- Use unname() to remove names from mapped vectors.
- Use utf8ToInt() to convert characters in a text string to integers.

Sequential Runs of Base Types

Here we present a bit of code that finds the beginnings and ends of sequential runs of integers and has many potential uses. In this example we will find the beginning and end positions of RNA bases that are paired or unpaired. Whereas in protein-coding genes DNA bases are grouped in triplets called codons (which are translated into amino acid sequences by ribosomes via messenger RNA), the bases in RNA molecules form complex 3D structures important for their enzymatic functions. Variation in RNA sequences between species has played an extremely important role in molecular phylogenetic studies. The analysis of such sequence data can be very sophisticated and different evolutionary models may need to be applied to groups of bases with different evolutionary properties, such as those that are paired and those which are not, and the programs used to conduct the analyses need to be fed this information usually in the form of lists of positions of each base type.

Below is an example RNA sequence and a corresponding string in which bases known to be paired (P) or unpaired (U) are indicated, and we extract the positions of pairing and unpairing bases using the familiar combination **which**, **unlist** and **strsplit** functions.

```
rnasq<-"UAUAUUUCAAAUAUAAAGUAAUUUAGAAAGUAAGAUCAAAGA
    UUAAU"
types<-"PPPPPUUUUPPPPPPPUPPUUUPPPPUUUUUUUUUPPPPPP
    UUUUU"
U<-which(unlist(strsplit(types,"")))=="U")
P<-which(unlist(strsplit(types,"")))=="P")
```

Now to find the groups of paired and unpaired we introduce two new functions, **split** and **diff** in a neat combination

```
Pgroups<-split(P,cumsum(c(1,diff(P) != 1)))
Pgroups
$`1`
[1] 1 2 3 4 5
$`2`
[1] 10 11 12 13 14 15 16
$`3`
[1] 18 19
$`4`
[1] 23 24 25 26
$`5`
[1] 37 38 39 40 41 42
```

Our code found five runs of consecutive pairing types, in this case pairing. How was this accomplished? **diff** tells us the numerical difference between consecutive numbers in our vector starting with the difference between the second element and the first.

```
diff(P)
[1] 1 1 1 1 5 1 1 1 1 1 2 1 4 1 1 1 11 1 1 1 1 1
```
So the result of diff is a vector with one fewer elements than the vector passed to it, which is why we use **c** to add an initial value of 1 to the list. What we get then is a vector of the same length as U
```
c(1,diff(P))
[1] 1 1 1 1 1 5 1 1 1 1 1 2 1 4 1 1 1 11 1 1 1 1 1
```
Then in the same **c** function call we use not equal to 1 (!=1) to see where the value of the next element is not a consecutive number. We see above that the sixth element differs from the fifth by five not by one, so there is an end to the sequence of Ps after the fifth element.
```
c(1,diff(P) !=1)
[1] 1 0 0 0 0 1 0 0 0 0 0 1 0 1 0 0 0 1 0 0 0 0 0
```
This tells us that the second to fourth elements differ from the first and each other in steps of 1.

The next neat trick is to calculate the cumulative sum (**cumsum**) to find the blocks of a given base type
```
cumsum(c(1,diff(P) != 1))
[1] 1 1 1 1 1 2 2 2 2 2 2 3 3 4 4 4 4 5 5 5 5 5 5
```
These are the five groups of 'P' in our vector P, which you recall is the list of positions of all the 'P' characters in the types string. These group numbers can be considered factors (we don't have to specifically make them into factors using **as.factor**). Therefore, passing this list of factors to the function **split** returns the positions of each pairing base in discrete groups.

Exercise Box 24.2

1) A number of phylogenetic programs such as RAxML will want the list of bases to which a given evolutionary model has to be applied given in a format such as
DNA, unpairing bases = 6-9, 17-17, 20-22 ...
Modify the code given above to get the list of unpairing base clusters and then convert that to the above format. Hint: there are many ways of doing this, but using **range** might help.

Sometimes we may want to generate random DNA sequences of a given base composition, perhaps in order to test software or to see the influence of random sequences on the outcome of an analysis. This line of code does just that.
```
randomDNA<-paste(sample(c("A","C","G","T"),10000,rep=TRUE,
    prob=c(0.4,0.1,0.1,0.4)), collapse="")
```

Tool Box 24.2

- Use cumsum() to return the cumulative sum of a numeric vector.
- Use diff() to return differences between consecutive values of a numeric vector.
- Use prob= in sample to specify probabilities of sampling each component of a vector.
- Use split() to divide a vector into a list-based on a specified criterion.

Downloading DNA Sequences from GenBank

Previously we installed the package **ape** (see Chapter 23). Now we will use the **read.GenBank** function, which uses the site http://www.ncbi.nlm.nih.gov/ from where it downloads sequence(s) by supplying it with the unique GenBank Accession number. In this example we will supply that of a tropical African parasitic wasp of the genus *Monilobracon* whose accession number is AY529648. This will not work if your computer is not connected to the Internet.
library(ape) # use install.packages("ape") if you have not done so in Chapter 23
Monilobracon1<-read.GenBank("AY529648",as.character=TRUE)
Monilobracon1
$AY529648
 [1] "t" "t" "t" "c" "g" "a" "a" "c" "g" "c" "a" "c" "a" "t"
     "t" "g" "c" "g" "g" "t" "c" "c" "a" "c" "g" "g" "a" "t"
     "c" "c" "a" "a" "t" "t" "c" "c" "c" "g" "g" "a" "c" "c"
 [43] "a" "c" "g" "c" "c" "t" "g" "g" "c" "t" "g" "a" "g" "g"
     "g" "t" "c" "g" "t" "t" "a" "t" "g" "c" "a" "t" "t" "a"
     "a" "a" "a" "a" "a" "c" "t" "g" "c" "t" "t" "a" "t" "a"
 [85] "t" "a" "t" "a" "t" "t" "t" "t" "t" "g" "t" "t" "g" "t"
     "a" "t" "a" "t" "g" "c" "t" "g" "c" "g" "t" "g" "t" "t"
     "g" "t" "g" "t" "a" "t" "g" "t" "g" "t" "g" "c" "g" "c"
[127] "g" "a" "t" "t" "c" "a" "t" "a" "a" "t" "t" "t" "a" "t"
     "a" "a" "a" "t" "a" "t" "t" "t" "g" "a" "a" "t" "a" "t"
     "a" "c" "a" "g" "a" "t" "g" "a" "a" "t" "c" "a" "c" "a"
[169] "a" "a" "a" "a" "t" "a" "t" "a" "c" "a" "c" "a" "c" "a"
     "a" "c" "a" "a" "t" "a" "c" "a" "c" "a" "c" "a" "c" "a"
     "c" "a" "c" "t" "c" "t" "t" "a" "c" "a" "t" "a" "c" "a"
[211] "c" "a" "a" "a" "g" "t" "a" "t" "a" "t" "a" "t" "a" "c"
     "t" "a" "g" "c" "g" "c" "a" "t" "a" "t" "t" "t" "t" "a"
     "a" "a" "a" "a" "a" "a" "g" "c" "t" "g" "a" "a" "t" "g"
[253] "t" "n" "c" "g" "t" "c" "a" "a" "t" "t" "t" "t" "t" "g"
     "t" "g" "t" "g" "c" "a" "a" "a" "a" "a" "g" "a" "a" "a"
     "g" "t" "g" "t" "a" "t" "a" "t" "a" "c" "a" "t" "t" "t"
[295] "t" "t" "c" "g" "a" "a" "t" "a" "t" "a" "t" "a" "t" "g"
     "t" "a" "t" "a" "t" "a" "t" "t" "a" "t" "t" "c" "t" "t"
     "c" "t" "t" "t" "a" "a" "a" "a" "c" "a" "a" "a" "a" "w"
[337] "a" "t" "a" "c" "a" "c" "a" "c" "t" "t" "a" "t" "a" "a"
     "t" "t" "t" "g" "a" "c" "g" "t" "c" "a" "t" "t" "t" "t"
     "a" "a" "a" "t" "t" "a" "a" "c" "a" "a" "t" "a" "t" "g"
[379] "t" "g" "t" "g" "a" "a" "g" "a" "a" "a" "a" "a"
     "t" "g"
attr(,"species")
[1] "Monilobracon_sp._NL827"
Our vector Monilobracon1 is a list as we can see by using the function **mode**.
mode(Monilobracon1)
[1] "list"
Using the function **str** we can see that the sequence has 392 bases

Working with DNA Sequences and Other Character Data

```
str(Monilobracon1)
List of 1
 $ AY529648: chr [1:392] "t" "t" "t" "c" ...
 - attr(*, "species")= chr "Monilobracon_sp._NL827"
```

Note that the number of bases you see on a line in your R output depends on the width of your R console window. To just look at the first 100 bases

```
Monilobracon1$AY529648[1:80]
 [1] "t" "t" "t" "c" "g" "a" "a" "c" "g" "c" "a" "c" "a" "t"
     "t" "g" "c" "g" "g" "t" "c" "c" "a" "c" "g" "g" "a" "t"
     "c" "c" "a" "a" "t" "t" "c" "c" "c" "g" "g" "a" "c" "c"
[43] "a" "c" "g" "c" "c" "t" "g" "g" "c" "t" "g" "a" "g"
     "g" "g" "t" "c" "g" "t" "t" "a" "t" "g" "c" "a" "t" "t"
     "a" "a" "a" "a" "a" "a" "c" "t" "g" "c" "t"
```

```
s<-paste(Monilobracon1$AY529648,collapse="")
s
[1] "tttcgaacgcacattgcggtccacggatccaattcccggaccacgcc
    tggctgagggtcgttatgcattaaaaaactgcttatatatttttgttg
    tatatgctgcgtgttgtgtatgtgtgcgcgattcataatttataaatat
    ttgaatatacagatgaatcacaaaaatatacacacaacaatacacacaca
    cactcttacatacacaaagtatatatactagcgcatatttaaaaaaagct
    gaatgtncgtcaattttttgtgtgcaaaagaaagtgtatatacat
    ttttcgaatatatatgtatatattattcttctttaaaacaaaawataca
    cacttataatttgacgtcattttaaattaacaatatgtgtgaagaaaatg"
```

The function '**attributes()**' is used to extract the name(s) and "species" of the downloaded sequence

```
attributes(Monilobracon1)[1]
$names
[1] "AY529648"
attributes(Monilobracon1)[2]
$species
[1] "Monilobracon_sp._NL827"
```

Several sequences can be downloaded from GenBank with a single command. For example if we have a list of accessions numbers

```
all_Monilos<-c("MH234998","MH260676","AY296646","AY529647",
    "AY532320", "AY529649","AJ296046")
Mons<-read.GenBank(all_Monilos,as.character=TRUE)
```

The result here actually comprises a mixture of different gene fragments from different species of *Monilobracon*.

```
data<-cbind(attr(Mons,"species"),names(Mons))
data
     [,1]                                  [,2]
[1,] "Monilobracon_sp._CCDB-27844-E10"    "MH234998"
[2,] "Monilobracon_sp._CCDB-27844-E10"    "MH260676"
[3,] "Monilobracon_sp._NL785"             "AY296646"
[4,] "Monilobracon_sp._NL823"             "AY529647"
[5,] "Monilobracon_sp._NL832"             "AY532320"
[6,] "Monilobracon_sp._NL832"             "AY529649"
[7,] "Monilobracon_sp._RDB-2000"          "AJ296046"
```

Many (but not all) sequence analysis programs require sequence data in something called fasta format, which consists of two parts, a name and any other data on one line, then a greater than sign followed by the sequence (sometimes split into convenient length blocks for display). The function **write.dna** in **ape** allows us to save our downloaded sequences in fasta and other formats, and has some useful arguments. Run the following and then open it from your working directory to see what it has produced.

write.dna(Mons,file="Monilobracon.fas",format="fasta",nbcol=6,
    colsep=" ",colw=10)

The argument 'colw' specifies the number of nucleotides in a column, 'nbcol' the number of columns and 'colsep' the character to put between columns. There is a **read.dna** function in **ape** too, and **read.fasta** and **write.fasta** functions in the **seqinr** library, which are easier.

Translating DNA to Amino Acids

The **Biostrings** library has a very useful function, translate, which translates protein-coding DNA or RNA to their corresponding amino acid sequences. Because there are multiple genetic codes we must specify which. The standard code is designated 'SGC0' but *Monilobracon* is an insect and the sequence is mitochondrial so we need to use the invertebrate mitochondrial code SGC4.
SGC4<-getGeneticCode("SGC4")
The sequence Mons$MH260676 from above is a cytochrome oxidase subunit 1 barcode, which we can concatenate thus:
barcode<-paste(Mons$MH260676,collapse="")
barcode
[1] "ctcgaataaataatataagattttggttattaattccttctattatttat
    tattattaagaggaattttaaatattggagtaggtactggatgaacagt
    ttaccctccattatcttcattaattgggcatagaggatatcagttgat
    ttagcaattttttctttacatatagctggtatttcttctattataggttc
    tattaattttatttctacaattttaaatatacgtttatttttttaaaatt
    agatcaattaactttgttaatttgatcatttttttaacaacaattttat
    tattattatctttaccagttttagcaggaggaattactatattattaa
    cagatcgtaatttaaatacttcattttttgattttctggtgggggg
    gatcctgttttatttcaacatttattt"
dna1<-DNAString(barcode) # we have to use the function **DNAString** to tell
    Biostrings what type of object our sequence is
We will assume that we do not know which codon position the first base is, so we will try all three using **substr**
dna2<-DNAString(substr(barcode,2,nchar(barcode)))
dna3<-DNAString(substr(barcode,3,nchar(barcode)))
Now we use the translate function (it turns out that the third base in our sequence is the start of a codon).
translate(dna3,genetic.code=SGC4,no.init.codon=TRUE)

```
140-letter "AAString" instance
seq: RMNNMSFWLLIPSIILLLLSGILNIGVGTGWTVYPPLSSLIGHSGMSVDLA
    IFSLHMAGISSIMGSINFISTILNMRLFFLKLDQLTLLIWSIFLTTILLLLSL
    PVLAGGITMLLTDRNLNTSFFDFSGGGDPVLFQHLF
```
The output is in the form of single letter IUPAC amino acid abbreviations available at bioinformatics.org/sms2/iupac.html (accessed 21 April 2020).

Exercise Box 24.3

1) The translation tables for the various genetic codes are available at ncbi.nlm.nih.gov/Taxonomy/Utils/wprintgc.cgi#SG3 (accessed 21 April 2020); choose the *Mold, Protozoan, and Coelenterate Mitochondrial Code* and make it into a table. Hints: (i) you probably have to hand edit it a bit; and (ii) use **textConnection**.
2) Download the cytochrome oxidase subunit 3 sequence for the sea anemone *Metridium senile lobatum* from GenBank, its accession number is JF833002. Hints: you probably want to **unlist** it and maybe **unname** it.
3) Convert the DNA sequence to its equivalent amino acid sequence using the translation table. Hints: (i) the first base is a first codon position; and (ii) you might want to combine codon bases into strings of three and use **which**.

Tool Box 24.3

- Use attributes() to show the names and species of the downloaded GenBank sequence.
- Use DNAString in BioStrings to convert a character string to a DNA sequence object.
- Use genetic.code= in **translate** to specify which genetic code to use.
- Use mode() to get the type or storage mode of an R object.
- Use read.dna in ape to read a fasta format sequence from disc.
- Use read.GenBank in ape to download DNA sequences from GenBank.
- Use translate() in **Biostrings** to convert a coding DNAsequence to single letter amino acid sequence.
- Use write.dna in ape to save a fasta format sequence to disc.

Prettifying a Table

A frequent requirement when creating matrices of morphological characters is to add new columns or re-order columns, or split the table into blocks of columns to make visually locating individual data points easier. In a large block of data that are hand-generated and edited such as a morphological data matrix, editing individual values can be a nightmare and carries a high risk of introducing errors because you can get confused between rows and columns. Suppose we want to split our columns into nice, easily comprehended groups of five, each group separated by a space. What we essentially want to do is copy for each row, five characters and then insert a space character. Easy to define, but how to do it?

First we want to know which numbers are whole multiples of five, i.e. dividing by five gives an integer result. But whilst we know that an integer is a number followed by a decimal point and an infinite number of zeros, computers cannot, each system has a very finite limit to its precision. If you create a vector declared as an integer it is easy, but if you want to test whether say 435 is an integer multiple of 15, your computer will do a floating point calculation and come up with 29.00000000000. It won't show you all the zeros, but 435000000000000001/15000000000000000
[1] 29

So testing whether a number is truly an integer is not so straightforward and the answer produced could depend on the precision of the programming language or chip. R therefore does not have any way of showing whether a value is truly an integer: it cannot. So to test, say, whether elements of a series are whole number multiples of another integer (say five) we need to write a function to do this, essentially just one line of code. These all basically ask whether the test value does not differ from an integer at a level of accuracy determined by your software and computer. For most practical purposes in biology these are probably satisfactory.

R has a built-in function **all.equal**, which compares two numbers, in this case the test number and nearest whole number to it using **round**. Playing with the number we see that our particular computer can distinguish 8.00000012 from 8 but it cannot distinguish 8.00000011 from 8.
all.equal(8.00000012,round(8.00000012))
[1] "Mean relative difference: 1.5e-08"
all.equal(8.00000011,round(8.00000011))
[1] TRUE

This is not terribly impressive and we can improve it using exactly equals, ==, by writing a short function such as
is.whole<-function(a) a%%1==0 # %% means modulo, i.e. remainder after division
or
is.whole<-function(a) round(a)==a
is.whole(7.0000000000000004)
[1] TRUE
is.whole(7.0000000000000005)
[1] FALSE

Here we see our computer has a limit around 10^{-16}.

Going back to our pretty presentation of a table
Xenarcha<-"00000001000000101110111100000000?01?10000100 0?0?0100"
Colastes<-"00000001000001100110101100000000?01?110101000?0?0100"
Gnamptodon<-"00000001000000100100100100100111?01?0?0001?00?0?0101"
wasps<-rbind(Xenarcha,Colastes,Gnamptodon)
wasps
          [,1]
Xenarcha "00000001000000101110111100000000?01?100001000?0?0100"

```
Colastes  "00000001000001100110101100000000?01?110101000
   ?0?0100"
Gnamptodon "000000010000001001001001001000111?01?0?0001?
   00?0?0101"
```
It is clear that there is a lot of scope for making silly mistakes in editing values in such a table and it is not always convenient to do this in a spreadsheet. Therefore we will insert spaces every five characters using **is.whole** to test whether our current position is a whole multiple of five, and **ifelse** to copy a character value or a character value followed by a space, depending on the **is.whole** TRUE or FALSE output.

As good practice, we will first make a back-up copy of 'wasps' under a different name; this is not necessary but it helps us keep track if errors occur.

```
newwasps<-wasps
for(i in 1:length(wasps)){
    newdata<-NULL
    for(j in 1:nchar(wasps[i])){
        ifelse(is.whole(j/5),newdata<-paste(newdata,substr(wasps[i],j,j),
            "",collapse="",sep=""),newdata<-paste(newdata,substr(wasps[i],
            j,j), collapse="",sep="")) # end ifelse
    } # end for j loop
    if(i==1) ONE<-TRUE
    if(i==2) TWO<-TRUE
    newwasps[i]<-newdata} # end i loop
newwasps
[,1]
Xenarcha   "00000 00100 00001 01110 11110 00000 000?0 1?100
   00100 0?0?0 100"
Colastes   "00000 00100 00011 00110 10110 00000 000?0 1?110
   10100 0?0?0 100"
Gnamptodon "00000 00100 00001 00100 10010 01000 111?0 1?0?0
   001?0 0?0?0 101"
```

Easy Ways to Extract Taxon Names from a Phylogenetic Matrix

Given a matrix, we might just be interested in the taxon names, for example, from the online file 'Tuatara.txt', which we will use here,

```
tuatara<-readLines("Tuatara.txt")
tuatara
[1] "mammal 0000011001" "bird 1111111002" "lizard 1000101112"
    "turtle 1000101000"
[5] "alligator 1111101002" "Tuatara 1000(0,1)01002"
```

Since the first character following the name is going to be a space (or possibly a tab) we could loop through each line and use **strsplit** and then take the first element of the resulting list. Our code could be something like

```
names<-NULL
for(i in 1:length(tuatara))
names<-c(names,unlist(strsplit(tuatara[i]," "))[1])
names
[1] "mammal" "bird" "lizard" "turtle" "alligator" "Tuatara"
```
Alternatively, we can use **lapply** (list apply) with the function **strsplit**, the primitive R function **[**, which has to be inside simple quotes (single or double), and the argument 1 meaning apply the function to the rows.
```
names<-unlist(lapply(strsplit(tuatara,split=" "),"[",1))
names
[1] "mammal" "bird" "lizard" "turtle" "alligator" "Tuatara"
```

Replacing Specified Ambiguity Codes with a Question Mark

Some phylogenetic packages cannot accept explicit polymorphisms such as the '(0,1)' and these would need to be replaced with a single character, normally a '?'. In the Tuatara data there happens to be one such ambiguous character, in this case meaning that tuataras excrete both urea and uric acid. To make the substitution using **gsub** we must remember that the '(' and ')' are metacharacters and so need to be preceded (escaped) by \\
tuatara[6]<-gsub("*\\(.*?\\)","?", tuatara[6])
In between the escaped opening and closing round bracket characters we write '.*?', which stands for replace all contents, with a question mark, which is the next argument in the **gsub** call.

Tool Box 24.4

- Use == to test if two things are exactly equal to one another.
- Use all.equal() to test if values are equal.
- Use round() to round values to a specified number of decimal places.

25 Spacing in Two Dimensions

Summary of R Functions Introduced
   rect

There are many interesting questions in biology that revolve around the spacing of individuals, for example in territoriality, or spatial clumping of genotypes. Here we give a very brief demonstration from basics of looking at the randomness of spacing of a sedentary, but not immobile animal, a European sea anemone. We will test for spatial structure using nearest neighbour distances. This is a huge area of statistics so any reader wishing to work in the area will need to consult more specialized sources. Baddeley *et al.* (2016) provides a very comprehensive, R-based, coverage.

Brace and Quicke (1986) examined seasonal changes in inter-individual spacing of the common intertidal sea anemone, *Actinia equina*, that occurs on rocky shores in western Europe. Here we digitized the x and y coordinates of part of the pattern of distributions from their Fig. 3A using the program PlotDigitizer (plotdigitizer.sourceforge.net/ accessed 22 April 2020) though there are numerous alternatives. It should be emphasized that for spatial data a buffer zone around the studied points is needed to reduce edge effects – these may be artefactual or represent biological features. The raw digitizer output is accurate to 5 decimal places which is rather extreme given that the positions of the measurements was probably only accurate to about 3 mm,

data=read.csv("ActiniaXYdata.csv",header=TRUE)

```
head(data)
         X         Y
1 11.23853  47.37443
2 15.02294  47.26027
3 19.03670  47.48858
4 19.26605  44.74886
5 25.91743  44.17808
6 31.76605  46.46119
```

so we will use the function **round** to make these a bit more presentable

data$X<-round(data$X,1);data$Y<-round(data$Y,1) # multiple R commands on one line separated by a semicolon – or we can just use data<-round(data,1)
pdd<-c(4.9,1.3,5.3,1.6,3.9,3.8,1.8,3.5,3,2.9,3,3.9,3,3.3,2.1,4.5,3.2,4.2,6.1,3.8,4, 4,3.8,1.5,3.9,4.3,4.1,3.9,1.8,3.2,4.2,3.2,1.9,4,2,3.8,1.5,5.9,3.9,4.1,3,3.6, 3.2,3.8,3.9,2.5,5.5,3,1.5,3.9,2.5,4,1.2,4,3.9,3.7,1.8,4.4,3.2,3.9,3.7,3.7, 4.4,3.9,4.6,4.2,5,4,4.6,5,3.7,3,4.1,1.5,4,3.2,3,1.7,4,3,3.1,3.8,4.2,5,2,4, 4.1,3.9,3.1,3,1.6,3.7,3.5,3,3.8,3.9,4,4.9)
plot(data,pch=19,xlim=c(0,50),ylim=c(0,50),col="darkred",cex=pdd/1.6, xlab="X coordinate (cm)",ylab="Y coordinate (cm)") # cex is adjusted to scale diameter of anemones correctly.

R has the useful built-in function **rect**, which takes for its position and size the following coordinates in order (xleft, ybottom, xright, ytop). The border presence/absence (bty=TRUE or FALSE), border colour, line type, and the fill, shading, colour and transparency (density) can all be specified.
rect(0,0,50,50,lty=4,lwd=0.6)

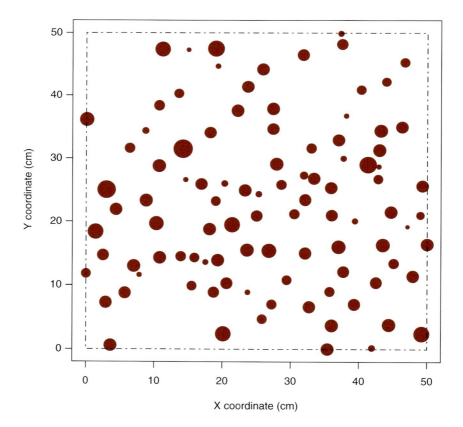

A cumbersome way to find the distance from each anemone to its nearest neighbour (nearest neighbour distance or NND) is to loop through each individual, and then from it calculate the distances to all other individuals. The smallest value is the NND. We will write some code to do this and employ our user-defined function 'hypotenuse' to calculate the Pythagorean distance between the centres of each individual.

```
hypotenuse<-function(x1,y1,x2,y2) sqrt((x1-x2)^2+(y1-y2)^2)
```

Because this is a one step calculation we can put it on a single line and dispense with the curly brackets and return value. We can check it is correct with the standard 3, 4, 5 right angle triangle

```
hypotenuse(0,0,3,4)
[1] 5
```

The following code steps through each anemone twice in a nested pair of loops. The focal anemones are in the *i* loop, and then the distance to all others is calculated in the *j* loop.

```
N<-nrow(data)
distances<-matrix(nrow=N,ncol=N) # the cells of the matrix are automatically
    filled with NAs if nothing is specified
for(i in 1:N){
    for(j in 1:N){
    distances[i,j]<-hypotenuse(data$X[i],data$Y[i],data$X[j],data$Y[j])
    }}
```

This operation on just the 98 times 98 x and y values took approximately a quarter of a second, so if this is the approximate size of your dataset then there is not much point spending time writing much faster code.

Our NNDs for each anemone are therefore the smallest non-zero values in each row of the 'distances' matrix.

```
NNDs<-NULL
for(i in 1:N) NNDs<-c(NNDs,min(distances[i,which(distances[i,]>0)]))
```

This line looks complicated so we will go through it starting at the innermost part:

1. 'which(distances[i,]>0)' asks which positions in row number *i* of our vector 'distances' are non-zero.
2. 'distances[i,which(distances[i,]>0)]' then gives all of the non-zero values in row *i*.
3. 'min(distances[i,which(distances[i,]>0)])' finds the minimum value, i.e. the NND for the anemone represented by the data in row *i*.
4. 'c(NNDs, min(distances[i,which(distances[i,]>0)]))' in each step in the for *i* loop concatenates the NND into the vector NNDs.

```
hist(NNDs,col="pink")
```

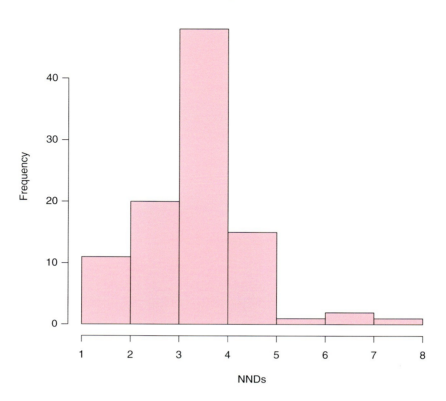

Histogram of NNDs

Clark and Evans (1954) invented a simple NN index, usually called R, which is the ratio of the observed mean nearest neighbour distance to the expected, and showed that the expected mean NND is given by

$$R_E = \frac{1}{2\sqrt{(\rho)}}$$

where ρ is the density. Within our area of 2500 cm^2 we have 98 anemones so $\rho = 98/2500$ per cm$^2 = 0.0392$
RE<-1/(2*sqrt(0.0392))
RE
[1] 2.525381
so
R<-mean(NNDs)/RE
R # index of dispersion
[1] 1.334161
If the anemones were distributed at random then R would be closer to unity. Values of R less than one indicate clumping and values greater than one indicate

overdispersion. Our observed value is quite a lot larger than one so we would want to know if this is a significant difference. A one sample t-test would be appropriate as we know the expected mean NND precisely
t.test(NNDs,mu=RE)

```
    One Sample t-test
One Sample t-test
data: NNDs
t = 7.6925, df = 97, p-value = 1.201e-11
alternative hypothesis: true mean is not equal to 2.525381
95 percent confidence interval:
    3.151536 3.586992
sample estimates:
mean of x
3.369264
```

So our observation points are highly significantly overdispersed. Ignoring the aggressive behaviour of the anemone, can you think of a simple explanation based on the plot of the anemone positions (see p. 298)?

An alternative way of testing for random spacing relies on the fact that we know the standard error, s_r, of NNDs for n randomly distributed individuals with overall density ρ (Clark and Evans, 1954)

$$s_r = \frac{0.26136}{\sqrt{n\rho}}$$

SR<-0.26136/sqrt(N* 0.0392)
SR
[1] 0.1333469

The standard normal deviate, z is then given by

$$z = \frac{mean(NND) - r_E}{s_r}$$

z<-(mean(NNDs)-RE)/SR
z
[1] 6.328475

Since the absolute value of z ($|z|$) is much greater than 1.96 we can see that the anemones are highly significantly overdispersed. Smith (2016) discusses several other methods that can be used to detect spatial non-randomness.

Weiss (1981) presented a simple graphical way of examining a list of NNDs to see if the coordinates were randomly dispersed, overdispersed or clumped, or a combination of these at different scales. We want to plot

$$-\log(1 - (R/(N+1)))$$

where R is the rank order of the NND, against NND squared (x axis)
plot(sort(NNDs)^2,-log(1-((1:N)/(N+1))),ylab="Weiss Index",xlab=
    "NND^2",pch=18,main="Weiss plot of anemone spacing")
lines(sort(NNDs)^2,-log(1-((1:N)/(N+1))),col="red")

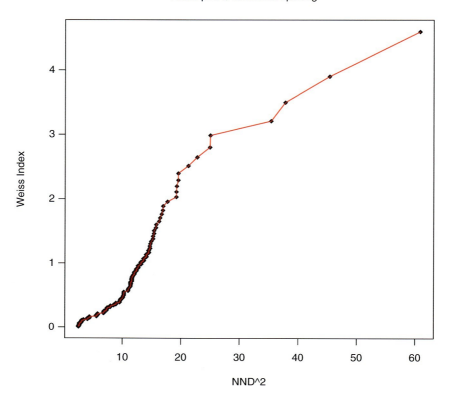

Weiss plot of anemone spacing

The concavity towards the left-hand side indicates overdispersion. There are at least two possible reasons for this, can you think of any?

Exercise Box 25.1

1) Create a set of randomly distributed data points (artificial anemones, for example 150) within a plot area, and make a Weiss plot to confirm that the random points in a plane produce a straight line.
2) Because we are randomizing things we ought to do a number of replicates.
3) With the real data above take into account the pedal disc areas to calculate actual physical separation of the individuals and repeat the Weiss index plot.

Tool Box 25.1

- Use rect() to draw a rectangle.

26 Population Modelling Including Spatially Explicit Models

Summary of R Packages Introduced	ceiling
beepr	expand.grid
simecol	rainbow
	return
Summary of R Functions Introduced	runif
beep in "beepr"	system
cat	

R can be used for some quite fast modelling jobs but its speed is nowhere near that of a compiled programming language such as C++. Here we will show how user-defined functions can be used to perform highly repetitive jobs efficiently, and this section also demonstrates various mathematical functions. The first example shows how a vector can be incremented and the calculated points plotted on a graph as the simulation proceeds. The second example runs a loop, and each time passes values to a user-defined function, and receives back multiple values from that function, which it then stores for plotting later. The third example is necessarily more complex and shows how R code can be used to carry out spatially explicit analyses. Finally, a simple example shows how R can be used to teach how evolution takes place, even in the absence of natural selection due to genetic drift and population bottle-necking.

Example 1 – Ricker Population Growth Model, Plotting as You Go

This is a nice example that lets you watch the simulation progressing in real time. Ricker (1954) introduced a simple, discrete time interval equation to show how the population of a species N_t changes in response to two variables, r_0, its intrinsic rate of increase and K is the carrying capacity for that species,

$$N_{t+1} = N_t e^{r_0}(1 - N_t/K)$$

As the species population N_t approaches the carrying capacity, N_t/K approaches 1 so $1 - N_t/K$ approaches zero, and this reduces the overall rate at which it can reproduce. Under these conditions, the resources needed to reproduce at the maximal (unconstrained rate, r) are no longer available.

Here we will use some code based on Soetaerd and Herman (2009) that will model the organism's population assuming that the limiting carrying capacity is 1, and so N_t will be treated as a population density rather than a number of individuals, and we will start each time with a density between 0 and 1 randomly chosen using the **runif** function. Note that therefore we do not have to enter a carrying capacity since $N_t/1 = N_t$. What is neat about this code is that the graph grows as the program steps through increasingly large values of r.

The code involves a **for** loop and a user-defined function '**ricker**', which is sent two numbers from the main body of the program each time it is called: the current number of individuals and the intrinsic rate of increase.

```
ricker<-function(Nt,r){ # for simplicity K is taken to be 1
    NtPlus1<-Nt*exp(r*(1-Nt))
    return(NtPlus1) # you can omit "return" and just write NtPlus1
    }
minimum_r<-0.001 # set the starting value of r
maximum_r<-5
r_sequence<-seq(minimum_r,maximum_r,0.001) # creates a vector of r-values
    that will be stepped through:
plot(NULL,xlim=range(r_sequence),ylim=c(0,5),xlab="r",ylab="Nt",
    main="Ricker logistic population model") # create a blank plotting field
    for(r in r_sequence){ # note that here we have presented a sequence of
        real numbers so we cannot use a colon to separate extremes
    Nt<-runif(1) # picks a random value for Nt between 0 and 1 to start with
    for(i in 1:200) Nt<-ricker(Nt,r) # runs the simulation for 200 generations
        ignoring the result, you can think of this as a burn-in phase
    for(i in 1:200){Nt<-ricker(Nt,r) # now calculates and plots the population
        size for each of the 200 generations after the burn in.
    points(r,Nt,pch=".",cex=1.5)} # plotting a full stop because there are
        going to be so many points
    } # end of r loop
```

Here is the result of running the code until all the 5000 values of r have been multiplied by 400 Ricker calculations – that's 2 million calculations and takes about 2 minutes on a laptop. During this time R has also plotted 1 million points so please be a little patient.

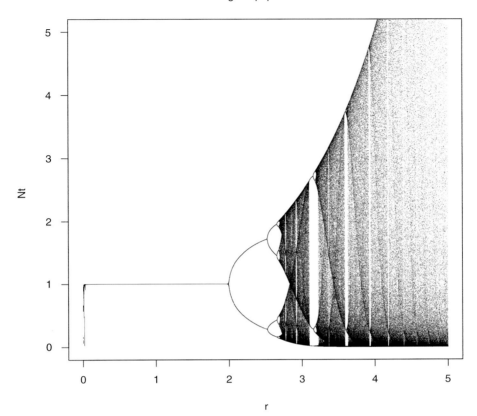

Ricker logistic population model

The graph shows that for values of r slightly larger than 0 (note exp(0)=1) up to 2, each simulation homed in on a stable population density of 1. For r between 2 and 2.5, the population alternates each generation between a value above 1 and a value below 1 (it is bistable), then as r increases there is a range of values where the population alternates between four fixed values, and so on until it becomes truly chaotic.

Writing R code can be done very efficiently, for example, the **ricker** function can be simplified to
ricker<-function(Nt,r) Nt*exp(r*(1-Nt))
and since the entire content of the function comprises only one line, we can omit {}. However, when you first start writing code and functions it is best to do so in quite a long-hand way, adding lots of comments to remind yourself, and other potential users, what the various bits of code do.

> **Information Box 26.1**
>
> *Getting sound notifications.* R is not really meant for sounds but it can sometimes be helpful (or fun) to get a sound out of it. This may be helpful for example, when a program is taking a long time (maybe running in the background while you are using your computer for something else) and you want to be prompted when it has finished its run. The simplest way is the **alarm** function, which is equivalent to **cat**('\a'): enter **alarm**(), and you should get hear a beep sound, but unfortunately this does not work on all platforms! Note, sound must be turned on, of course.
>
> The package **beepr** allows you to get a small range of sound prompts, specifically 'ping', 'coin', 'fanfare', 'complete', 'treasure', 'ready', 'shotgun', 'mario', 'wilhelm', 'facebook' and 'sword'. beep(3) gives you a fanfare, etc. Try:
> i<-1
> while(i < 500000000){i<-i+1}; beep(8)
> If you are out of the room you might want to have the notification repeated. You could try
> cat("press [escape key] to stop"); i<-1; while(i<2) beep(9)
> On at least some MacOSX systems there is actually a way to get your computer to speak using the **system** function
> system("say Program finished!")

Example 2 – Host–Parasitoid Population Modelling – Discrete Time Version

Host–parasitoid population dynamics were simulated using Beddington *et al.*'s (1975) modification of Nicholson–Bailey discrete generation equations with density dependence (equations for host, *N*, and parasitoid, *P*).

$$N_{t+1} = N_t e^{\left[r\left(1-\frac{N_t}{K}\right) - \alpha P_t\right]}$$

$$P_{t+1} = cN_t \left[1 - e^{(-\alpha P_t)}\right]$$

where *t* is the generation, *r* is the host's intrinsic rate of increase, *K* is the host's carrying capacity, α the *per capita* parasitoid searching efficiency and *c* the number of parasitoids completing development per host.

To write R code to do this we will have to set some initial values and then we can explore the effects of these and other variables on the outcomes.

Nt<-1000
Pt<-20
alpha<-0.002
r<-3
c<-1
K<-2000

Now we will build the structure leaving some details blank.

The body part of the program will be a loop cycling once per generation, and the numbers of hosts (Nt) and parasitoids (Pt) will be passed to a function to calculate the numbers for the next generation. The numbers of each in every generation will be remembered by concatenating them onto a list using 'c' and the previous generation values will be substituted by the new ones, etc.

```
hosts<-Nt # starting these two vectors with the initial host and parasitoid
    populations
parasitoids<-Pt
number_of_generations<-1000
for(i in 1: number_of_generations){
    # THE MAIN BODY TEXT WILL GO HERE
}
```

In each generation the body text will have to pass 'Nt', 'Pt', 'alpha' and 'r' to the function which we will call 'Beddington' (see Beddington et al., 1975), which in the working version of the program will be placed above the body text. Our 'Beddington' function will then pass back the next generation host and parasitoid values. A single function can only pass back one R object but this can be a list or dataframe or matrix, so passing multiple values back in one go is not a problem.

Here is the function to calculate the next generations. It is good practice not to use the same variable names as in the rest of the program, but it is not essential. Six parameters are passed to it. Inside the function these are referred to in order as 'fNt', 'fPt', 'fr', 'falpha', 'fK' and 'fc', corresponding to the current host and parasitoid populations ('fNt' and 'fPt') and the parameters 'r', 'alpha', 'K' and 'c'. The function **exp** raises the base of natural logarithms to the power passed to it

```
Beddington<-function(fNt,fPt,fr,falpha,fK,fc){
    NtPlus1<-fNt*exp((fr*(1-(fNt/fK)))-(falpha*fPt))
    PtPlus1<-fc*fNt*(1-(exp(-falpha*fPt)))
    answer<-c(NtPlus1,PtPlus1)
    answer
}
```

The body text therefore simply has to pass the necessary parameters to 'Beddington' 1000 times, receive back 'answer', remember the values of answer 'hosts' and 'parasitoids' and then reset Nt to NtPlus1 and Pt to PtPlus1 before going around the *i* loop again. At the end we will plot the results. The whole piece of code will therefore be:

```
Nt<-1000
Pt<-20
alpha<-0.002
r<-3
c<-1
K<-2000 # we put the user-defined constants first to make editing them easier
Beddington<-function(fNt,fPt,fr,falpha,fK,fc){
    NtPlus1<-fNt*exp((fr*(1-(fNt/fK)))-(falpha*fPt))
    PtPlus1<-fc*fNt*(1-(exp(-falpha*fPt)))
    answer<-c(NtPlus1,PtPlus1)
    answer # this passes the value of "answer" to where the function was
        called
} # end Beddington function
```

```
hosts<-Nt  # starting these two vectors with the initial host and parasitoid
    populations
parasitoids<-Pt
number_of_generations<-1000
for(i in 1: number_of_generations){
    answer<-Beddington(Nt,Pt,r,alpha,K,c)
    hosts<-c(hosts,answer[1])
    parasitoids<-c(parasitoids,answer[2])
    Nt<-answer[1]
    Pt<-answer[2]} # end of i loop
plot(1:length(hosts),hosts,ylim=c(0,max(hosts)),col="green",pch=20,
    xlab="Generations",ylab="Population size",main="Parasitoid-Host
    Population Dynamics")
points(1:length(hosts),parasitoids,col="red",pch=20)
# there is no room for a legend so we have to insert the legend manually with
    a bit of trial and error for positions
points (number_of_generations/4.5,max(hosts)*1.07,col="green",pch=20,xpd=NA)
text(number_of_generations/3.5,max(hosts)*1.07,"Hosts",xpd=NA)
points(number_of_generations/2,max(hosts)*1.07,col="red",pch=20,xpd=NA)
text (number_of_generations/1.65,max(hosts)*1.07,"Parasitoids",xpd=NA)
```

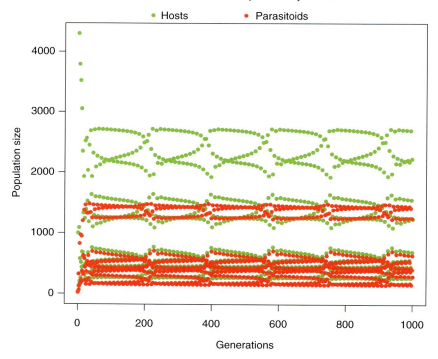

And/or we can plot parasitoid and host population sizes against one another
plot(hosts,parasitoids,pch=20,xlab="Host population size",ylab="Parasitoid
    population size",main="Parasitoid-Host Phase Plot",cex=0.5)

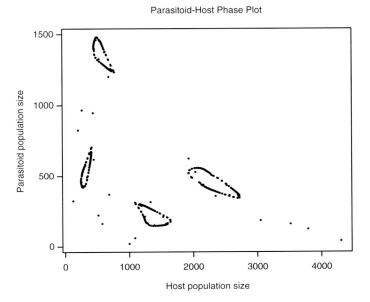

To see how the relative host parasitoid populations change we can use
plot(hosts,parasitoids,pch=20,xlab="Host population size",ylab="Parasitoid
    population size",main="Parasitoid-Host Phase Plot",cex=0.5)
lines(hosts,parasitoids)

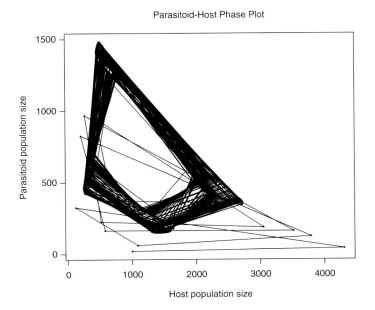

and if the points were thought intrusive, we could add 'col="white"' to the plot command so nothing appears in the plot area, and then run the **lines** command.

The particular combination of parameters we used here results in very complicated dynamics. Changing 'alpha' to 0.005 results in chaos, alpha = 0.0015 leads to stable equilibrium.

> **Exercise Box 26.1**
>
> 1) Explore the effects of changing parameters. Try reducing and increasing the value of '*r*' and playing with alpha values. Above what '*r*' value does the parasitoid go to extinction?
> 2) Set up the plot area to produce four plots in the same window. Place the population size versus generation top left, phase plot top right, parasitoid versus host population track bottom left and a plot of host–parasitoid ratio versus host population size bottom right.

Example 3 – Spatial Host–Parasitoid Model

Spatially explicit models are inevitably more complicated and usually involve square grids of cells and in each generation or time step the program not only calculates what happens in each cell but also the effects of each cell on its neighbours, so as to simulate things like migration, competition for resources, etc. Here we just give a feel of how this can be done. There are numerous books going into a lot more detail, and in particular Soetaerd and Herman (2009) provide examples with the coding worked in R.

We use the 'Beddington' function we wrote about above to calculate the numbers of hosts and parasitoids in successive generations. We need to think of the sequence of operations we want to perform, and then we write code to do each and, hopefully, when we run the entire lot together it will work. Most likely you will have to do some debugging.

A summary of what we might want to do might be as follows:

1. Define grids for host and parasitoid populations.
2. Use the 'Beddington' function from above to calculate host and parasitoid numbers for each grid square.
3. Write a function to migrate hosts and parasitoids randomly to adjacent cells.
4. Include a plot function to observe progress of runs.
5. Define population dynamic, generation number and migration parameters.
6. Start generations loop.
7. Step through all cells doing the calculation of how many hosts and parasitoids will exist in the next generation – use the 'Beddington' function for this.
8. Migrate a proportion of individuals to adjacent cells stepping through matrix cell by cell – need to write new function for this.
9. Remember host and parasitoid populations at selected generations for future plotting.
10. End of generations loop, repeat until desired number of generations reached.
11. Create beautiful plots of selected generations using colour palette and add.

We need to define a grid (we'll make it a square for simplicity) of Z rows and Z columns. Let Z = 10 to start with. We need two matrices with these dimensions to represent the host and parasitoid populations.
Z<-30
H<-matrix (NA,nrow=Z,ncol=Z) # number of hosts
P<-matrix (NA,nrow=Z,ncol=Z) # number of parasitoids
We will run the model for G generations
G<-100
and we need to include the Beddington function we wrote above.
Beddington<-function(fNt,fPt,fr,falpha,fK,fc){
    NtPlus1<-fNt*exp((fr*(1-(fNt/fK)))-(falpha*fPt))
    PtPlus1<-fc*fNt*(1-(exp(-falpha*fPt)))
    answer<-c(NtPlus1,PtPlus1)
    answer}

The more fiddly part of spatial analyses like these is the need for hosts and/or parasitoids to migrate between cells each generation. What happens at the edges? To solve this, we will write the migration function that wraps the edges of the plot area such that if individuals migrate off the top-most row, they appear in the bottom-most row and if they migrate off the area left, they reappear in cells on the right, and vice versa. Clearly there are three categories of cells, corner ones, edge ones and central ones, and to do the job properly we need to create special cases for each. We check whether each cell is central, edge or corner and then apply the migration to adjacent cells accordingly.

For a central cell of coordinates (a, b) we want migrating individuals to move randomly to the eight surrounding cells. These are shown below (row first, then column).

a-1, b-1	a-1, b	a-1, b+1
a, b-1	a, b	a, b+1
a+1, b-1	a+1, b	a+1, b+1

For the left edge, the neighbours are

a-1,Z	a-1, 1	a-1, 2
a, Z	a, 1	a, 2
a+1, Z	a+1, 1	a+1, 2

and so on for right, top and bottom margins. Finally, for corners, the neighbours follow the pattern for the top left corner.

Z, Z	Z, 1	Z, 2
1, Z	1, 1	1, 2
2, Z	2, 1	2, 2

It is clear that we could write a load of sets of operations to deal with each of these alternatives, but that would be unbelievably ugly, long-winded and slow, so we will therefore automate it by the trick of adding and subtracting 1 to column and row numbers, and then if we get 0 (i.e. we started in row or column 1 and 'moved' left or up), we will then replace the 0 with Z, thus wrapping in one direction. If we get Z + 1 (i.e. we started in column or row Z and right or down) we will replace Z + 1 with 1 thus jumping to the first row or column.

We need to divide the number of migrating individuals randomly between the eight cells surrounding our focal cell. An easy way to calculate how many to move to each of these is to create a vector with a length of the number of individuals that have to be moved and then to divide it into eight random chunks by dividing at seven random places. The number in each chunk will then be added to the surrounding eight cells. Here is a function to do that:

```
random_asign_to_neighbours<-function(to_Add_f,Number_moving,neighbours){
    breaks<-sort(sample(1:Number_moving,7,replace=TRUE))
    cell<-rep(NA,8) # because we are going to assign values to numbered
        positions in the vector "cell" we need to define it first
    for(i in 2:7){
        cell[i]<-breaks[i]-breaks[i-1]} # end i loop
    cell[1]<-breaks[1]
    cell[8]<-Number_moving-breaks[7]
    for(x in 1:8){
        to_Add_f[neighbours[x,1],neighbours[x,2]]<-to_Add_f
        [neighbours[x,1],neighbours [x,2]]+cell[x]}
    to_Add_f # same as return(to_Add_f), the last variable will be returned
} # end of random_asign_to_neighbours function
```

To observe what is going on each generation we create a small plotting function

```
plot_generation<-function(X,Z){ # X receives the host or parasitoid matrices
mcol<-max(X) # to colour each plotted symbol in no particular way
    for(j in 1:Z){
        for(k in 1:Z){
            points(j,k,col=round(10*X[j,k]/mcol),pch=15)
        }} # end j and k loops
    } # end plot_generation function
```

We will start by seeding each host matrix cell with 100 hosts and 10 parasitoids respectively and define various other parameters

```
H[1:Z,1:Z]<-100
P[1:Z,1:Z]<-10 # we could use P<-matrix(10,nrow=Z,ncol=Z)
```

and setting some initial population dynamics parameters

```
alpha<-0.0025
r<-2.1 # intrinsic rate of host population increase
c<-1
K<-1000 # carrying capacity
```

Define proportion of individuals in each cell to migrate to adjacent cells proportion_to_migrate<-0.1 # 10% of individuals in each cell will migrate to neighbouring cells

neighbours<-matrix(NA,nrow=8,ncol=2) # hold the relative coordinates of neighbouring cells

We wish to save/remember how many parasitoids and hosts there were in each cell (that will be 20,000 values in this case) at various times in the run to see how things change. The easy way to do this is to write the values of all the cells in a grid into a single vector of length Z^2 and to do this we will use the R function **expand.grid**, and we can store all these vectors as rows in a table by binding the new one to the old list each generation. At the end we can recreate the grid for each generation as we please using the function **matrix**. Here we define the holding matrices:

save_runs<-c(10,30,60,100) # these will be used to make snapshots after these numbers of generations; we plot 10 and 100 at the end

hold_H<-matrix(NA,nrow=Z*Z,ncol=length(save_runs)) # these will store the four host population snapshots

hold_P<-matrix(NA,nrow=Z*Z,ncol=length(save_runs)) # ditto for parasitoids

count<-0 # we will use this to set the column in the hold matrices to fill

par(mfrow=c(3,2)) # set graphics parameter for plotting spatial arrays during simulation

Here we will start the main body of the simulation program, looping for each generation and within each loop, passing host and parasitoid numbers for each cell to the 'Beddington' function, then calculating how many individuals will migrate to each cell, plot the spatial patterns, and if the number of generations 'g' is in our list of those results we want to save for later (save_runs) saving them to hold_H and hold_P.

```
for(g in 1:G) { # stepping through generations
    for(ba in 1:Z){ # stepping through rows
        for(bb in 1:Z){ # stepping through columns
            answer<-Beddington(H[ba,bb],P[ba,bb],r,alpha,K,c)
            H[ba,bb]<-answer[1]
            P[ba,bb]<-answer[2]
        } # end bb loop
    } # end ba loop
```

For migration we will create matrices (tomove_Host and tomove_Parasitoid) that have the numbers of individuals from each cell that we want to migrate, and then we remove them from the cells leaving only resident, non-migrating individuals. We need to deal with whole numbers of hosts and parasitoids so we use the function **round** to set numbers to the nearest integer.

tomove_Host<-round(H* proportion_to_migrate)

tomove_Parasitoid<-round(P* proportion_to_migrate)

H<-round(H-tomove_Host) # subtracts the values of cells in tomove_Host from corresponding cells in H, leaving just the residents

P<-round(P-tomove_Parasitoid) # ditto for parasitoids

We will create two matrices of zeros (to_add_Host and to_add_Parasitoid) that will store the final movement places of all individuals after we have stepped through all the cells containing numbers of individuals to migrate (tomove_Host and tomove_Parasitoid). For each cell in the 'tomove . . .' matrices we need to create a list of their eight neighbouring cells. The row and column locations are ba and bb respectively. After having done that, we wrap the object for cells on the borders, connecting left with right and top with bottom, and then pass numbers to move, the locations of the neighbouring cells and the 'to_add . . .' matrices to our function 'random_asign_to_neighbours'.

```
to_add_Host<-matrix(0,nrow=Z,ncol=Z)
to_add_Parasitoid<-matrix(0,nrow=Z,ncol=Z)
for(ba in 1:Z){ # stepping through rows
    for(bb in 1:Z){ # stepping through columns
        neighbours[1,1:2]<-c(ba-1,bb-1)
        neighbours[2,1:2]<-c(ba-1,bb)
        neighbours[3,1:2]<-c(ba-1,bb+1)
        neighbours[4,1:2]<-c(ba,bb-1)
        neighbours[5,1:2]<-c(ba,bb+1)
        neighbours[6,1:2]<-c(ba+1,bb-1)
        neighbours[7,1:2]<-c(ba+1,bb)
        neighbours[8,1:2]<-c(ba+1,bb+1)
        neighbours[which(neighbours==0)]<--999 # minus 999 is an
            impossible holding value
        neighbours[which(neighbours==Z+1)]<-1
        neighbours[which(neighbours==-999)]<-Z
        to_add_Host<-random_asign_to_neighbours(to_add_
            Host,H[ba,bb], neighbours)
        to_add_Parasitoid<-random_asign_to_neighbours(to_add_
            Parasitoid, P[ba,bb],neighbours)
    } # end bb loop
} # end ba loop
```

Having now calculated the migrating host and parasitoid numbers to be added to each cell, we add them to the existing residents and plot the new numbers using our function **plot_generation** to add the coloured points. Note that including these plots slows down the runs a lot and is there for de-bugging purposes. It is better to comment (#) them out in a real run.

```
H<-H + to_add_Host
P<-P + to_add_Parasitoid
plot(NULL,xlim=c(1,Z),ylim=c(1,Z),axes=F,xlab="",ylab="",main=paste
    ("generation =",g)) # the default sep argument in paste is " "
mtext("Host")
plot_generation (H,Z)
plot(NULL,xlim=c(1,Z),ylim=c(1,Z),axes=F,xlab="",ylab="",main=paste
    ("generation =",g))
plot_generation (P,Z)
mtext("Parasitoid")
```

Finally, we wish to store the populations of hosts and parasitoids in each cell at selected times (numbers of generations) through the simulation run for later consideration.

```
if(g %in% save_runs){
    count<-count+1
    hold_H[,count]<-unlist(expand.grid(H)) # you must use unlist because
        expand.grid returns a list of length 1
    hold_P[,count]<-unlist(expand.grid(P))
    } # end if g
} # end G loop
```

We have now completed our simulation and all that remains is to create a nice plot! The R base package contains four predefined colour palettes ('gray', 'heat.colors', 'rainbow' and 'terrain.colors') and we will use 'rainbow' here. If you type 'rainbow(256)' you will see the list of colours in hexadecimal *rgbt* format. The colours go almost full circle from red (FF0000FF) through the spectrum to violet and then back to a similar red (FF0006FF). To avoid ambiguity, we are going to curtail the list at the 220th value, a deep pink-violet (FF00DDFF). Because we will access colours separately for each point, we will create a function, 'getcoloursrainbow', to obtain the colour for each point.

```
getcoloursrainbow<-function(x){
    purplef<-220
    f<-unlist(as.list(rainbow(256))) # rainbow is a built-in colour palette
        comprising 256 shades
    mn<-0
    mx<-max(x,na.rm=T)
    cor<-(x-mn)/(mx-mn)
    ans<-f[(1+ceiling(purplef*(cor)))] # ceiling returns the integer
        immediately above a real number, cf. floor and round
    ans} # end function getcoloursrainbow
```

Now the plots: to make things easier we will get the colours for the population values that are held in Z*Z long columns in 'hold_H' and 'hold_P' first. And then we will back convert the list of colours to a Z by Z matrix using **matrix**({*colour list*}, nrow=Z). Then we plot each point in its appropriate colour, ranged between the minimum and maximum population size.

```
par(mfrow=c(2,2))
point_colours<-getcoloursrainbow(hold_H[,1])
point_colours2<-matrix(point_colours,nrow=Z)
point_colours3<-getcoloursrainbow(hold_P[,1])
point_colours4<-matrix(point_colours3,nrow=Z)
plot(NULL,xlim=c(1,Z),ylim=c(1,Z),axes=F,xlab="",ylab="",main="Host")
for(j in 1:Z){
    for(k in 1:Z){
        points(j,k,col=point_colours2[j,k],pch=15)
    }} # end j and k loops
plot(NULL,xlim=c(1,Z),ylim=c(1,Z),axes=F,xlab="",ylab="",main="Parasitoid")
```

```
for(j in 1:Z){
    for(k in 1:Z){
        points(j,k,col=point_colours4[j,k],pch=15)
}} # end j and k loops
point_colours<-getcoloursrainbow(hold_H[,4])
point_colours2<-matrix(point_colours,nrow=Z)
point_colours3<-getcoloursrainbow(hold_P[,4])
point_colours4<-matrix(point_colours3,nrow=Z)
plot(NULL,xlim=c(1,Z),ylim=c(1,Z),axes=F,xlab="",ylab="",main="Host")
for(j in 1:Z){
    for(k in 1:Z){
        points(j,k,col=point_colours2[j,k],pch=15)
}} # end j and k loops
plot(NULL,xlim=c(1,Z),ylim=c(1,Z),axes=F,xlab="",ylab="",main="Parasitoid")
for(j in 1:Z){
    for(k in 1:Z){
        points(j,k,col=point_colours4[j,k],pch=15)
}} # end j and k loops
```

As a last tweak, we will add a colour scale bar to the middle of the four plots. The whole display window has x and y coordinates based on the values in the most recent plot, i.e. if we have completed the fourth plot, all points in the plotting area will be defined by the axes of that plot. We know the right-hand edge of that is 1 and its left-hand edge is 30, so we can make a guess that the middle between the plots is x = –7 (i.e. 7 below the upper row of plots) and y = 45 (i.e. 15 beyond the right edge of the plot). Then we apply some judgement to choose where to start our colour scale on the left, which will have 220 increments, and for each colour we draw a coloured vertical line, moving incrementally to the right until we have drawn all of them.

```
centreX<--7 # minus 7; you do not need to put a space between the two -s
centreY<-45
LX<-8
lowY<-centreY-2
highY<-centreY+2
increment<-0.08
line_colour<-getcoloursrainbow(1:220)
for(i in 1:220){
    X<-centreX-LX+(i*increment)
    lines(c(X,X),c(lowY,highY),col=line_colour[i],xpd=NA)}
text(xpd=NA,-24,centreY,"Population = 0") # xpd=NA can go anywhere
text(xpd=NA,15,centreY,"Maximum population")
```

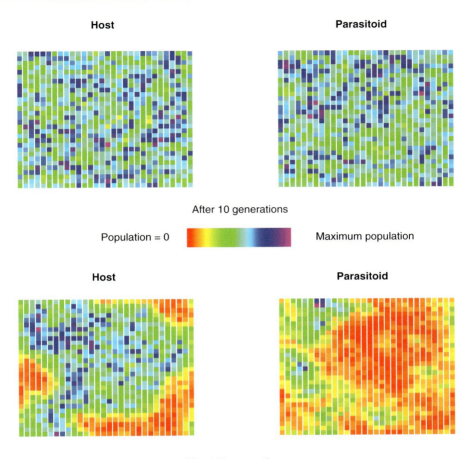

After 100 generations

It is not quite perfect, the colour bar could be a little more central, and you would probably want to add text for each pair of plots to indicate how many generations had passed. If you do simulations with far larger grids, you will need to reduce the size of square (pch = 15) that you plot, etc.

Here we have created a spatial simulation in R by hand to show the logic of producing a moderately complicated piece of code. The R-package **simecol** (Petzoldt and Rinke, 2007) provides some utility functions to speed up running cellular automata models in R. We can see that whereas early in the simulation there was only limited spatial structure and limited ranges in the population sizes of the host and parasitoid, after 100 model generations, large zones were either host-rich or host-poor with hosts and parasitoids having been driven to extinction in some cells.

> **Exercise Box 26.2**
>
> 1) Add a function to calculate overall host–parasitoid ratio across generations and then after completing the model run, plot them on a grid.
> 2) Write a function to calculate a measure of host and/or parasitoid clumpedness each generation and plot it against generation. There are a huge number of functions that could give a good indication. One simple, easy and quick way that we suggest might be to calculate the variance in numbers of individuals in the cells of the array, though this is also likely to reflect randomness. This result could be compared with a null expectation created by distributing the individuals randomly across all cells in a Monte Carlo procedure (see Chapter 16).

> **Tool Box 26.1**
>
> - Use expand.grid() to save all the values of a grid as a vector.
> - Use matrix() to convert expand.grid() object back to a 2D matrix.
> - Use rainbow() to show the hexadecimal code for the colours.
> - Use return() to specify what should be returned from a function call.

Example 4 – Genetic Drift, a Program Aimed at Teaching Students about Evolution

By definition, evolution is any change of allele frequencies in a population over time, and whilst many people think of it as being due to natural selection, purely random processes also lead to evolution and can have profound biological consequences such as when new populations are founded by only a small number of individuals, or when only a small proportion of individuals in a generation succeed in reproducing, or when dramatic events cause major population crashes just short of extinction, or, as in the case of inbreeding, when populations are fairly consistently small. In all these cases, random effects will have a strong influence on gene frequencies in subsequent generations. Here we will consider a situation in which only 10% of individuals reproduce in each generation. Starting with an overall population of 1000 individuals with equal frequencies of five alleles (200 individuals each of red, green, blue, turquoise and yellow), each generation, 100 individuals will be selected at random to reproduce and provide the starting gene frequencies for the next generation.

First, we will write the body of the program, but when you run it you must 'run' the plotting function first then the body. The first two lines define the plotting area as being for 3 × 2 separate plots (two columns and three rows), and that each has smaller margins all around than the default so as not to waste space. The initial gene frequency vector ('alleles') comprises 200 of each allele type. We will then randomize their order using the **sample** with **replace**=F so that the new version of 'alleles' contains precisely the same numbers of each allele. The vector 'plotlist' contains a list of the generations whose allele frequencies we want to plot graphically, with an **if** statement in the loop checking if the current generation is one that we included in our 'plotlist'. In each generation represented by the *i* loop we randomly sample 100 individuals from the population of 1000, and replicate each of their genotypes 10 times to get back to the total of 1000 to start the next generation.

par(mfrow=c(3,2)) # to create a two columns of three rows plotting area
par(mar=c(2,2,2,2))
alleles<-c(rep("red",200),rep("blue",200),rep("green",200),rep("yellow",200),
    rep("turquoise",200))
alleles<-sample(alleles,1000,replace=F)
plotlist<-c(1,5,10,25,50,100)

Next, we will write a small function, 'plotpoints', to plot, for visual impact only, the genotypes of individuals whenever we choose to call it, in this case at selected generations. When called from the body of the program, 'plotpoints' receives a vector of the genotypes in the current cycle ('all'), and, just to help explain the plot, the generation number ('gen').

plotpoints<-function(all,gen){
    plot(NULL,axes=F,xlim=c(0,100),ylim=c(0,100),ylab="",xlab="",main="")
    for(j in 1:length(all)){ # stepping through vector of genotypes to plot
        x<-sample(1:1000000,1)/10000 # randomly selecting and scaling
            it to axis length of 100 to minimize the probability of points pre-
            cisely overlapping
        y<-sample(1:1000000,1)/10000
        points(x,y,pch=15,col=alleles[j])} # end *j* loop
    text(xpd=NA,50,110,paste("generation = ",gen,sep=""),cex=0.8)
        # adding the generation number as a label centred above each plot
    } # end plotpoints function

Now the body of the program; we'll just run the simulation for 100 generations.
for(i in 1:100){
    if(i %in% plotlist) plotpoints(alleles,i) # we don't need to use {} as only
        one operation to be performed
    breeders<-sample(alleles,100,replace=F)
    alleles<-rep(breeders,10)} # end *i* loop

Within a few minutes you will see the genotype distribution after each of the six selected generations. In the run shown here all five genotypes survived in the population until generation 25, and indeed at that time 'red' was doing rather well, but by generation 50 'blue' had been lost, and by generation 100 'red' and 'turquoise' had also been lost. Readers might like to explore the effect of varying the proportion reproducing, and/or what happens with smaller population sizes.

Exercise Box 26.3

1) Here we chose to illustrate the results in a pictorial way, but to be more scientific write code to run many simulations, recording the allele frequencies for each generation and in each simulation, and then plot the means and standard deviations for generation.

2) We have been analysing a situation with haploid organisms, would diploids show the same? Write code to model a diploid organism and re-run the analyses, keeping the total number of allele copies the same.

Tool Box 26.2

- Use ceiling() to return the integer immediately above a real number.
- Use floor() to return a number immediately below a real number.
- Use round() to obtain the closest integer to a real number.
- Use runif(n) to obtain *n* random numbers between 0 and 1.

27 More on *apply* Family of Functions – Avoid Loops to Get More Speed

Summary of functions introduced	proc.time
	rnorm

This short chapter only really matters to those readers who need to process very large datasets, or who need to perform loop-type operations on largish datasets but perhaps in a nested fashion. It is reasonable to say that if you are only doing some calculation or data manipulation a few times and each time takes only a second or so, it is no big deal. But if, using standard loops each iteration takes several seconds or minutes, then waiting for the output can be at best, a bit frustrating, or at worst maybe take a day or so, or more.

The message is: avoid loops, and we are fortunate that R provides some tools that can massively increase calculation speed when it comes to otherwise slow looping calculations. The main set of these belong to the 'apply' set of functions. These can take a bit of getting used to but with large projects they are basically essential.

We encountered these functions earlier in Chapters 8, 15, 22 and 24. There are several apply family functions: **apply**, **lapply** (= list apply), **sapply**, **tapply** and **mapply**, but we won't consider the last one, which has little use in biology.

First, let us look at how long it takes to process a big calculation using a loop with many steps/iterations. In this example we will generate 100,000 values of a vector 'a' drawn at random from a normal distribution using **rnorm** (see Chapter 30):

```
a<-rnorm(100000)
```

This generates a list of 100,000 values of 'a' that can each be accessed using the standard subscript format a[*i*] where *i* is a number between 1 and 100,000. We will time the whole processing time using a function, **proc.time**, that starts the system clock. Now let us loop through all values of 'a' just adding 1 to each of them:

```
start_time<-proc.time() # start computer clock
for(i in 1:10000000) a[i]<-a[i] + 1 # loop through the vector
proc.time()-start_time # the computer clock returns a new value to proc.
    time
 user  system elapsed
4.352   0.504   4.817
```
i.e. about 5 seconds on our laptops. But for this operation of adding one to each of the values in vector 'a' we do not need to loop at all, we can simply use a < –a + 1. Now doing this again without the loop we get
```
a<-rnorm(10000000)
start_time<-proc.time() # record computer clock time
a<-a+1
proc.time()-start_time
 user  system elapsed
0.085   0.009   0.088
```
This is approximately 50× faster than using the loop!

Using *apply*

The function **apply** takes the vector to be operated on, an argument specifying rows or columns (1 or 2 respectively) and the function (either a built-in function such as **mean** or **max**, or a user-defined one, e.g. function(x) x > 5). Here we apply mean and then x > 5 to a small dataset that we will call d, created here:
```
a<-c(1,2,3,4)
b<-c(0,6,3,12)
c<-c(0,2,3.1,1000)
d<-rbind(a,b,c)
d
  [,1] [,2]  [,3]  [,4]
a   1    2   3.0     4
b   0    6   3.0    12
c   0    2   3.1  1000
    apply(d,1,mean) # the 1 means by row, there are 3 rows so 3 means
      a       b       c
  2.500   5.250 251.275
apply(d,2,mean) # the 2 means apply the function to columns, there are 4
    means
[1] 0.3333333 3.3333333 3.0333333 338.6666667
```
The next example is a user-defined function that calculates how many values in a column are less than 10
```
apply(d,2,function(x) length(x[x<10]))
[1] 3 3 3 1
apply(d,1,function(x) x>5) # the 1 in this case means by row so the answers
    are given as rows × columns
```

```
         a     b     c
[1,] FALSE FALSE FALSE
[2,] FALSE  TRUE FALSE
[3,] FALSE FALSE FALSE
[4,] FALSE  TRUE  TRUE
```
apply(d,2,function(x) x>5) # the 2 in this case means by column so the answers are given as columns × rows
```
    [,1]  [,2]  [,3]  [,4]
a FALSE FALSE FALSE FALSE
b FALSE  TRUE FALSE  TRUE
c FALSE FALSE FALSE  TRUE
```
The functions **sapply** and **lapply** are very similar. **lapply** returns a list whereas **sapply** returns a vector or matrix.
sapply(c(1,2,3,4,5),function(x) x^2) # you could replace c(1,2,3,4,5) by 1:5
```
[1]  1  4  9 16 25
```
lapply(1:3,function(x) x^2) # gives the same answers as **sapply** but in the form of a list
```
[[1]]
[1] 1

[[2]]
[1] 4

[[3]]
[1] 9
```
To see the same result as **sapply** use **unlist**
unlist(lapply(1:3,function(x) x^2))
```
[1] 1 4 9
```

Using *tapply* to Calculate Values Based on Factors

This function calculates values of a column in a dataframe for each factor listed in another column. For this we will again use the first few lines of the fish parasite load data from Wiriya *et al.* (2013) that we explored in Chapter 6.
Tilapia<-5
Mystus_singaringan<-c(1,2,2,2)
Osteochilus_vittatus<-c(26,30,11,2,28,6,19,12,4,4,18,13,29,21,9,8,21,19,28)
Henicorhynchus_siamensis<-c(46,22,33,28,50,42,31,4,13,8,11,8,18)
First we need to combine these into a dataframe with the first column, the species, as a factor, and the second, the values in the vector. Therefore the species names will be repeated as many times as the length of the corresponding vector. The logical way is to create a vector of names and a vector of values. We will use the function **lengths**, which has to be applied to lists.
names<-c("Tilapia","Mystus","Osteochilus","Henicorhynchus")
wize<-lengths(list(Tilapia,Mystus_singaringan,Osteochilus_vittatus,
    Henicorhynchus_siamensis))
fishnames<-c(rep(names,wize))

```
metacercaria<-c(Tilapia,Mystus_singaringan,Osteochilus_vittatus,
    Henicorhynchus_siamensis)
df<-data.frame(fishnames,metacercaria)
head(df,7)
```

```
    fishnames  metacercaria
1     Tilapia             5
2      Mystus             1
3      Mystus             2
4      Mystus             2
5      Mystus             2
6  Osteochilus            26
```

Now use **tapply** to summarize things for each species, for example, the mean for each species

tapply(df$metacercaria,df$fishnames,mean)

```
Henicorhynchus       Mystus    Osteochilus       Tilapia
      24.15385      1.75000       16.21053       5.00000
```

or

tapply(df$metacercaria,df$fishnames,range)

```
$Henicorhynchus
[1]  4 50

$Mystus
[1] 1 2

$Osteochilus
[1]  2 30

$Tilapia
[1] 5 5
```

Exercise Box 27.1

1) Use the dataset YouthRisk in the **Stat2Data** package, which provides data from the US Centers for Disease Control and Prevention (CDC). Write code that calculates how many 17- and 18-year-olds were in the sample, how many individuals in each age group, how many in each age group smoke, how many in each age group are female. Hint: you should use **apply**, **tapply** and **table**.

2) Write code to determine how many females in each age group smoke and what proportion of total females that is. Hint: use **subset**.

Tool Box 27.1

- Use apply() to carry out functions on all rows or columns in an array.
- Use lapply() to apply a function to a vector and return a list of individual elements.
- Use proc.time() to get/show the processing time.
- Use rnorm() to obtain numbers randomly picked from a normal distribution.
- Use sapply() to apply a function to a vector and return a vector of individual elements.
- Use tapply() to apply a function to each of the values specified by a factor in a dataframe.

28 Food Webs and Simple Graphics

> *Summary of Packages Introduced*
> cheddar
>
> *Summary of Functions Introduced*
> attr
> PlotCircularWeb in "cheddar"
> PlotWebByLevel in "cheddar"
> names

Food webs are fundamental in much of ecology and there has been a steady increase in studying their structure and properties over the past 50 years, nowadays often utilizing molecular methods too (Hrcek et al., 2011). First we will create code to draw our own food web, then we will introduce the package **cheddar**. The reason for learning how to produce our own is not just to improve programming skill and logical thinking, it also means we are in a position to customize our diagrams in ways that perhaps are not available in pre-written packages.

A Parasitoid *foodweb* Example

A food web shows the trophic links between a set of members of a community, and additionally might indicate the strengths of these links. In this example from Thailand, 22 braconid parasitoid wasps, representing a total of 9 species were associated with 22 lepidopteran hosts representing a total of 11 species using DNA barcoding. What we aim to do is to represent parasitoid species as a line of rectangles along the top of the plot, where the width of each rectangle represents the number of rearings of that parasitoid species. Similarly, the host species will be represented by a line of rectangles across the lower part of the figure with the widths of each rectangle representing the number of parasitized hosts. Then we will illustrate the number of rearings of each parasitoid from each host by lines of various widths connecting parasitoid host rectangles. Names can be added, and ideally we will write code so that further associations can be added with minimal inconvenience.

Food Webs and Simple Graphics

The vectors 'parasitoid' and 'host' represent the recovered associations in order. Of course we would probably read the data from file as a table but here we will just enter the vectors:

```
parasitoid<-c("Bracon1","Bracon1","Bracon1","Bracon1","Bracon1","agathidine1",
    "agathidine1","agathidine1","agathidine1","Cotesia1","Cotesia1",
    "Cotesia1","Cotesia1","Cotesia2","Cotesia3","Cotesia3","Cotesia3",
    "Cotesia4","Aleiodes1", "Aleiodes2","Macrocentrus1")
host<-c("Nymphalidae2","Erebidae1","Erebidae1","Erebidae1","Erebidae1",
    "Erebidae2", "Erebidae2","Noctuid1","Lycaenidae1","Hesperidae1",
    "Hesperidae1","Hesperidae1", "Hesperidae2","Noctuidae2","Geometridae1",
    "Geometridae2","Erebidae1", "Erebidae1","Erebidae1","Erebidae3","Erebidae4")
```

Use levels to determine how many species of parasitoid and host we have in our sample:

```
l<-levels(as.factor(parasitoid))
m<-levels(as.factor(host))
```

We start by creating a blank sheet on which the plot will build up. The width of the plot can be the number of associations, which is the number of either parasitoid rearings or parasitized hosts (they are the same, the lengths of the vectors parasitoids and hosts must be identical). The height of the y axis can be any convenient number; we chose 0 to 5.

```
plot(NULL,axes=F,xlab="",ylab="",xlim=c(0,length(parasitoid)),ylim=c(0,5))
```

Since our parasitoid–host association data are in no particular order, we will calculate how many individuals of each species of parasitoid 'P', and how many individuals of each host species there are, 'H'. We can do this by parsing through each list using length(which{*vector*} == "each of the levels"), which we stored above in the vectors 'l' and 'm'.

```
P<-NULL
H<-NULL
for(i in 1:length(l)) P<-c(P,length(which(parasitoid==l[i])))
for(i in 1:length(m)) H<-c(H,length(which(host==m[i])))
S<-cumsum(P)
S1<-c(0,S)
T<-cumsum(H)
T1<-c(0,T)
for(i in 1:length(l)) rect(S1[i],3.9,S1[i+1],5.7,col="red",lwd=6,border="white")
for(j in 1:length(m)) rect(T1[j],0.7,T1[j+1],2,col="green",lwd=6,border="white")
midT<-T1[1:length(T)]+(T-T1[1:length(T)])/2
midS<-S1[1:length(S)]+(S-S1[1:length(S)])/2
for(i in 1:length(l)) text(midS[i],4.1,l[i],srt=90,cex=0.8,xpd=NA,adj=0,font=3)
for(i in 1:length(m)) text(midT[i],1.9,m[i],srt=270,cex=0.8,xpd=NA,adj=0,
    font=3)
for(k in 1:length(l)){ # stepping through each different parasitoid
pp<-which (parasitoid==l[k])
np<-length(pp)
hostsofparasitoid<-host[pp]
links<-length(levels(as.factor(hostsofparasitoid)))
for(q in 1:links){
```

```
noRearings<-length(which(hostsofparasitoid==levels(as.factor
    (hostsofparasitoid))[q]))
temph<-which(m==levels(as.factor(hostsofparasitoid))[q])
lines(c(midS[k],midT[temph]),c(3.8,2.2),col="grey50",lwd=noRearings*2,
    lend=3)
} # end q loop
} # end k loop
points(midT,rep(2.2,length(T)),pch=15,col="blue")  # putting the squares at
    the end of the connecting lines
points(midS,rep(3.8,length(S)),pch=15,col="blue")
```

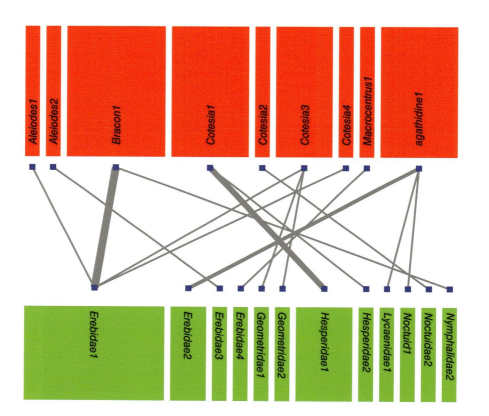

Foodweb and Community Packages

An earlier R package with the obvious name 'foodweb' (Perdomo et al., 2012) no longer functions on various systems (Mac and PC) because of the following error.
```
Loading required package: rgl
Error in dyn.load(file, DLLpath = DLLpath, ...)
```

Food Webs and Simple Graphics 329

There are complex work-arounds that can be found on the web, but we will not go into those. Luckily there is a new package **cheddar** (Hudson *et al.*, 2013) that has many functions for analysis and graphing food webs and many other community features.

install.packages("cheddar")
library(cheddar)
options(stringsAsFactors=F) # this line is necessary because **cheddar** does not automatically recognize character strings as factors

Cheddar has several associated vignettes, the first two of which are essential for learning how to use most of the functions: 'CheddarQuickstart', 'Community', 'PlotsAndStats', 'Collections' and 'ImportExport'. As explained in the last of these vignettes the input to cheddar is called a community and that in turn is created from three R dataframes. We will use our same host parasitoid data as above with vectors 'parasitoids' and 'hosts'.

properties<-data.frame(title="ThaiWasps")
properties

```
        title
1 ThaiWasps
```

To create our nodes dataframe we need a list of all the unique host and parasitoid names, a column specifying their relationship, in this case host or parasitoid, and optionally we have included a 'colour' column with hosts coded green, and parasitoids red.

nodes<-data.frame(node=unique(c(host,parasitoid)),type=c(rep("host",length(unique (host))),rep("parasitoid",length(unique(parasitoid)))),colour=c(rep("green", length(unique(host))), rep("red",length(unique(parasitoid)))))
head(nodes)

```
          node type  colour
1 Nymphalidae2 host   green
2    Erebidae1 host   green
3    Erebidae2 host   green
4     Noctuid1 host   green
5  Lycaenidae1 host   green
6  Hesperidae1 host   green
```

For the links file we need to specify all the unique links and add a column called 'weight' to specify the number of occurrences of each link type (zero is not permitted)

all.records<-as.data.frame(cbind(host,parasitoid)) # this is just our data as a dataframe
unique.links<-as.data.frame(unique(all.records)) # **unique** finds unique rows in a dataframe
head(unique.links)

```
         [,1]            [,2]
[1,] "Nymphalidae2"  "Bracon1"
[2,] "Erebidae1"     "Bracon1"
[3,] "Erebidae2"     "agathidine1"
[4,] "Noctuid1"      "agathidine1"
```

```
[5,] "Lycaenidae1"  "agathidine1"
[6,] "Hesperidae1"  "Cotesia1"
```
To obtain the frequencies of the different combinations of host and parasitoids we will first combine them using **paste** to make single character strings for each row of our raw data

hp_pairs<-paste(all.records$host,all.records$parasitoid)
hp_pairs
```
[1] "Nymphalidae2 Bracon1" "Erebidae1 Bracon1" "Erebidae1
    Bracon1" "Erebidae1 Bracon1" "Erebidae1 Bracon1"
    "Erebidae2 agathidine1" "Erebidae2 agathidine1"
[8] "Noctuid1 agathidine1" "Lycaenidae1 agathidine1"
    "Hesperidae1 Cotesia1" "Hesperidae1 Cotesia1" "Hesperidae1
    Cotesia1" "Hesperidae2 Cotesia1" "Noctuidae2 Cotesia2"
[15] "Geometridae1 Cotesia3" "Geometridae2 Cotesia3"
    "Erebidae1 Cotesia3" "Erebidae1 Cotesia4" "Erebidae1
    Aleiodes1" "Erebidae3 Aleiodes2" "Erebidae4 Macrocentrus1"
```
From this vector we can use **table** to obtain number of occurrences of each combination

weight<-as.vector(table(hp_pairs)) # using as.vector just gives us the numbers
weight
```
[1] 1 4 1 1 2 1 1 1 3 1 1 1 1
```
and we add these as an extra column ('weight') to the 'unique.links' dataframe and give it the required column names

unique.links$weight<-weight
colnames(unique.links)<-c("resource","consumer","weight")

Our three dataframes are now ready and we combine them into an object that **cheddar** will recognize, as follows:

thaiFW<-list(nodes, properties,unique.links)
names(thaiFW)<-c("nodes","properties","trophic.links")
attr(thaiFW,"class")<-c("Community","list")

For this illustration we want to plot three food webs side by side in a wide plot window, how to do this varies according to system. On a Mac machine we use the function **quartz** to set width and height in inches (e.g. **quartz-**(width=11,height=4)); on a Windows machine use the function **windows** (e.g. **windows**(11,4)). Depending on your machine you may have to adjust these dimension numbers.

par(mfrow=c(1,3))

For the plotting we will use two **cheddar** functions **PlotCircleWeb** and **PlotWebByLevel**. We only have two levels but cheddar can analyse and plot multiple trophic levels (see the vignettes):

1. a simple food web showing only the links.

PlotCircularWeb(thaiFW,show.nodes.as="labels",node.labels="node",
    col=NPS(thaiFW)$colour,main="Thai host-parasitoid food
    web",xpd=NA)

2. a web with the thickness of the links according to the number of occurrences of each link.

PlotCircularWeb(thaiFW,show.nodes.as="labels",node.labels="node",
    col=NPS(thaiFW)$colour,link.lwd=TLPS(thaiFW)$weight,main="Thai
    host-parasitoid food web",xpd=NA)

3. the same plot in triangular food web format.
PlotWebByLevel(thaiFW,pch=21,bg=NPS(thaiFW)$colour,link.lwd=TLPS
(thaiFW)$weight, main="Thai host-parasitoid food web",xpd=NA)

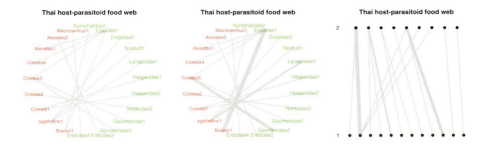

If we want to save the nodes, properties and trophic.links files to disc, we should save them using **write.csv** to the same directory as files called properties.csv, nodes.csv and trophic.links.csv. However, because the objects we have created are dataframes we need to remove the line numbers by including 'row.names=F' in the write.csv call. To read them from the directory use the function **LoadCommunity**, which searches the specified directory for files of those names. If they are in the current working directory use **LoadCommunity(getwd())**.

Tool Box 28.1 ✗

- Use attr() to specify attributes of an object.
- Use PlotCircularWeb() in **cheddar** to plot circular community links.
- Use PlotWebByLevel() in **cheddar** to plot community links as horizontal tiers.
- Use names() to specify the names of parts of a complex R object.

29 Adding Photographs

Summary of R Packages Introduced
  jpeg
  tiff

Summary of R Functions Introduced
  as.raster
  dev.cur
  names
  rasterImage
  read.jpeg in "jpeg"
  read.tiff in "tiff"

In this chapter we show how to insert photographic files into plots. The R base package does not have any tools for handling photographs, but recently Simon Urbanek has created the CRAN packages **jpeg** (which seems to work with most versions) and **tiff** (which allows importing and exporting some of the standard .tiff file types), and displaying them on the console. Inevitably because there are various graphics devices used by different R versions on different systems there will probably have to be some fiddling with code to get the perfect output. It may be easier on a MacOSX, and PCs do not support some graphics features of .tiff files. If you have difficulties you may find some helpful parameters in Appendix 1. In particular, the functions **dev** and **dev.size** and the 'mar=' and 'pin=' arguments in **par**, which allow you to manipulate the sizes of the plotting area and margins.

As a demonstration, we saved the image of an Oriental pied hornbill (credit Gossipguy 2012) from Wikimedia Commons (commons.wikimedia.org/wiki/File:Oriental_Pied_Hornbill.jpg, accessed 22 April 2020) and saved it as 'Oriental_Pied_Hornbill.jpg' in the online data files.

install.packages("jpeg")
library(jpeg)
and/or
install.packages("tiff")
library(tiff)
and then read the image in, e.g.
imgjpg<-readJPEG("Oriental_Pied_Hornbill.jpg",native=TRUE)

Adding Photographs

Here we will create a blank plot and then place copies like tiles over the area. So as not to distort the image we note its dimensions (22.56 wide by 12.7 high) and will create the plot so that it precisely accommodates 4 rows by 3, which involves changing the plotting area using par(pin=c({*width in inches i.e. default 6.004166*},{*height in inches i.e. 5.43 the default*})). You can find the current default of **par** arguments by typing their name as follows,

par("mar")
[1] 5.1 4.1 4.1 2.1

The R function **rasterImage** may not be available on all devices – you can check whether it is on your computer by using the **exists** function, i.e. **exists**("rasterImage"), which will return TRUE if it is available.

We want the height to width ratio of our plotting area to be 4*12.7 to 3*22.56 while maintaining the current width at the default value 6.001 inches. Therefore, we wish to set a new height of ((4*12.7)/(3*22.56)) *6.004, which we set in plot.

par(mfrow=c(1,1))
plot(c(0,3*22.56),c(0,4*12.7),type="n",xlab="",ylab="")

and now we can plot all 12 copies of the hornbill photo in an array

for(i in 1:3){
    for(j in 1:4){
        rasterImage(imgjpg,(i-1)*22.56,(j-1)*12.7,i*22.56,j*12.7)
}} # end *i* and *j* loops

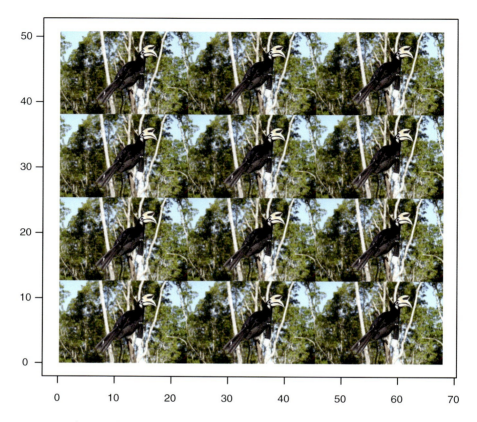

Obviously, you do not have to include the axes, you can use the functions **points** and **lines** to overlay data on such a plot, and 'par(new=TRUE)' (see Chapter 6) to superimpose another graph on top of your background.

Exercise Box 29.1

1) Take any jpg and create a coloured margin frame around it.
2) Tile your picture 5 × 4 to give 20 replicates.
3) Alter the margins so the tiles fit nicely.

There is no answer to this, you have to do it all by yourself.

Tool Box 29.1

- Use dev.cur() to get the name of the plotting device (platform dependent).
- Use pin= in **par** to set plot dimensions within par() function in inches.
- Use polygon() to draw polygons made out of straight line sections.
- Use rasterImage() to create tiling pattern of images in plot screen.
- Use readJPEG() in package **jpeg** to import .jpeg files.
- Use readTIFF() in package **tiff** to import .tiff files.

30 Standard Distributions in R

> *Summary of R Functions Introduced*
> d..., p..., r... and q... to go with norm, binom, nbinom, chisq, gamma, exp, lnorm, pois, t, unif, weibull
> curve

There are a number of in-built probability distributions, including uniform, binomial, negative binomial, normal, log-normal, logistic, exponential, Chi-squared, Poisson, gamma, Fisher's F, Student's t, Weibull and others. These are used to generate *p*-values from test statistics, to generate random values from a distribution or to generate expected distributions (Table 30.1).

The Normal Distribution

To obtain 100 numbers drawn randomly from a normal distribution with mean 20 and standard deviation 4 we use **rnorm**.
```
head(rnorm(100,20,4),20)
 [1]  15.860557  22.433278  17.663737  22.433839  20.511562
      18.590710  15.221031  16.901198  28.289713  21.385006
[11]  23.710613  15.656098  20.126080  20.546819  23.035948
      11.264748   9.739163  24.369990  25.105256  17.542089
```
Which we can visualize as a histogram
```
hist(rnorm(100,20,4))
```
or to get a smoother outline, take a far larger sample size.
```
hist(rnorm(1000000,20,4))
```

Table 30.1. R functions associated with selected in-built distributions and examples.

Distribution	Probability density function	Cumulative probability	Get random numbers from distribution	Quantiles
normal	**dnorm**	**pnorm**	**rnorm**	**qnorm**
binomial	**dbinom**	**pbinom**	**rbinom**	**qbinom**
negative binomial	**dnbinom**	**pnbinom**	**rnbinom**	**qnbinom**
Chi squared	**dchisq**	**pchisq**	**rchisq**	**qchisq**
gamma	**dgamma**	**pgamma**	**rgamma**	**qgamma**
exponential	**dexp**	**pexp**	**rexp**	**qexp**
lognormal	**dlnorm**	**plnorm**	**rlnorm**	**qlnorm**
Poisson	**dpois**	**ppois**	**rpois**	**qpois**
Student's t	**dt**	**pt**	**rt**	**qt**
uniform	**dunif**	**punif**	**runif**	**qunif**
Weibull	**dweibull**	**pweibull**	**rweibull**	**qweibull**

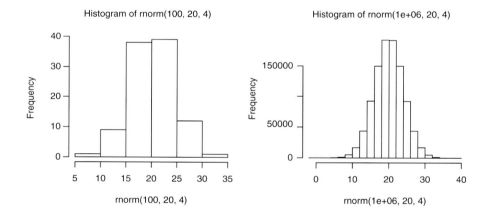

To plot a smooth curve to overlay you create a vector of x values (xv) and get y values (yv) for each x value using the function **dnorm**

```
hist(rnorm(1000000,20,4))
xv<-seq(-5,45,0.1)  # from −5 to 45 in steps of 0.1
yv<-dnorm(xv,mean=20,sd=4)*2000000
lines(xv,yv,col="red",lwd=2)
```

Standard Distributions in R

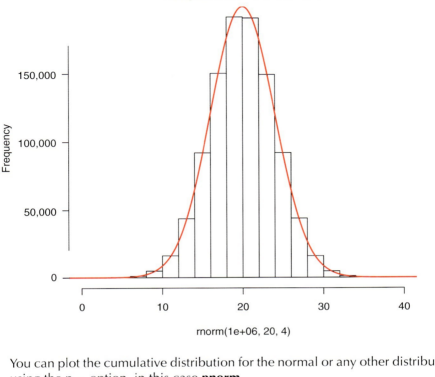

You can plot the cumulative distribution for the normal or any other distribution using the p... option, in this case **pnorm**

```
xv<-seq(-20,20,0.1) # from -20 to 20 in steps of 0.1
yv<-pnorm(xv,mean=0,sd=5)
plot(xv,yv,col="red",lwd=2,type="l",main="Cumulative values of the normal
    distribution",xlab="X",ylab="Cumulative P") # type= "l" tells R to plot lines
mtext("mean =1; S.D. =5")
```

Ninety-five per cent of normally distributed observations lie within 1.96 standard deviations of the mean.

```
lower95<-pnorm(-1.96*5,mean=0,sd=5)
upper95<-pnorm(1.96*5,mean=0,sd=5)
lines(c(-1.96*5,-1.96*5),c(0,lower95),col="blue",lwd=3,lty=2,lend=2)
lines(c(1.96*5,1.96*5),c(0,upper95),col="blue",lwd=3,lty=2,lend=2)
xrange<-1.96*5 - -1.96*5 # 19.6, same as xrange<-1.96*5--1.96*5
x_increment<-xrange/1000
for(i in 7:993){ # not from 1 to 1000 because we don't want to obliterate the
    blue dashed lines
    x<--1.96*5+(i* x_increment)
    lines(c(x,x),c(0,pnorm(x,mean=0,sd=5)),col="grey80",lwd=0.5)
    } # end i loop
rect(-1.5,0.152,7.5,0.275,col="white",bty="n")
text(3,0.25,"95% of",cex=0.8)
text(3,0.21,"observations",cex=0.8)
```

```
text(3,0.17,"are in grey zone",cex=0.8)
lines(xv,yv,col="red",lwd=2) # redrawing the red line that has been slightly
    covered by grey
```

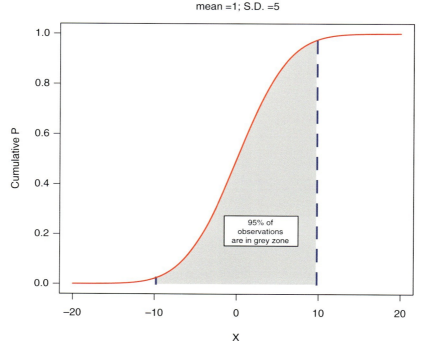

Exercise Box 30.1

1) Compare the normal and binomial distributions for a range of sample sizes such as 10, 100, 1000 and 10,000. Hints: (i) create a binomial data vector first using $p = 0.5$ and calculate its mean and standard deviation to pass to **rnorm**; (ii) use 'add=TRUE' in the **hist** function to allow superimposing histograms on top of one another; and (iii) use 'rgbt=' colours with transparency to show overlap.

Student's t Distribution

The t distribution is closely related to the normal distribution and represents how accurately we know the mean when we have a small sample, which is typically taken to mean that n is <30. Specifically when sample sizes are small, the estimated standard deviation is likely to be biased and usually an underestimate of the population standard deviation.

In exactly the same way as above we can plot a histogram of a random sample of values of t (green; in this example for 10 degrees of freedom), the cumulative distribution curve (blue) and indicate the areas of the distribution representing values of t that occur in 5% or fewer of cases (2.5% at both the negative and positive extremes). Appropriate code is:

```
sample_size<-10000
degf<-10
x<-rt(sample_size,degf)
hist(x,col="grey30",ylab="Probability and cumulative probability",xlab="t",
    prob=TRUE,ylim=c(0,1),main="Histogram of t with 10 D.F.")
curve(dt(x df=degf),lwd=2,col="green",add=TRUE)
curve(pt(x,df=degf),lwd=2,col="blue",add=TRUE)
lines(c(qt(0.975,degf),qt(0.975,degf)),c(0,0.975),col="red",lty=2,lwd=2)
lines(c(-100,qt(0.975,degf)),c(0.975,0.975),col="red",lty=2,lwd=2)
text(2.65,0.5,paste("P=0.05,critical (2-tailed) t = ",round(qt(0.975, degf),4),
    sep=""),col="red",srt=270,cex=0.8)
lines(c(qt(0.025,degf),qt(0.025,degf)),c(0,0.025),col="red",lty=2,lwd=2)
lines(c(-100,qt(0.025,degf)),c(0.025,0.025),col="red",lty=2,lwd=2)
```

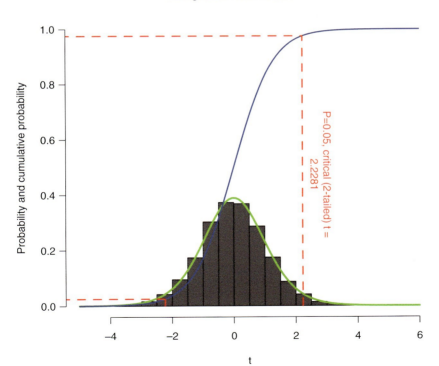

Critical values of t for various numbers of degrees of freedom are given in Table 30.2 for one- and two-tailed tests. Normally people are simply asking the question whether two samples or values differ significantly (e.g. $p < 0.05$ level) and are not predicting beforehand which will be larger – in this case we are applying a two-tailed test so the value of t has to be in the 2.5% at either end of the distribution for the result to be deemed significant. Occasionally, very occasionally, we might be asking specifically whether one value is significantly larger than the other based on a specific hypothesis – in that case we are interested in whether the calculated t value is in the 5% at a specific end of the distribution. It is often difficult to publish the results of a one-tailed test unless you have a really strong prediction.

Table 30.2. Critical values of t for one- and two-tailed tests.

Degrees of freedom	One-tailed		Two-tailed	
	$p = 0.05$	$p = 0.01$	$p = 0.05$	$p = 0.01$
1	6.3138	31.8205	12.706	63.6567
2	2.92	6.9646	4.3027	9.9248
3	2.3534	4.5407	3.1824	5.8409
4	2.1318	3.7469	2.7764	4.6041
5	2.0150	3.3649	2.5706	4.0321
6	1.9432	3.1427	2.4469	3.7074
7	1.8946	2.9980	2.3646	3.4995
8	1.8595	2.8965	2.3060	3.3554
9	1.8331	2.8214	2.2622	3.2498
10	1.8125	2.7638	2.2281	3.1693

To obtain these values and others you need only enter
round(qt(0.95,1:10),4)
round(qt(0.99,1:10),4)
round(qt(0.975,1:10),4)
round(qt(0.995,1:10),4)

Lognormal Distribution

The lognormal distribution is often considered a good approximation to the relative abundances of species. It is a right skewed distribution with possible x values ranging from zero to plus infinity. The x values are the logarithms of the x values of a normal distribution which extends from minus to plus infinity. To demonstrate this, we will put all the arguments including x and y calculations in a single plot line

```
plot(seq(0,10,0.01),dlnorm(seq(0,10,0.01)),type="l",lwd=2,xlab="X",ylab=
    "Lognormal (x)")
```

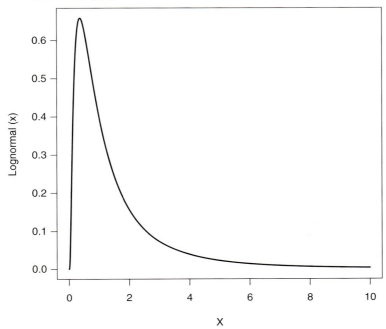

Logistic Distribution

There are several families of the 'S'-shaped distributions depending on how many parameters are specified. A simple two-parameter version is:

$$y = \frac{e^{ax+b}}{\left(1+e^{ax+b}\right)}$$

```
xv<-seq(-10,10,0.1)
a<-1
b<--0.5 # changing beta to negative makes the function start at 1 and
    decrease to zero
P<-a/(a+ exp(a+b*xv))
plot(xv,P,ylab="Probability",xlab="X",type="l",lwd=2)
```

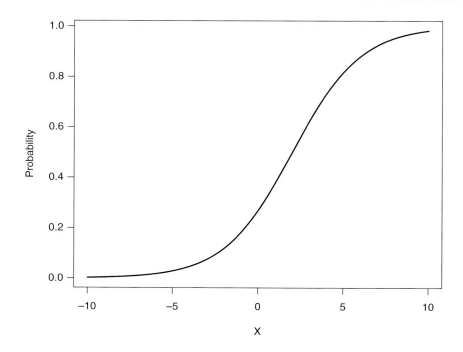

Poisson Distribution

The Poisson distribution is appropriate when the mean is equal to the variance. This applies particularly to count data when the mean is small. It assumes the logarithm of the expected values (mean) can be modelled into a linear form by some unknown parameters. When the mean of count data is low, there are likely to be many zero values, a few 1s, fewer 2s, etc., and the distribution left-skewed. If we only have 0s and 1s in our data, the mean will be between 0 and 1 and so will the variance, asymptotically reaching 0.25; if our counts range from 0 to say 2, then the mean will be between 0 and 2 and the variance will asymptotically reach 1.0, and as the mean increases, the asymptotic maximum variance increases ever more. This is an important distribution in statistics because it describes the frequency distribution of rare events (where there will be lots of zeros) and the error distribution of count data where the mean is low.

We will look at the effect of varying the mean from 1 to 10 on the distribution. To avoid overlapping lines, x values are incremented by a small amount (inc) for each value of the mean

```
inc<-0.085
p<-c("red","blue","green","purple","brown","pink","seagreen2","violet",
     "orange", "turquoise") # a vector of colours for each mean value
plot(NULL,xlim=c(0,20),ylim=c(0,0.4),xlab="X",ylab="Poisson (x)",main="Poisson
     distributions with means 1:10")
```

Standard Distributions in R

```
for(i in 1:10){
        for(j in 0:20){
        lines(c(j+(i-1)*inc,j+(i-1)*inc),c(0,dpois(j,i)),lwd=3,col=p[i])
        }}
par(new=TRUE)
for(i in 1:10) lines((0:20)+(i-1)*inc,dpois(0:20,i),col=p[i])
legendLabels=c(paste("mean =",1:10)) # we want a space after = so collapse
        not used
legend("topright",inset=0.05,legend=legendLabels,col=p,lty=1,cex=0.8)
```

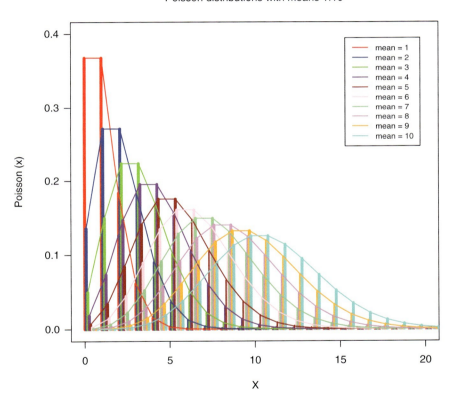

Poisson distributions with means 1:10

Gamma Distribution

This distribution is specified by two parameters (alpha and beta) and often closely resembles strongly skewed data with a long tail of values to the right – either starting at infinity on the left or zero on the left depending on alpha (<1 or >1). The **dgamma** function takes the form 'dgamma(x, alpha, beta)'. Because of the infinity issue, we cannot plot the value when x is zero, so we start our demonstration x vector at a small positive number.

```
par(mfrow=c(2,2))
xv<-seq(0.001,5,0.01)
plot(xv,dgamma(xv,0.5,0.1),type="l") # type= "l" tells R to plot lines not points
plot(xv,dgamma(xv,0.5,5),type="l")
plot(xv,dgamma(xv,1.5,0.5),type="l")
plot(xv,dgamma(xv,1.5,5),type="l")
```

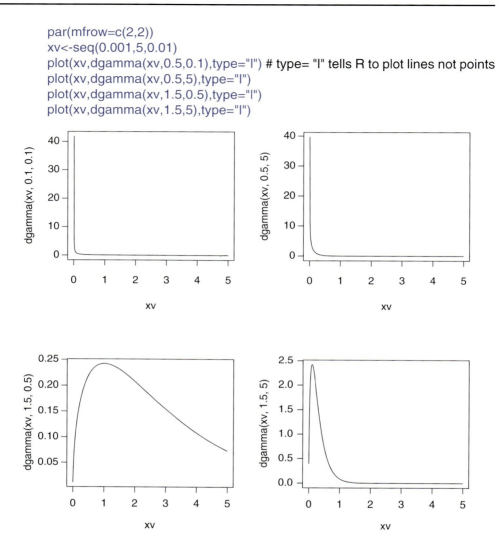

The Chi-squared Distribution

The Chi-squared distribution is a special case of the gamma distribution and is the sum of the squares of k randomly picked values of a standard normal distribution where k is the number of degrees of freedom. A standard normal distribution has a mean of zero and a standard deviation of one. In sampling from a Chi-squared distribution we therefore need to specify the number of degrees of freedom, which we put as the second argument in the function **rchisq**; here d.f. = 5

```
par(mfrow=c(1,2))
sample_size<-10000
x<-rchisq(sample_size,5)
hist(x,col="grey30",ylab="Frequency")
```

We used a large sample size to obtain a rather smooth and accurate distribution. The y axis here is not very helpful though (see left-hand figure), as it depicts the number of random Chi-squared values sampled. More helpful would be to see relative probabilities, which we could do by rescaling the y axis, but the function **hist** has an easier built-in option; to get probabilities we add the argument 'prob=TRUE'.
hist(x,prob=TRUE,col="grey30",ylab="Probability")

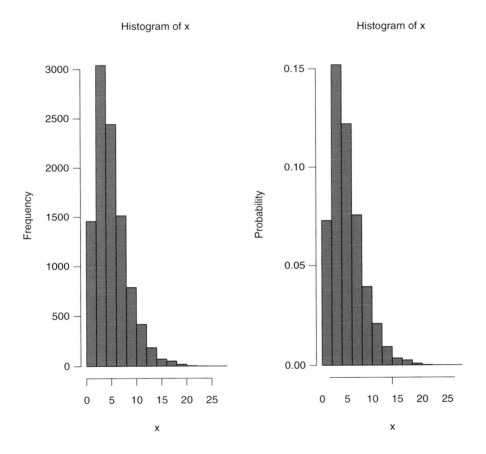

To examine how the Chi-squared distribution changes with different numbers of degrees of freedom we can first plot a histogram as above, and we introduce a new R function to plot curves on top of the histogram; the function **curve** is an independent plotting function and acts like **plot** if presented with an appropriate data vector. It is simpler than creating a vector of x values and plotting lines. However, like **plot** it wipes out previous plots unless you add the argument 'add=TRUE' in which case it plots the curves on top of an existing plot.

```
par(mfrow=c(1,1))
hist(x,prob=T,col="grey30",ylab="Probability")
curve(dchisq(x,df=2),lwd=2,col="green",add=TRUE)
curve(dchisq(x,df=5),lwd=2,col="blue",add=TRUE)
curve(dchisq(x,df=10),lwd=2,col="red",add=TRUE)
leg_text<-c("d.f. = 2","d.f. = 5","d.f.=10")
legend("topright",lwd=2,col=c("green","blue","red"),leg_text)
```

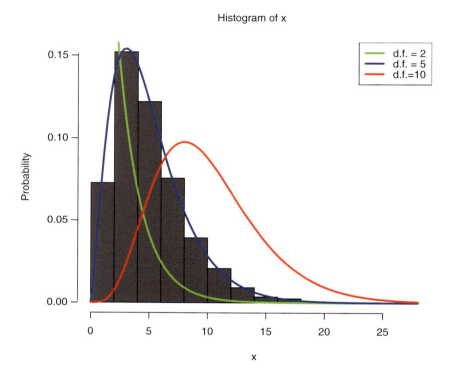

As should be intuitive since Chi-squared is the sum of the squares of samples from a standard normal distribution, as the number of samples added (i.e. the degrees of freedom) increases, so too does the peak value of this sum. This is why, for a Chi-square value to be significant at any given *p*-value, e.g. 0.05, it will have to be larger, the more degrees of freedom there are. Critical values, which Chi-squared must be above if it is to be a significant result at the 0.05 and 0.01 levels, can be found in Table 30.3.

Standard Distributions in R

Table 30.3. Critical values of Chi-squared for two values of significance.

Degrees of freedom	Probability = 0.05	Probability = 0.01
1	3.84	6.64
2	5.99	9.21
3	7.82	11.34
4	9.49	13.28
5	11.07	15.09
6	12.59	16.81
7	14.07	18.48
8	15.51	20.09
9	16.92	21.67
10	18.31	23.21

Going back to the histogram of Chi-squared values for degrees of freedom = 5, this means that 95% of results will give Chi-squared values to the left of $x = 11.07$. You can use similar code to plot all the other in-built statistical distributions.

Tool Box 30.1 ✖

- Use add=T to allow plotting over the existing histogram.
- Use curve() to plot a curve using histogram values.
- Use prob=T in **hist** to plot probabilities rather than numbers.

31 Reading and Writing Data to and from Files

Summary of R Packages Introduced
  openxlsx
  pdftools
  readxl
  tibble

Summary of R Functions Introduced
  as_tibble in tibble
  openxlsx
  read
  read.delim
  scan
  write.csv
  write.table

It is often most convenient to put your data and code files for a given project in a specific project folder and then set R to always look there to read or write data. See Appendix 2 for details of how to set or change the working directory. After selecting the location, all subsequent read and write commands will automatically go there until you start a new session. Alternatively, you can specify a path to the relevant directory in the read or write statement, e.g.

txt<-readLines("/Users/donald/Desktop/books/testgrab.txt")

which just reads in plain text, or

newdata<-read.table(file="/Users/donald/Desktop/DougChestersCo1/Data_ Release_1/ ChestersDataSumR.txt",header=T) # this will obviously only work on DQ's computer

which reads data saved in a table format. In this case the first row of the table has the names of the column headers so we also pass 'header=TRUE' or 'header =TRUE' to the **read.table** command so that the first line of the input is treated differently. It is vital to remove spaces between the elements of column names; if you leave spaces in them, then **read.table** or **read.csv** will error, because they will treat each element as a possible column header and therefore expect more columns of actual data than there are.

All writing functions require you to specify the object that you want to save as well as the name of the file that you want to save it to, either as a file path, e.g.

write(file="/Users/donald/Desktop/DougChestersCo1/newdata.txt")

or, if you have set a working directory then just

write(x,file="newdata.txt") # if you just want the contents of 'x' saved – suitable for text

The basic file-writing function **write** is suitable for saving single vectors of text, however, as noted in Chapter 20, it saves by default, numbers in groups of five, e.g.

write(1:27,file="test.write.txt")
readLines("test.write.txt")
[1] "1 2 3 4 5"   "6 7 8 9 10"   "11 12 13 14 15" "16 17 18 19 20" "21 22 23 24 25" "26 27"

As mentioned previously, the function **scan** usefully overcomes this annoyance.

scan("test.write.txt",quiet=TRUE) # 'quiet' suppresses reporting how many items were read
[1] 1 2 3 4 5 6 7 8 9 10 11 12 13 14 15 16 17 18 19 20 21 22 23 24 25 26 27

The function **write** is really unsuited to saving arrays, dataframes or tabular data because it transposes rows and columns, loses column names and needs you to specify the number of columns to be saved. Therefore, to save a dataframe or table we should use **write.table** or **write.csv**. When these files are subsequently read using **read.table** or **read.csv**, R will recognize that the contents are a dataframe. However, the default of **write.csv** is 'row.names=TRUE', so you need to change that to 'row.names=FALSE' if the first column is data.

When reading a table it is important to know whether the first line of the saved file is column names or not and to specify this using the header argument as appropriate. The available read functions are:

readLines for plain text – note the lower case r and capital L (header is not applicable)

read.table for tab-delimited tables (header is FALSE by default)

read.csv for comma-delimited tables (header is TRUE by default)

read.delim when variables are separated by some user-specified character such as space (spaces), tab, comma, colon, semicolon or almost any single byte character (header=TRUE by default)

scan for reading a given type of input including 'logical', 'integer', 'numeric', 'character' and 'list', specified by the argument 'what='; elements are separated by a space; entry from terminal stops when you enter a blank line; useful when number of items in each row are not constant or for inputting data direct from terminal (header is not applicable).

Appending Data to an Existing File

By default, R writes a file of a given name over any pre-existing one of that name in that location. Sometimes you may be progressively calculating output and want to save each time around, and to do that you would set 'append=TRUE', e.g.

write(file="newdata.txt",append=TRUE)

Using *read.delim* with Non-tab Separator

read.delim is basically like **read.table**. The default separator is the tab character, \t. By default it assumes header=TRUE. However, we can use it to read data where elements are consistently separated by another (single byte) character. The data file 'Panamanian_cycads_colon.txt' is a single line list of the *Zamia* species occurring in Panama, each species name being separated by a semicolon. It looks like 'Zamia cunaria;Zamia dressleri;Zamia elegantissima; ....'. Using **read.delim** we can specify this separator and input the list of species
zamias<-read.delim("Panamanian_cycads_colon.txt",sep=";",header=F)
zamias
```
                V1               V2              V3
V4          V5             V6              V7         V8
V9     V10
1 Zamia cunaria Zamia dressleri Zamia elegantissima Zamia
fairchildiana Zamia hamannii Zamia imperialis Zamia
ipetiensis Zamia lindleyi Zamia manicata Zamia nana
            V11            V12        V13
V14            V15            V16            V17
1 Zamia nesophila Zamia neurophyllidia Zamia obliqua
Zamia pseudomonticola Zamia pseudoparasitica Zamia
skinneri Zamia stevensonii
```
The result is a list with elements given default names V1 ... V17
zamias[6]
```
              V6
1 Zamia imperialis
```
To convert to a standard set of plant names we need to **unlist** the list and remove the names using **unname**.
unlist(unname(zamias))
```
[1] Zamia cunaria Zamia dressleri Zamia elegantissima
    Zamia fairchildiana Zamia hamannii Zamia imperialis
    Zamia ipetiensis
[8] Zamia lindleyi Zamia manicata Zamia nana Zamia
    nesophila Zamia neurophyllidia Zamia obliqua Zamia
    pseudomonticola
[15] Zamia pseudoparasitica Zamia skinneri Zamia
    stevensonii
17 Levels: Zamia cunaria Zamia dressleri Zamia
    elegantissima Zamia fairchildiana Zamia hamannii
    Zamia imperialis Zamia ipetiensis Zamia lindleyi
    Zamia manicata ... Zamia stevensonii
```

Choosing a File to Read Interactively

The function **file.choose**() can be used with any read function. It opens the working directory as a window showing the file names or icons, then you just browse and click on the one you want, e.g. **read.csv**(**file.choose**(),header=TRUE).

Using Excel for Data Entry

Microsoft Excel® is almost universally used for entering data. Here is an example of what some data entry might look like

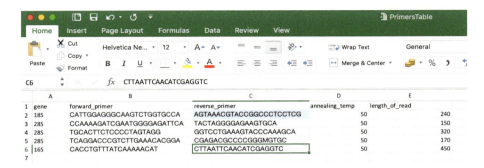

Note that the multi-word column names were entered as single entities by using an underscore to separate words; you can also use a full stop or just remove spaces. If you leave spaces in column names, then 'read.table' or 'read.csv' will error, because they will be expecting more columns of data than there really are.

In its normal saving mode, i.e. as an Excel spreadsheet (.xlsx), Excel places a huge amount of information about dates, fonts, cell sizes, borders, etc., in the file which R cannot usefully handle and you will get some sort of error message.

test<-readLines("PrimersTable.xlsx")

There were 25 warnings (use warnings() to see them

and the top few rows of it look like:

head(test,3) # this shows the top three lines in the file test

```
[1] "PK\003\004\024"
[2] "?\x889L\xec3\025\xc8sbg⌣|\xc81!\xf5\xf9\032USh9i\
    xb0b\x9er:\"y_dl\xc0\xf3D\x9b\xbf\023\xfd|-N\x9c\
    xc8R\"4\022\xf82\xcfG\xc7%\xa0\xf5\177Z\xb44\xf1˝y\
    xc47\të\xc8\xf0\u0242\x8b\037\xa8\xde\001"
```

Clearly this looks nothing like the data that we entered into Excel. One option is to save data from Excel either as tab-delimited text which has the default file name suffix '.txt', or as comma separated text with the file ending '.csv'. Both give the same.

test<-read.table("PrimersTable.txt",header=TRUE)
test

```
    gene           forward_primer              reverse_primer
annealing_temp length_of_read
1   18S  CATTGGAGGGCAAGTCTGGTGCCA AGTAAACGTACCGGCCCTCCTCG
            50              240
2   28S  CCAAAAGATCGAATGGGGAGATTCA      TACTAGGGGAGAAGTGCA
            50              150
3   28S     TGCACTTCTCCCCTAGTAGG  GGTCCTGAAAGTACCCAAAGCA
            50              320
```

```
4    28S    TCAGGACCCGTCTTGAAACACGGA     CGAGACGCCCCGGGMGTGC
              50                 170
5    16S    CACCTGTTTATCAAAAACAT         CTTAATTCAACATCGAGGTC
              50                 450
```

However, to read and write R tables, matrices and dataframes to and from Excel there is the **openxlsx** package

install.packages("openxlsx")

```
also installing the dependency 'zip'
```

library(openxlsx)

Here we will read open-source data from Pitt *et al.* (2013) on various biological parameters of pelagic organisms. The XLSX file is supplied in our online Appendix or can be downloaded from https://doi.org/10.1371/journal.pone.0072683.s003.

pelagic1<-read.xlsx("https://doi.org/10.1371/journal.pone.0072683.s003")

The default for **read.xlsx** is to read only the first sheet, however the Pitt *et al.* file has six sheets, so to obtain other sheets we need to specify them with the argument 'sheet=', e.g.

pelagic_respiration<-read.xlsx("https://doi.org/10.1371/journal.
    pone.0072683.s003",sheet=2) # note drop the .XLSX

The *readxl* Function and Tibbles

Another package, **read_xl**, also allows XLSX files to be read, but instead of creating a dataframe it creates an object called a tibble. Tibbles are a type of data frame with some differences, for example, they do not change variable names and types but do report more errors which can be an advantage. The package **tibble** contains a function **as_tibble** (with **as.tibble** as an alias) that will convert ordinary dataframes to tibbles.

install.packages("readxl")
library(readxl)
tun<-read_excel("Tunjai_regeneration.xlsx")
head(tun)

```
# A tibble: 6 x 9
  Species      Regeneration_guild   Size_categorical Shape
Coat    Coat_mm   `%MC`  establishment `establishment%`
   <chr>        <chr>                <chr>            <chr>
<chr>   <chr>     <chr>  <chr>         <chr>
1 Archidend… Pioneer_exclusion     I              R
Tn      7.000000… M      ME            11.25
2 Artocarpu… Late-successional_…   I              R
M       0.13      M      ME            36.25
3 Callerya_… Late-successional_…   L              R
Tk      1.34      M      HE            41.57
4 Cinnamomu… Late-successional_…   I              R
Tn      0.08      L      ME            21.25
```

```
5 Diospyros… Late-successional_… I            O
M       0.31        L      HE       43.13
6 Diospyros… Late-successional_… I            O
M       0.289999… L       ME       17.5
```

Whether tibbles will catch on we are not sure. Each column has length 1,
```
tun[1]
# A tibble: 19 x 1
   Species
   <chr>
 1 Archidendron_clyperia
 2 Artocarpus_dadah
 3 Callerya_atropurpurea
 4 Cinnamomum_iners
 5 Diospyros_oblonga
 6 Diospyros_pilosanthera
 7 Garcinia_cowa
 8 Garcinia_hombroniana
 9 Garcinia_merguensis
10 Lepisanthes_rubiginosa
11 Litsea_grandis
12 Microcos_paniculata
13 Morinda_elliptica
14 Pajanelia_longifolia
15 Palaquium_obovatum
16 Peltophorum_pterocarpum
17 Scolopia_spinosa
18 Sandoricum_koetjape
19 Vitex_pinnata
```

so to access cell values you need to use the argument 'range=' or call columns by name
```
tun$Species
 [1] "Archidendron_clyperia"  "Artocarpus_dadah"
     "Callerya_atropurpurea"  "Cinnamomum_iners"
 [5] "Diospyros_oblonga"      "Diospyros_pilosanthera"
     "Garcinia_cowa"          "Garcinia_hombroniana"
 [9] "Garcinia_merguensis"    "Lepisanthes_rubiginosa"
     "Litsea_grandis"         "Microcos_paniculata"
[13] "Morinda_elliptica"      "Pajanelia_longifolia"
     "Palaquium_obovatum"     "Peltophorum_pterocarpum"
[17] "Scolopia_spinosa"       "Sandoricum_koetjape"
     "Vitex_pinnata"
```

To view all rows of a tibble use **print**({tibble name},{length of the tibble}).

Reading PDF Files for Data Mining

The package **pdftools** allows you to import text elements of a PDF file. We will demonstrate with the Thai National Parks list of wildlife at Doi Inthanon National Park, Thailand – the PDF of this is in the online Appendix (www.cabi.org/openresources/45349)

install.packages("pdftools")
library(pdftools)
pdf_file<-("Wildlife-at-Doi-Inthanon-National-Park.pdf")
pdftext<-pdf_text(pdf_file)

The first few lines of 'pdftext' look like

[1] "18/05/2019
    Wildlife at Doi Inthanon National Park\n   Doi Inthanon
    National Park (/doiinthanonnationalpark)\n   Wildlife\n
    Have you been to Doi Inthanon National Park and seen
    species not listed here? Please let us know
    (/contactus) in case, it will be much appreciated!\n
    Birds (Class: Aves)\n   Total 526 species\n
    Non-passerines\n   Grebes (Order:
    Podicipediformes, Family: Podicipedidae)\n
    Tachybaptus ruficollis (/species/littlegrebe),
    Little grebe\n   Cormorants (Order:
    Pelecaniformes, Family: Phalacrocoracidae)\n
    Microcarbo niger (/species/littlecormorant), Little
    cormorant\n   Bitterns, herons and
    egrets (Order: Ciconiiformes, Family: Ardeidae)\n
    Ardea alba (/species/easterngreategret), Great egret

This is where a lot of the methods presented in Chapters 21, 22 and 24 in particular will come in handy.

Writing Graphics Directly to Disc

Normally after having created a graphics output, the window or frame can be selected and then saved. However, you can choose to define an output file that will be saved directly to the working directory without appearing in a graphics window using the functions **png**, **jpeg**, **tiff**, **bmp** or **pdf**. As an example we will make a default plot of the function **sin**. To save this as a .png file we first create a call to the function **png** specifying the dimensions (in pixels is the default), then we make the plot and then we have to turn the device off if we want the next output to appear on the screen, e.g.

png("My sine graph",bg="wheat",width=600,height=600)
plot(seq(1,10,by=0.01),sin(seq(1,10,by=0.01)),pch=10,col="darkgreen",
    type="l",lwd=3)
dev.off()

The .png file is now saved to R's working directory on the computer.

Tool Box 31.1

- Use append= to save progressive output to a file rather than overwriting existing file content.
- Use col.names= to specify whether to write columns to file when writing a table, or to specify whether a file being read has column names.
- Use file= to specify the name of a file to be written to.
- Use ncol= to set the number of columns.
- Use quiet=TRUE to suppress reporting how many items were read.
- Use read.csv() to tell R where to read a comma delineated table.
- Use read.delim() to tell R where to read a file with elements separated by a specified character.
- Use read.table() to tell R where to find a table.
- Use readLines() to tell R where to find text.
- Use row.names= as for col.names but for row names.
- Use t() to transpose a table.
- Use the package openxlsx to use the read.xlsx() to read xlsx files as dataframes.
- Use the package readxl to use the read excel() to read Excel files as tibbles.
- Use write() to tell R where to save text.
- Use write.table() to tell R where to save a table.
- Use bmp(), png(), jpeg(), tiff(), pdf() to have graphics saved directly to disc.
- Use dev.off() to make subsequent graphics as a window or frame.

Appendix 1: Summary of Graphical Parameters

The following are some explanations of parameters that can be passed to the functions **plot** or **par**, the latter dealing with global graphic settings until reset. There are numerous more that are not listed below, but they are rather seldom used – see Crawley (2012; pp. 848–850) for more details. Many of the arguments used by **plot** can also be applied to **barplot**, **boxplot**, **legend** and some to **lines**, **text**, **points**, **hist**, etc.

Arguments Passed Directly to *par* Function

par(bty= …) draw a box around the plot is the default, if bty="n" no box is drawn
par(fin=c({*width of plot window,height of plot window*}) in inches.
par(mfrow=c({*rows of plots*},{*columns of plots*}) sets the arrangement of multiple plots within the window, each new plot filling any remaining gaps or cycling over to start a new plot window:
    par(mfrow=c(1,1)) is the default
    par(mfrow=c(1,2)) gives two plots side-by-side (one row) in the plot window
    par(mfrow=c(2,1)) gives two plots one above the other (two rows)
    par(mfrow=c(2,3)) gives a block of six plots (two rows of three).
par(mar=c({*bottom,left,top,right*})) sets the width of the four margins around the plot in the plot window in lines. The default values are 5.1, 4.1. 4.1 and 2.1 respectively.
par(mai=c({*bottom,left,top,right*})) same as **par**(mar…) except the margin widths are in inches (how archaic!).
par(new=TRUE) plot another plot on top of the existing one without erasing.
par(pin=c({*width of plot area within the plot window,height ditto*})) in inches.
par(plt=c({*width of plot area within the plot window,height ditto*})) as a fraction of the plot window dimensions.
To see what the current settings of all the **par** parameters are, use par(). To see the value of a particular parameter enter par("{*name of setting*}"), e.g. **par**(fin=…).

Arguments Applied Directly to the *plot* Function as well as in Some Others

adj= text justification also used in **text**. 0 = left, 0.5 = centred, 1 = right.
bg= the background colour of the plot area or legend or fill of characters 20 to 25.

cex= sets the sizes of points and text characters with the default being cex=1 for most things. cex is subdived into cex.axis, cex.lab, cex.main and cex.sub for these specific items.
col= sets the colour of the plotted symbols. col is subdivided into font.axis, font.lab, font.main and font.sub for the colours of the axes, labels, main heading and subheading respectively.
family= allows choice of serif, sans-serif fonts: see the vignette https://cran.r-project.org/web/packages/svglite/vignettes/fonts.html
fg= the colours of axes and boxes, default black.
font= sets the font for text (1=plain, 2=bold, 3= italics, 4=bold italics); font is subdived into font.axis, font.lab, font.main and font.sub.
lac= the orientation of axis numbers.
new={TRUE or FALSE} true means that a plot command will cause the new plot to appear on top of the previous one if the plot window is still open.
pch= there are 26 defined geometric symbols, and you can also use pch= various other keyboard symbols such as '.'.
srt= sets the rotation of character strings in degrees (0 = horizontal); particularly in text commands for manually labelling long names below barplot columns.
sub= an optional subtitle to plot.
type= "l" this tells a plot to draw lines between the coordinates instead of the default points.
xlab= and **ylab=** labels for the *x* and *y* axes respectively.
xlim= and **ylim=** the upper and lower bounds of the *x* and *y* axes respectively given as c({*lower*},{*upper*}).
xpd={TRUE or FALSE} true allows things to be drawn or plotted outside of the margins of the plot area. Useful for annotating things for publication.

Arguments for the *lines* Function

lty= line type. 'lty=1' is a normal solid line, values 2 to 6 give different types of dashing.
lend= the shape of the ends of a line – the default is rounded (0 or 'round'). Use 'lend=2' for square.
lwd= line width.

Having Multiple Graphics Windows Open at the Same Time

Whenever you call a graphics function such as using **plot**, **hist**, **barplot**, etc., your computer will automatically open its appropriate graphics device. When you create a new plot with a subsequent plot command the first plot is erased. This is normally just fine because you are probably tweaking the parameters and exploring your data. You can however open more than one graphics window at the same time using **dev.new**().

Appendix 1

Macintosh-specific Graphics

The graphics device on MacOS is called Quartz and there is a function **quartz** which can be used to open a new quartz graphics window, and allows you to set some basic parameters for it such as width and height (in inches). It is not normally that necessary because the default parameters serve for most things. However, there can be occasions when you want to have multiple windows open at the same time either for comparison or for experimentation. Every time you enter the **dev.new**() command a new window will open in addition to previously open ones. Each window is numbered sequentially and you can select which one you want your next graphics output to appear in using dev.set(). For example if you have four graphics windows open (numbered 1 to 4) and you want to plot add something on to the output in window number 2, then enter **dev.set**(2) before your plotting command.

Using the *layout* Function

layout is a much more versatile alternative to 'par(mfrow=...)' but is a bit more complicated. It allows you to define a plot space with any arrangement of numbers of figures in rows, columns, etc.
layout(matrix(c(1,2,3,4),2,2)) # is equivalent to par(mfrow=c(2,2)) with the plot windows numbered 1 to 4 by row
layout(matrix(c(1,1,2,3),2,2,byrow=TRUE)) # gives a single wide plot at the top (called number 1), and numbers 2 and 3 on the lower row
layout.show(3) will show you what plotting areas your layout call has defined; the number in the round brackets is the total number of separate areas.

Using the *split.screen* Function

This function is roughly equivalent to **layout**, and allows you to specify a number of variously sized plots on one screen. First create a four column matrix with the number of rows equal to the number of separate plots. Each row comprises the coordinates on a 0:1 scale of the left x value, right x value, lower y value and upper y value, for each separate plot, e.g. for three plots we can write
corners<-matrix(c(0,.6,.6,1,.65,1,0,1,0,.6,0,.65),byrow=TRUE,nrow=3)
split.screen(corners)
[1] 1 2 3
screen(1)
plot(0:10,10:0,xlim=c(0,10),ylim=c(0,10),col="red",type="l")
screen(2)
plot(0:10,10:0,xlim=c(0,10),ylim=c(0,10),col="blue",pch=15)
screen(3)
plot(sample(1:100),sample(1:100),xlim=c(0,100),ylim=c(0,100),col=c("blue","green"),pch=0)

Appendix 2: General Housekeeping R Functions and Others Not Covered in the Main Text

General Housekeeping Functions

help.search("{a key word}")) searches help pages for that word.
ls show the names of all objects currently in R's memory.
ls.str to show the contents of each of the objects in memory.
ls.str(pat="{*some character string in object name*}") shows those objects whose names contain the pattern.
ls("package:{*package name*}") shows all function names in that named package.
objects shows the names of the variables you have created (same as **ls**).
rm({*object name*}) remove that object from memory.
rm(list=**ls**()) removes all objects from memory, a possible alternative to restarting R.
search() shows which libraries and/or dataframes you have attached.

Setting or Changing the Working Directory

R will normally look for data files in, or save files to, its working directory. To find out what this is at the moment on your computer use the function **getwd**(), which returns the path of the current working directory, e.g.
getwd()
[1] "/Users/donaldquicke/Desktop/Books/R by DQBBRKW"
There are several ways to change the directory. There are pull-down menus once R has been started. On Windows: File > Change directory; On Mac computers via the menu bar 'Misc > Change working directory' or on older versions 'Tools > Change the working directory'. We normally do this every time we start up R, and depending on what we are doing may change it during a session several times as we swap between various projects. There is also a command line function to set the directory, **setwd**() into which you need to enter the path, e.g.
setwd("C:/Users/Rachel/Downloads/wetransfer-126312/DonaldData") # for a PC-user
setwd("/Users/donaldquicke/Downloads") # for a MacOSX-user

Appendix 2   361

Finding What Files Are in a Directory

To list the files in a directory use the **list.files** function,
my_files<-list.files(path="C:/Users/Rachel_R_data")
or on a MacOSX system
my_downloads<-list.files("/Users/donaldquicke/Downloads/")
or for the working directory on either system
list.files(path=getwd()) # or just list.files(getwd())
There is also the function **list.dirs** or **list.dirs**(path = ".") to get a list of subdirectories within the working directory. Both can be very useful if, for example, you have forgotten exactly what you called a file or perhaps cannot remember which subdirectory you placed a file in.

Graphical Functions and Parameters

par(ask=T) pauses after presenting one graph until you enter a carriage return.
screen(3) for example, prepares R to output the next plot in the third screen, in this case the second one in the lower row of three, screen one being the upper undivided plotting area.
split.screen split.screen(c(2,1)) splits the screen into two rows of plot areas (see Appendix 1).
split.screen(c(1,4),screen=2) tells R to use the current second screen (lower one) and split that into four plot areas. There are now five plot areas, which are specified in sequence using screen().

Interaction with User

R is not a very interactive language but there may be occasions when you want your program to ask you to make a choice.
askYesNo("{*some yes/no question*}") e.g. a<-askYesNo("Do you want to continue? Y/N") returns a=TRUE if you type Y or y and FALSE if you type N or n
readline(prompt="Enter your choice: ") reads a line of text that can then be used in if statements for example to choose what function to use

Mathematical Functions

abs # returns the absolute value of x (i.e. with any minus sign removed).
choose(n,p) # the number of unordered ways that p can be chosen from n.
colMeans(x) # returns a vector containing the means of all the columns in a matrix or dataframe. Similar functions are **colSums**, **rowMeans** and **rowSums**.
cumprod(x) # returns a vector of the cumulative products of the elements of x up to each value, e.g. x=c(2,1,3,4,5), cumprod(x)=c(2,2,6,24,120).

combn(x,n) # returns a vector of all different combinations of *n* items from list *x*.
factorial(x) # returns the factorial value of *x* (i.e. $x*(x-1)*(x-2) \ldots 2*1$).
floor(x) # returns the largest integer less than *x*.
GCD(x) # returns greatest common denominator (devisor) of two or more elements; needs library(DescTools) activated.
is.data.frame(x) # returns TRUE if *x* is a dataframe.
LCM(x) # returns the least common multiple of two or more elements; needs library(DescTools) activated.
median(x) # returns the median value of a vector of numbers.
quantile(x) # returns a vector containing the minimum, lower quartile, median, upper quartile and maximum of *x*.
tan(x) # returns the tangent of *x* radians.

Writing Concatenated Data Straight to File (in the Working Directory) Using *cat*

The function **cat** can be used to save concatenated text efficiently, e.g.
cat(c(0,6,8,4,3),"\n","egg, caterpillar, pupa, adult","\n",4*21,"\n",letters)
0 6 8 4 3
egg, caterpillar, pupa, adult
84
a b c d e f g h i j k l m n o p q r s t u v w x y z
and it outputs the result to the console. However, if you include 'file= "filename"' as an argument, it will create a text file (in this case called "filename") in the working directory (use **getwd**() to find what that is if you do not know). Note that it accepts '\n' for line breaks, and performs calculations (4*21 = 84).

Troubleshooting Package Installation

In installing packages you might encounter a number of error messages; this can be annoying and often arises if you have moved to a new computer, sometimes different platforms or whether the computer is 32 or 64 bit might be causing problems. Some packages require a package called 'rJava' and we have found this can be a headache on some systems. Some error messages can be ignored, but others cannot. If you see
"tar: Failed to set default locale", find out what your current system locale is. Here is what one of us got when running R on a new German Mac computer:
Sys.getlocale()
[1] "C" # this, or "c" is the default for the computer language C and reflects North American usage – also known as "POSIX". The solution (from the ever helpful Stack Overflow website) is to enter
system('defaults write org.R-project.R force.LANG en_US.UTF-8')

and then restart R and reinstall the package you were attempting to load. We suggest you use '?{name of package}' to make sure it has loaded OK. If you get a message saying that no information was found, one possibility might be that you have to go to package manager and tick the relevant box.

We obviously cannot cover all options here, but googling, for example, Stack Overflow and/or 'R program' plus some generic part of the error message may provide useful help pages. Rarely, it may be necessary to update some files or in a worst case scenario, completely re-install R.

Appendix 3: Some Useful Statistical and Mathematical Equations

The same terminology is used for all examples: n_i is the number of observations in the ith sample, N is the total sample size, x and y are the raw values of the two samples, bar x ($\bar{x}$) and bar y ($\bar{y}$) are the mean values of x and y respectively, μ is the population mean, s is the sample standard deviation, σ is the population standard deviation, d is the difference between paired values.

Logical Mathematical Operators

```
>   # greater than
<   # less than
>=  # greater than or equals
<=  # less than or equals
!=  # does not equal
```

Descriptive Statistics

Variance of a sample, s^2 (= sample standard deviation, s squared)

$$s^2 = \frac{\sum(X - \bar{X})^2}{N - 1}$$

Variance (σ^2 = population standard deviation (σ) squared)

$$\sigma^2 = \frac{\sum(X - \mu)^2}{N}$$

Standard error of the mean (SEM or S.E.) of a sample. If you are dealing with the whole population, replace s with σ.

$$SEM = \frac{s}{\sqrt{N}}$$

Appendix 3

Distributions

Normal distribution
The basic equation is very simple

$$y = e^{-x^2}$$

However, the version that we work (the standardized normal distribution) with looks rather more complicated because it has been scaled so that the mean is zero and the area under the curve (the integral from − infinity to + infinity) is one.

$$f(x) = \frac{1}{\sigma\sqrt{2\pi}} e^{-\frac{1}{2}\left(\frac{x-\mu}{\sigma}\right)^2}$$

Correlation Coefficients

Pearson's product moment correlation coefficient r is

$$r = \frac{\sum_{i=1}^{n}(x_i - \bar{x})(y_i - \bar{y})}{\sqrt{\sum_{i=1}^{n}(x_i - \bar{x})^2 (y_i - \bar{y})^2}}$$

Spearman's non-parametric rank correlation statistic rho (r_s)

$$r_s = 1 - \frac{6\Sigma d_i^2}{n(n^2 - 1)}$$

where n is the number of observations and d_i is the difference in the ranks of the observations.

Statistical Tests

G-test

$$G = 2\Sigma observed.\log_e \frac{observed}{expected}$$

t-test (single sample)

$$t = \frac{|\bar{x} - \mu|}{\frac{s}{\sqrt{n}}}$$

t-test (two sample)

$$t = \frac{|\bar{x}_1 - \bar{x}_2|}{\sqrt{\frac{(s_1)^2}{n_1} + \frac{(s_2)^2}{n_2}}}$$

where the vertical lines mean the absolute value (i.e. ignore the sign of the subtraction).
t-test (paired)

$$t = \frac{\Sigma d}{\sqrt{\frac{n(\Sigma d^2) - (\Sigma d)^2}{n-1}}}$$

because the observations are paired, n is the same for both samples.

Logarithms and Exponents

The default logs in R are to base e (where e is an irrational number close to 2.71828182845904).
For other bases specify base by adding a second argument, e.g. for base 7, e.g. log(x,7). Remember that if log(x)=a then x=exp(a).

Logistic Functions

These may be simple with just two parameters, e.g.

$$y = \frac{e^{a+bx}}{1+e^{a+bx}}$$

or have more, such as this with three parameters

$$y = \frac{a}{1+be^{-cx}},$$

Weibull and Gompertz Equations

The next two distributions are used for modelling mortality as well as age-specific growth (for comparisons and details see Juckett and Rosenberg, 1993; D.L. Wilson, 1994). These are the Weibull equations, which have up to four fit parameters (a, b, c and d) with the basic form

$$y = a - be^{-(cx^d)}$$

and the three-parameter form is the Gompertz equation, which has a longer lag phase at the beginning

$$y = ae^{be^{cx}}$$

The survivorship function derived from the Weibull equation is $y = \exp(-(ax)^b)$, which is sigmoidal starting high and declining, whereas its age specific hazard function $h = ax^{(a-1)}$ increases with age.

Trigonometric Functions

Base R includes the standard set of trigonometric functions: cos, sin, tan, cosh, sinh, tanh, acosh, ashinh and atanh plus a few others. Note that in R the angles passed to a trigonometric function are in radians. There are 2*pi radians in a circle, i.e. 360° (approx. 6.283185).
There is a special R constant for π
pi
[1] 3.141593
Therefore, 1 degree = 6.283185/360 radians, i.e.
2*pi/360
[1] 0.01745329
For a right-angled triangle the sine of one of the angles is the length of the opposite side divided by the hypotenuse, the cosine is the length of the side adjacent to the angle divided by the hypotenuse and the tangent is the length of the opposite side divided by the adjacent one, or

tan (x) = sin(x)/cos(x)

R has the corresponding functions, **sin**, **cos** and **tan** which take angles in radians as their arguments.

Convert Radians and Degrees Functions

```
radian2degree<-function(radian) radian*180/pi
degree2radian<-function(degree) degree*pi/180
```

Bibliography

Akaike, H. (1973) Information theory and an extension of the maximum likelihood principle. In: Petrov, B.N. and Csáki, F. (eds.) *2nd International Symposium on Information Theory, Tsahkadsor, Armenia, USSR, September 2–8, 1971, Budapest: Akadémiai Kiadó.* pp. 267–281. Republished in Kotz, S. and Johnson, N.L. (eds) (1992) *Breakthroughs in Statistics, I.* Springer-Verlag, New York, pp. 610–624.

Atuo F., O'Connell T.J., Saud, P. and Wyatt, C. (2018) Are oil and natural gas development sites ecological traps for nesting killdeer? *Wildlife Biology* 2018(1). DOI: 10.2981/wlb.00476 (accessed 20 April 2020).

Baddeley, A., Rubak, E. and Turner, R. (2016) *Spatial Point Patterns: Methodology and Applications with R.* CRC Press, Boca Raton, Florida.

Basset, Y., Eastwood, R., Sam, L., Lohman, D.J., Novotny, V. *et al.* (2013) Cross-continental comparisons of butterfly assemblages in rainforests: implications for biological monitoring. *Insect Conservation and Diversity* 6, 223–233. DOI: 10.1111/j.1752-4598.2012.00205.x.

Beall, G. (1942) The transformation of data from entomological field experiments. *Biometrika* 29, 243–262. Available at: https://www.jstor.org/stable/pdf/2332128.pdf (accessed 30 April 2020).

Beckerman A.P., Childs, D.Z. and Petchey, O.L. (2017) *Getting Started with R: an Introduction for Biologists, 2nd edition.* 240pp. Oxford University Press, Oxford, UK.

Beddington, J.R., Free, C.A. and Lawton, J.H. (1975) Dynamic and complexity in predator–prey models framed in difference equations. *Nature* 255, 58–60.

Blondel, J., Perret, P. and Maistre, M. (1990) On the genetical basis of the laying-date in an island population of blue tits. *Journal of Evolutionary Biology* 3, 469–475. DOI: 10.1046/j.1420-9101.1990.3050469.x

Bolker, B.M. (2008) *Ecological Models and Data in R.* Princeton University Press, Princeton, New Jersey, USA.

Bonferroni, C.E. (1936) Teoria statistica delle classi e calcolo delle probabilità. *Pubblicazioni del R Istituto Superiore di Scienze Economiche e Commerciali di Firenze* 8, 3–62.

Booth, R.G. (2012) Coccinellidae Latreille, 1807. In: Duff, A.G. (ed.) *Checklist of Beetles of the British Isles. 2nd edition.* Pemberley Books, Iver, UK.

Brown, J.H. and Maurer, B.A. (1989) Macroecology – the division of food and space among species on the continents. *Science* 243(4895), 1145–1150.

Brace, R.C. and Quicke, D.L.J. (1986) Seasonal changes in dispersion within an aggregation of the anemone, *Actinia equina*, with a reappraisal of the role of intraspecific aggression. *Journal of the Marine Biological Association of the United Kingdom* 66, 49–70. DOI: 10.1017/S0025315400039631

Butcher, B.A., Smith, M.A., Sharkey, M.J. and Quicke, D.L.J. (2012) A turbo-taxonomic study of Thai *Aleiodes (Aleiodes)* and *Aleiodes (Arcaleiodes)* (Hymenoptera: Braconidae: Rogadinae) based largely on COI bar-coded specimens, with rapid descriptions of 179 new species. *ZooTaxa* 3457, 1–232. Available at: biotaxa.org/Zootaxa/article/view/zootaxa.3457.1.1 (accessed 20 April 2020).

Chantanaorrapint, S. (2010) Ecological studies of epiphytic bryophytes along altitudinal gradients in Southern Thailand. *Archive for Bryology Special Volume* 7, 1–102. Available at: https://hss.ulb.uni-bonn.de/2010/2049/2049.pdf (accessed 20 April 2020).

Clark, P.J. and Evans, F.C. (1954) Distance to nearest neighbour as a measure of spatial relationships in populations. *Ecology* 35, 445–453. Available at: www.jstor.org/stable/1931034 (accessed 20 April 2020).

Clutton-Brock, T.H. and Iason, G.R. (1986) Sex ratio variation in mammals. *The Quarterly Review of Biology* 61, 339–374. Available at: www.jstor.org/stable/2826773 (accessed 20 April 2020).

Conway, R. and Maxwell, W.L. (1962) A queuing model with state dependent service rate. *Journal of Industrial Engineering* 12, 132–136.

Crawley, M.J. (2002) *Statistical Computing: An Introduction to Data analysis Using S-Plus.* John Wiley & Sons, Chichester, UK.

Crawley, M.J. (2007) *The R Book.* John Wiley & Sons, Chichester, UK.

Dalgaard, P. (2006) *Introductory Statistics with R.* Springer Verlag, Berlin.

Day, A.J., Hawkins, A.J.S. and Visootiviseth, P. (2000) The use of allozymes and shell morphology to distinguish among sympatric species of the rock oyster *Saccostrea* in Thailand. *Aquaculture* 187, 51–72. DOI: 10.1016/S0044-8486(00)00301-X

Dunn, P.K. and Smyth, G.K. (2018) *Generalized Linear Models with Examples in R.* Springer, New York.

Dytham, C. (2011) *Choosing and Using Statistics, 3rd edition.* Wiley-Blackwell, Chichester, UK.

Feigl, P. and Zelen, M. (1965) Estimation of exponential survival probabilities with concomitant information. *Biometrics* 21, 826–838. Available at: www.jstor.org/stable/2528247 (accessed 20 April 2020).

Garzón, M.J. and Schweigmann, N. (2015) Thermal response in pre-imaginal biology of *Ochlerotatus albifasciatus* from two different climatic regions. *Medical and Veterinary Entomology* 29, 380–386. DOI: doi.org/10.1111/mve.12128

Geffeney, S., Brodie, E.D., Jr, Ruben P.C. and Brodie E.D., III (2002) Mechanisms of adaptation in a predator-prey arms race: TTX-resistant sodium channels. *Science* 297, 1336–1339. DOI: 10.1126/science.1074310 (accessed 20 April 2020).

Gotelli, N.J. and Ellison, A.M. (2004) *A Primer of Ecological Statistics.* Sinauer Associates, Sunderland, Massachusetts.

Hector, A. (2016) *The New Statistics with R: an Introduction for Biologists.* Oxford University Press, Oxford, UK.

Herndon, T., Ash, M. and Pollin, R. (2014) Does high public debt consistently stifle economic growth? A critique of Reinhart and Rogoff. *Cambridge Journal of Economics* 38, 257–279. Available at: academic.oup.com/cje/article/38/2/257/1714018 (accessed 20 April 2020).

Heumann, C., Schomaker, M. and Shalabh (2016) *Introduction to Statistics and Data Analysis: with Exercises, Solutions and Applications in R.* Springer International Publishing, Gland, Switzerland.

Hill, M.O. (1973) Diversity and evenness: a unifying notation and its consequences. *Ecology* 54, 427–432. Available at: www.jstor.org/stable/1934352 (accessed 20 April 2020).

Holcomb, W.L., Chaiworapongsa, T., Luke, D.A. and Burgdorf, K.D. (2001) An odd measure of risk: use and misuse of the odds ratio. *Obstetrics and Gynecology* 98, 685–688. DOI: 10.1016/S0029-7844(01)01488-0

Holt, B.G., Lessard, J.-P., Borregaard, M.K., Fritz, S.A., Araújo, M.B. *et al.* (2013) An update of Wallace's zoogeographic regions of the world. *Science* 339, 74–78. DOI: 10.1126/science.1228282

Hrcek, J., Miller, S.E., Quicke, D.L.J. and Smith, M.A. (2011) Molecular detection of trophic links in a complex insect host-parasitoid food web. *Molecular Ecology Resources* 11, 786–794. DOI: 10.1111/j.1755-0998.2011.03016.x

Hudson, L.N., Emerson, R., Jenkins, G.B., Layer, K., Ledger, M.E. et al. (2013) Cheddar: analysis and visualisation of ecological communities in R. *Methods in Ecology and Evolution* 4, 99–104. DOI: 10.1111/2041-210X.12005

Inta, A., Shengji, P., Balslev, H., Wangpakapattanawong, P. and Trisonthi, C. (2008) A comparative study on medicinal plants used in Akha's traditional medicine in China and Thailand, cultural coherence or ecological divergence? *Journal of Ethnopharmacology* 116, 508–517. DOI: 10.1016/j.jep.2007.12.015.

Jenks, K.E., Chanteap, P., Damrongchainarong, K., Cutter, P., Passanan, C. et al. (2011) Using relative abundance indices from camera-trapping to test wildlife conservation hypotheses – an example from Khao Yai National Park, Thailand. *Tropical Conservation Science* 4(2), 113–131. DOI: 10.1177/194008291100400203

Jonkers, J. and Kučera, M. (2015) Global analysis of seasonality in the shell flux of extant planktonic Foraminifera. *Biogeosciences* 12, 2207–2226. DOI: 10.5194/bg-12-2207-2015

Juckett, D.A. and Rosenberg, B. (1993) Comparison of the Gompertz and Weibull functions as descriptors for human mortality distributions and their intersections. *Mechanisms of Ageing and Development* 69, 1–31. DOI: 10.1016/0047-6374(93)90068-3

Jutagate, T., Phomikong, P., Avakul, P. and Saowakoon, S. (2013) Age and growth determinations of chevron snakehead *Channa striata* by otolith reading. *Proceedings of the 51st Kasetsart University Annual Conference, Bangkok, Thailand, 5–7 February 2013.*

Kaplan, E.L. and Meier, P. (1958) Nonparametric estimation from incomplete observations. *Journal of the American Statistical Association* 53, 457–481. DOI: 10.2307/2281868

Kays, R., Dunn, R.R., Parsons, A.W., Mcdonald, B., Perkins, T., Powers, S., Shell, L., McDonald, J.L., Cole, H., Kikillus, H., Woods, L., Tindle, H. and Roetman, P. (2020) The small home ranges and large local ecological impacts of pet cats. *Animal Conservation.* DOI: 10.1111/acv.12563

Keszthelyi, S., Puskas, J. and Nowinszky, L. (2008) Changing of flight phenology and ecotype expansion of the European corn borer (*Ostrinia nubilalis* Hbn.) in Hungary. Part 1. Biomathematical evaluation. *Cereal Research Communications* 36(4), 647–657. DOI: 10.1556/CRC.36.2008.4.14

Kieschnick, R. and McCullough, B.D. (2003) Regression analysis of variates observed on (0, 1): percentages, proportions and fractions. *Statistical Modelling* 3, 193–213. DOI: 10.1191/1471082x03st053oa

Kim, T.-S., Cho, S.-H., Huh, S., Kong, Y., Sohn, W.-M. et al. (2009) A nationwide survey on the prevalence of intestinal parasitic infections in the Republic of Korea, 2004. *The Korean Journal of Parasitology* 47, 37–47. DOI: 10.3347/kjp.2009.47.1.37

Laohapensang, K., Rerkasem, K. and Kattipattanapong, V. (2004) Seasonal variation of Buerger's disease in northern part of Thailand. *European Journal of Vascular and Endovascular Surgery* 28, 418–420. DOI: 10.1016/j.ejvs.2004.05.014

Logan, M. (2010) *Biostatistical Design and Analysis Using R: A Practical Guide.* Wiley-Blackwell, Chichester, UK. DOI: 10.1002/9781444319620

Lomolino, M.V. (2000) Ecology's most general, yet protean pattern: the species-area relationship. *Journal of Biogeography* 27, 17–26. DOI: 10.1046/j.1365-2699.2000.00377.x

Longo, J.M. and Fischer, E. (2006) Efeito da taxa de secreção de néctar sobre a polinização e a produção de sementes em flores de *Passiflora speciosa* Gardn. (Passifloraceae) no Pantanal. *Revista Brasiliana Botanici* 29, 481–488. DOI: 10.1590/S0100-84042006000300015

Lynch, H.J., Thorson, J.T. and Shelton, A.O. (2014) Dealing with under- and overdispersed count data in life history, spatial, and community ecology. *Ecology* 95(11), 3173–3180. Available at: www.jstor.org/stable/43495231 (accessed 20 April 2020).

MacArthur, R.H. (1957) On the relative abundance of bird species. *Proceedings of the National Academy of Sciences of the U.S.A.* 43, 293–295. Available at: www.pnas.org/content/43/3/293 (accessed 20 April 2020).

MacArthur, R.H. and MacArthur, J.W. (1961) On bird species diversity. *Ecology* 42, 594–598. DOI: 10.2307/1932254

MacArthur, R.H. and Wilson, E.O. (1967) *The Theory of Island Biogeography*. Princeton University Press, Princeton, New Jersey.

Magurran, A.E. (2003) *Measuring Biological Diversity*. Blackwell Publishing, Oxford, UK.

Malone, J.C., Forrester, G.E. and Steele, M.A. (1999) Effects of subcutaneous microtags on the growth, survival, and vulnerability to predation of small reef fishes. *Journal of Experimental Marine Biology and Ecology* 237, 243–253.

Manly, B.F. (1985) *The Statistics of Natural Selection on Animal Populations*. Chapman & Hall, London.

Marsh, S.T., Brummitt, N.A., de Kok, R.P.J. and Utteridge, T.M.A. (2009) Large-scale patterns of plant diversity and conservation priorities in South East Asia. *Blumea* 54, 103–108. Available at: repository.naturalis.nl/document/564944 (accessed 20 April 2020).

McDonald, J.H. (2014) *Handbook of Biological Statistics, 3rd edition*. Sparky House Publishing, Baltimore, Maryland.

McKillup, S. (2011) *Statistics Explained: An Introductory Guide for Life Scientists, 2nd edition*. Cambridge University Press, Cambridge, UK.

Møller, A.P. (1988) Ejaculate quality, testes size and sperm competition in primates. *Journal of Human Evolution* 17, 479–488. DOI: 10.1111/j.1095-8312.1988.tb00812.x

Narum, S.R. (2006) Beyond Bonferroni: less conservative analyses for conservation genetics. *Conservation Genetics* 7, 783–787.

Nuraemram, K. (2011) Effects of forest fire on ant diversity in the dry dipterocarp forest at Lai Nan subdistrict, Wiang Sa district, Nan Province. Zoology major project, Chulalongkorn University.

Offenberg, J., Havanon, S., Aksornkoae, S., MacIntosh, D.J. and Nielsen, N.G. (2004) Observations on the ecology of weaver ants (*Oecophylla smaragdina* Fabricius) in a Thai mangrove ecosystem and their effect on herbivory of *Rhizophora mucronata* Lam. *Biotropica* 36. 344–351. DOI: 10.1646/03158

O'Hanlon, J.C., Holwell, G.I. and Herberstein, M.E. (2013) Data from: pollinator deception in the orchid mantis. *Dryad Digital Repository*. Available at: datadryad.org/stash/dataset/doi:10.5061/dryad.g665r (accessed 20 April 2020).

O'Hanlon, J.C., Holwell, G.I. and Herberstein, M. (2014) Pollinator deception in the orchid mantis. *American Naturalist* 183(1), 126–132. Available at: www.jstor.org/stable/10.1086/673858 (accessed 20 April 2020).

Paradis E. and Schliep K. (2018) Ape 5.0: an environment for modern phylogenetics and evolutionary analyses in R. *Bioinformatics* 35(3), 526–528. DOI: 10.1093/bioinformatics/bty633

Paternoster, R., Brame, R., Mazerolle, P. and Piquero, A.R. (1998) Using the correct statistical test for the equality of regression coefficients. *Criminology* 36, 859–866. DOI: 10.1111/j.1745-9125.1998.tb01268.x

Pitt, K.A., Duarte, C.M., Lucas, C.H., Sutherland, K.R., Condon, R.H. et al. (2013) Jellyfish body plans provide allometric advantages beyond low carbon content. *PLoS ONE* 8(8), e72683. DOI: 10.1371/journal.pone.0072683

Poonswad, P., Tsuji, A., Jirawatkavi, N. and Chimchome V. (1998) Some aspects of the food and feeding ecology of sympatric hornbills in Khao Yai National Park, Thailand. In: Poonswad, P. (ed.) *The Asian Hornbills: Ecology and conservation*, pp. 137–157, Thai studies in biodiversity no. 2. Biodiversity Research and Training Program, Bangkok, Thailand. Available at: pirun.ku.ac.th/~fforvjc/asianhornbillp137.pdf (accessed 20 April 2020).

Porta, M. ed. (2014) *Dictionary of Epidemiology, 6th edition*. Oxford University Press, New York.

Preston, F.W. (1962) The canonical distribution of commonness and rarity: Part I. *Ecology* 43, 185–215 and 410–432. Available at: www.jstor.org/stable/1931976 (accessed 20 April 2020).

R Development Core Team (2009) *R: A Language and Environment for Statistical Computing*. R Foundation for Statistical Computing, Vienna, Austria.

Ranjith, A.P., Quicke, D.L.J., Saleem, UK.A., Butcher, B.A., Zaldivar-Riveron, A. and Nasser, M. (2016) Entomophytophagy in an Indian braconid 'parasitoid' wasp (Hymenoptera): specialized larval morphology, biology and description of a new species. *PLoS ONE* 11(6), e0156997. DOI: 10.1371/ journal.pone.0156997

Ricker, W.E. (1954) Stock and recruitment. *Journal of the Fisheries Research Board of Canada* 11, 559–623. DOI: 10.1139/f54-039

Ridout, M.S. and Besbeas, P. (2004) An empirical model for underdispersed count data. *Statistical Modelling* 4, 77–89. DOI: 10.1191/1471082X04st064oa

Scheiner, S.M. (2003) Six types of species–area curves. *Global Ecology and Biogeography* 12, 441–447. DOI: 10.1046/j.1466-822X.2003.00061.x

Shmueli, G., Minka, T.P., Kadane, J.B., Borle, S. and Boatwright, P. (2004) A useful distribution for fitting discrete data: revival of the Conway-Maxwell-Poisson distribution. *Applied Statistics* 54, 127–142. DOI: 10.1111/j.1467-9876.2005.00474.x

Scott, J.A. (1972) Biogeography of Antillean butterflies. *Biotropica* 4, 32–45. Available at: www.jstor.org/stable/2989643 (accessed 20 April 2020).

Smith, T.E. (2016) *Notebook on Spatial Data Analysis*. Available at: seas.upenn.edu/~ese502/#notebook (accessed 20 April 2020).

Soetaerd, K. and Herman, P.M.J. (2009) *A Practical Guide to Ecological Modelling Using R as a Simulation Platform*. Springer, The Netherlands.

Sokal, R.R. and Rohlf, F.J. (1995) *Biometry: The Principles and Practice of Statistics in Biological Research, third edition*. W.H. Freeman, New York.

Sombatboon, K. (2014) Correlation between stress and health of bullfrog *Lithobates catesbeianus* in captivity. Senior student project, Chulalongkorn University.

South, A. (2011) rworldmap: A new R package for Mapping Global Data. *The R Journal,* 3(1), 35–43.

Srimuang, K., Watthana, S., Pedersen, H.Æ., Rangsayatorn, N. and Eungwanichayapant, P.D. (2010) Flowering phenology, floral display and reproductive success in the genus *Sirindhornia* (Orchidaceae): a comparative study of three pollinator-rewarding species. *Annales Botanici Fennici* 47, 439–448. Available at: http://www.sekj.org/PDF/anbf47/anbf47-439.pdf (accessed 20 April 2020).

Sukontason, K.L., Chaiwong, T., Piangjai, S., Upakut, S., Moophayak, K. and Sukontason, K. (2008) Ommatidia of blow fly, house fly, and flesh fly: implication of their vision efficiency. *Parasitology Research* 103, 123–131. DOI: 10.1007/s00436-008-0939-y

Suwannapong, G., Benbow, M.E. and Nieh, J.C. (2011) Biology of Thai honeybees: natural history and threats. In: Florio, R.M. (ed.) *Bees: Biology, Threats and Colonies*. Nova Science Publishers, Hauppauge, New York, pp. 1–98.

Teetor, P. (2011) *R Cookbook*. O'Reilly, Cambridge, UK.

Tunjai, P. and Elliott, S. (2012) Effects of seed traits on the success of direct seeding for restoring southern Thailand's lowland evergreen forest ecosystem. *New Forests* 43, 319–333. DOI: 10.1007/s11056-011-9283-7

van Emden, H.F. (2019) *Statistics for Terrified Biologists, 2nd edition*. John Wiley & Sons, Chichester, UK. 360pp.

Van Ngan, P., Gomes, V., Carvalhoz, P.S.M. and de A.C.R. Passos, M.J. (1997) Effect of body size, temperature and starvation on oxygen consumption of Antarctic krill *Euphausia superba*. *Revista Brasileira de Oceanografia* 45, 1–10. DOI: 10.1590/S1413-77391997000100001

Venables, W.N. and Ripley, B.D. (2002) *Modern Applied Statistics With S-PLUS, 4th edition*. Springer-Verlag, New York.

Verme, M.J. and Ozoga, J.J. (1981) Sex ratio of white-tailed deer and the estrous cycle. *Journal of Wildlife Management* 45, 710–715. Available at: www.jstor.org/stable/3808704 (accessed 20 April 2020).

Verhoef, J.M. and Boveng, P.L. (2007) Quasi-Poisson vs. negative binomial regression: how should we model overdispersed count data? *Ecology* 88, 2766–2772. Available at: digitalcommons.unl.edu/cgi/viewcontent.cgi?article=1141&context=usdeptcommercepub (accessed 20 April 2020).

Warton, D.I. and Hui, F.K.C. (2011) The arcsine is asinine: the analysis of proportions in ecology. *Ecology* 92(1), 3–10. DOI: 10.1890/10-0340.1 (accessed 20 April 2020).

Weiss, P.W. (1981) Spatial distribution and dynamics of populations of the introduced annual *Emex australis* in south-eastern Australia. *Journal of Applied Ecology* 18, 849–864.

White, T., van der Ende, J. and Nichols, T.E. (2019) Beyond Bonferroni revisited: concerns over inflated false positive research findings in the fields of conservation genetics, biology, and medicine. *Conservation Genetics* 20, 927–937. Available at: https://link.springer.com/content/pdf/10.1007/s10592-019-01178-0.pdf (accessed 20 April 2020).

Wilson, D.L. (1994) The analysis of survival (mortality) data: fitting Gompertz, Weibull, and logistic functions. *Mechanisms of Ageing and Development* 74, 15–33. DOI: 10.1016/0047-6374(94)90095-7 (accessed 20 April 2020).

Wilson, J.B. (1993) Would we recognise a broken-stick community if we found one? *Oikos* 67, 181–183. Available at: www.jstor.org/stable/3545108 (accessed 20 April 2020).

Wiriya, B., Clausen, J.H., Inpankaew, T., Thaenkham, U., Jittapalapong, S., Satapornvanit, K. and Dalsgaard, A. (2013) Fish-borne trematodes in cultured Nile tilapia (*Oreochromis niloticus*) and wildcaught fish from Thailand. *Veterinary Parasitology* 198, 230–234. DOI: 10.1016/j.vetpar.2013.08.008 (accessed 20 April 2020).

Wright, S.J and Muller-Landau, H (2006) The future of tropical forest species. *Biotropica* 38, 287–301. Available at: https://repository.si.edu/handle/10088/4183 (accessed 20 April 2020).

Zuur, A., Ieno, E.N. and Meesters, E. (2009) *A Beginners Guide to R*. Springer-Verlag, New York.

Web Resources

bioinformatics.org/sms2/iupac.html (accessed 22 April 2020)
coleoptera.org.uk/family/coccinellidae (accessed 23 April 2020).
cran.r-project.org/doc/contrib/Baggott-refcard-v2.pdf (accessed 22 April 2020)
cran.r-project.org/doc/contrib/Short-refcard.pdf (accessed 22 April 2020)
en.wikipedia.org/wiki/ISO_3166-1 (accessed 22 April 2020)
en.wikipedia.org/wiki/List_of_U.S._states_and_territories_by_median_age (accessed 16-04-20)
ncbi.nlm.nih.gov/Taxonomy/Utils/wprintgc.cgi#SG3 (accessed 22 April 2020)
plotdigitizer.sourceforge.net (accessed 22 April 2020)
rainforests.mongabay.com/03mammals.htm (accessed 22 April 2020)
rstudio.com/products/RStudio/#Desktop (accessed 22 April 2020)
rstudio.com/wp-content/uploads/2015/03/ggplot2-cheatsheet.pdf (accessed 22 April 2020)
stat.ethz.ch/R-manual/R-devel/library/datasets/html/00Index.html (accessed 22 April 2020)
RStudio at support.rstudio.com/hc/en-us/articles/200549016-Customizing-RStudio (accessed 22 April 2020)
tdwg.org (accessed 22 April 2020)
thainationalparks.com/khao-yai-national-park/wildlife (accessed 22 April 2020)
theplantlist.org/browse/A/Styracaceae/ (accessed 22 April 2020)
vincentarelbundock.github.io/Rdatasets/datasets.html (accessed 22 April 2020)
wikiwand.com/en/List_of_countries_by_southernmost_point (accessed 22 April 2020)

Index

R packages are in bold font and followed by (""); R functions are in bold font and followed by (); R arguments for functions are followed by =. Not all occurrences are listed for some very common entries where extra examples would not demonstrate any new features.

-1, remove intercept from model 50
-, minus indexing 14, 236, 263
: 15, 19, 23, 25, 30, 36, 39, 42, 51, 63, 67
; 19, 45, 47, 50, 78, 133, 135, 152, 175, 182, 195, 298, 306
! 18, 19, 221, 263, 267, 288, 289
? 22, 91, 118, 176, 210, 259, 279
?? 16, 279
() 14, 15, 193, 261
[, as a function 296
[[:punct:]] 269–271, 274
[] 15, 18, 78, 239
{} 15, 305, 319
*, in models 138, 145, 167, 169
/, nesting factors in **glm** 183, 184
\\ 259, 263, 272, 274, 296
\n 86, 87, 268, 354, 362
\r 86
\t 86, 274, 350
10, 45, 213, 221, 262, 300, 307, 352, 360
%/%, integer division 188
%%, modulus, modulo 188, 294
%in% 87, 88, 234, 271, 315, 319

^
    as power function 15, 19, 51, 90, 188, 189, 201, 208, 299, 324
    wildcard meaning beginning of string 268
+ in models 137, 138, 142
± 64
== 11, 18, 19, 23, 266, 294, 296, 327
>, prompt 6, 16
|
    logical OR 268
    in date format specification 228
~, tilde 70, 78, 99–101, 136, 137, 164
$
    in **aov**, **glm**, **lm** 98
    in **barplot** 52
    in selecting dataframe columns 33, 35, 88, 129, 273
    in **gsub**() 269
    in **Surv** 221, 222
abline() 122, 129, 146, 178, 190
abs() 133, 361
acp() in **amap** library 194
Actinia 297
ad.test() in **nortest** library 116, 118
adj= in **plot**(), **text**() 26, 327, 357

aggregate() 68, 69
AIC() 144, 145
Akaike's information criterion 144, 145
alarm() 306
alleles 107–110, 194, 195, 318, 319
all.equal() 294, 296
all.knots= in **smooth.spline**() 217
allozymes 194–197
amap("") 194
analysis of covariance (ANCOVA) 100, 148, 166–170, 225
analysis of variance (ANOVA) 74, 78, 99, 100, 113, 117–119, 126, 145, 155–165, 170, 225, 226
Anderson-Darling test see **ad.test**() in nortest library
angle=
  in **barplot**() 54, 56
  in **arrows**() 63
angular transformation see arcsine transformation
ants 53–56, 82, 114–116
anyNA, use with **Filter**() 221
aod("") 182
aov() 74, 78, 98, 101, 113, 156, 157, 162–165
ape("") 275, 276, 279, 290
Apis 188, 189
apostrophes 10, 252
append= in **write**() 349, 355
apply() 89, 178, 186, 216, 322–325
arcsine transformation 100, 157–163
arguments 2, 14, 18–22, 29, 32, 38, 39, 43, 44, 46, 48, 332, 333, 340, 344, 345, 349, 352, 353
arrows() 24, 26, 63, 69, 109
as.character() 35, 67, 69, 188, 271
as.data.frame() 35, 36, 82, 88, 177, 179, 192, 212, 221, 329
as.double(), same as **as.numeric**()
as.factor() 19, 51, 75, 76, 80, 170, 221, 289
as.matrix() 29, 30, 36, 72
as.numeric() 35, 36, 50, 51, 131, 170, 177, 188, 233, 252
as.POSIXlt() 228
as.table() 29
as_tibble(), **as.tibble**() 252
ASCII 64
asin() 158, 165
ask= in **par**() 361
askYesNo() 361

assigning values to a variable 14
asymptotic functions 135
at= see **axis**()
attach() 8, 33, 36, 80, 84
attr() 274, 330, 331
attributes() 277, 283, 291, 293
autoplot() in ggplot2 library 198
available.packages() 8
axes= in **plot**() 20, 48, 51, 67, 213, 248, 314–316, 319, 327
axis() 21, 22, 48, 51, 175

barplot() 52–56, 58–61, 63, 64, 149
bartlett.test() 99, 102
bar x ($\bar{x}$) and bar y ($\bar{y}$) 364
Beddington equation 306–310, 313
beep() in beepr library 306
beepr("") 306
beside= in **barplot**() 52, 53, 56, 58, 59, 63, 64
bg= in **plot**() 91, 357
binary response variable 172–186
binomial errors 100–102, 173, 184
binom.test() 117, 118
BiocManager("") 85, 284
Biostrings("") 275, 286, 292, 293
blue tits 228, 230
bmp() 354, 355
BOD dataset 27, 31
Bonferroni correction 77, 78, 138
brackets 9, 14, 15, 18, 33, 51, 80, 81, 251, 259, 261, 276, 296, 299, 359
Bracon 172–176
branch lengths in phylogenies 261, 276, 279
break command 15, 261, 262
breaks= in **hist**() 178, 201, 212, 217, 235
broken stick models 200, 211, 212, 215, 216
bryophytes 37–44, 47, 49, 52–53
bty= in **legend**(), **par**() 44, 51, 56, 59, 166, 298, 337, 357
built-in R datasets 27
butterflies 121–123, 200, 202, 208, 209, 278, 279, 282
byrow= in **matrix**() 63, 69, 221, 239

c() 14, 17, 19, 20–22, 24, 27, 28, 33, 34, 38, 39
c2s() in seqinr library 286

calculator mode 11
camera traps 60–62
cardinal bird 148
caret *see* ^
carriage return *see* \r
casefold() 271, 274
case sensitivity and character case
 13, 75, 227, 271
cat() 306, 362
cbind() 34, 54, 56, 63, 70, 82, 112, 123,
 160, 173, 175–177, 179, 183,
 189, 251, 291, 329
categorical explanatory variables
 155, 225
ceiling() 315
censored data 98, 147–149, 152, 218
cex= 39, 44, 48, 55, 67, 71–73, 114,
 166, 174, 175, 197, 202, 230,
 231, 237, 256, 298, 304, 309,
 319, 327, 338, 339, 343, 358
cex.axis= in **axis**() 51
cex.names= in **barplot**() 63, 64
change working directory 360
character(0) 266, 267
character classes 273
cheddar("") 329–331
ChickWeight dataset 138
Chi-squared tests, see **chisq.test**()
chisq.test() 100, 105, 110, 111, 118
choose() 361
circle.col= in **vennDiagram**() 89
circle drawing
 math approach 93
 using **symbols**() 91
circlize("") 64–69
class() 31–36, 139, 230
clear memory *see* **rm**()
clutch size 147
Clonorchis 107, 110
code= in **arrows**() 24, 26, 69
codons 288, 292, 293
coercing data types 24
col= in **arrows**(), **plot**(), **lines**(),
 points() 24–26, 29, 356
collapse= in **paste**() 190, 193, 237, 263,
 285, 286, 289, 291, 292,
 295, 343
colMeans() 177, 186, 206, 207, 361
col.names= in **write.csv**() 33, 36, 355
colnames() 33, 50, 58, 62, 88, 158, 177,
 195, 221, 253, 330

colour coding 24, 26, 39, 47, 53, 68
colour palettes 310, 315
coloured scale bars 316, 317
colours() or **colors**() 26, 30
colSums() 89, 177, 186, 216, 361
combn() 362
commas 6, 8, 14–16, 20, 32, 35, 51,
 195, 258, 261, 269
comp() in **seqinr** library 286
complement DNA bases 287
complement() in **Biostrings** library 275, 286
complete.cases() 255, 256
concatenation, *see also* **c**() 23, 47
consecutive numbers, runs of 288–289
contingency tables 100, 106, 107, 111
convert factors to other classes 34–36,
 50, 51
convert ragged array to matrix 28–30
convert types of variable in a matrix 31
Cook's distance 125–127, 141, 167
cor() 82
correlation coefficients 82, 365
cor.test() 82, 84
Corynopoma 82
cos() 367
count data 101, 147–154, 168–171,
 177, 342
Cox's Proportional Hazards 225, 226
CRAN xxii, 3, 8, 67, 85, 182, 242
creating lists *see* **c**()
cumsum() 67, 213, 217, 289, 362
curve() 339, 345–347

data()
data.frame() 35, 325
date *see* **Sys.date**()
debugging 266, 310
dchisq() 346
deer, white-tailed 117, 118
degrees of freedom 104, 133, 146, 181,
 339, 340, 344–346
 Chi-squared test 104, 112, 347
 contingency table 106, 107
degrees to radians 90, 367
demo() 6
density() 234–236
density= in **barplot**() 54
dependent variable *see* response variable
deprecated packages and functions
 8, 85

DescTools('''') 95, 100, 103, 110, 362
detach() 33, 82
dev.cur() 334
dev.new() 25, 26, 358, 359
dev.off() 354, 355
dev.set() 359
diff() 288, 289
dimnames= in matrix(), array() 36
dip() in diptest library 145
diptest('''') 145
dismo('''') 253
diversity() 16, 208–211, 217
diversity indices 200, 208–210
DNA data 268, 284–293
DNAString() in Biostrings library 286, 292, 293
DNAStringSet() in Biostrings library 286, 287
dnorm() 336
dose.p() in MASS library 176
downloading files from the www 8
dplyr('''') 226
dpois() 343
draw.ellipse() in plotrix library 93
draw.sector() in circlize library 67, 68
drop items from lists, dataframes, matrices, see - drop.tip() in ape library 281
dt() 336

e 10, 11
eaxis() in sfsmisc library 51
effect size 96, 137
ellipse() in car, plotrix libraries 93
else 24
erer('''') 234
error and warning messages
    Error in `[.data.frame`(dF, nameZSize) : undefined columns selected 244
    Error in !header : invalid argument type 139
    Error in gsub(..... : invalid 'pattern' argument 266
    Error in scan(file = file, ... 172
    Error in table(..... : all arguments must have the same length 29
    Error in xy.coords ... : ... 'x' and 'y' lengths differ 45
    Error: object '...' not found 34

Error: unexpected input in ... 10
In chisq.test(rbind(a, b)) : Chi-squared approximation may be incorrect 111
incomplete final line found on 85
tar: Failed to set default locale 362
The following objects are masked from 33, 275
There were ... or more warnings 351
Warning message, in bxp(list (stats = ... 73
Warning message: In bxp(.... : some notches went outside hinges ('box'): maybe set notch=FALSE 73
error bars 62, 63
escape characters 86, 296
escape key 6, 125, 306
European corn borer 234
exactly equals see ==
exists() 333
exiting from R 16
exiting a run or process 6
exp() 135, 146, 217, 307
explanatory variables 69, 70, 77, 78, 97, 100, 102, 119, 120, 123–125, 127, 130, 136–140, 144, 145, 155, 160, 171–186, 225
exponential notation for axes 51
exponential notation for numbers 51
expression() 108

F, see FALSE
F-test see var.test()
factor() 161
factorial() 362
faithful dataset in MASS library 108
FALSE 10, 14, 21, 24, 31, 36, 63, 139
false positive/negative 96
family=
    in glm() 148, 150, 152, 154, 169, 170, 173, 179, 181, 185
    in plot(), text() 26, 56
fasta format 292, 293
fg= in plot() 358
file.choose() 350
file= in read and write functions 355
fill= in legend() 53–55
Filter() 221, 255

Index

fish 69–73, 82, 134–135, 166–168, 324, 325
fisherfit() in **vegan** library 211
Fisher's alpha 201
Fisher's exact test 100, 111
fisher.test() 100, 111, 118
five columns default file writing 234, 349
Fligner-Killeen test see **fligner.test**
fligner.test() 95, 99
floor() 212, 213, 315, 362
fonts= in **mtext**(), **plot**(), **text**() 55, 56, 252, 327
font size see cex=
foodweb("") 326–331
food webs 326–331
for(){} 14, 23, 26, 68, 205, 206, 215, 221, 316, 323
force intercept through origin 135, 136, 168, 169
format() with **Sys.Date**() 227
fortify() 80
functions, user-defined 285, 299, 303, 304, 323

Gamma distribution 343–344
Gaussian distribution see normal distribution
GBIF 253, 255
GCD() 362
gbif() in **dismo** library 240, 256
GenBank 250, 258, 290–293
generalised linear models see **glm**
genetic codes 292, 293
getGeneticCode() in **Biostrings** library 292
getwd() 331, 360, 362
ggplot() in **ggplot2** library 84
ggplot2("") 59, 79, 80, 82, 84, 198, 235
ggplotly("") 79
ggpubr("") 79, 82–84, 99, 119, 235
ggqqplot() in **ggpubr** library 99, 119, 146
ggscatter() in **ggpubr** library 82, 84
ggsurvplot() in **survminer** library 226
github.com 8
glm() 148–150, 152, 154, 169, 170, 173, 179–182, 184–186
glm.nb() in **MASS** library 101
gobies 166–168
goodfit() in **vcd** library 149, 150

Gompertz 136, 366
Grammar of Graphics packages 79–84, 130, 198, 227, 235
graphical user interface see Rgui
graphics.off() 25, 26
gregexpr() 260, 266, 274
grep() 203, 205, 217, 251, 256, 262, 263, 265, 270, 272
grepl() 271
gridExtra("") 82
gsub() 24, 26, 233, 241, 252, 258, 262, 266–269, 274, 296
G-test 110
GTest() in **DescTools** library 110

h= in **abline**() 146
H_0 96, 105
HairEyeColour dataset 31
Hardy-Weinberg equilibrium 107–110
head() 49, 80, 84, 86, 140, 151, 177, 184, 195
header= in reading/writing files 32, 33, 49, 131, 140, 151, 158, 172, 177, 202, 211, 219, 232, 246, 247, 297, 348–350
heat.colors 68
help= in **library**() 7
help (?help) 78
help.search() 16, 360
heteroscedasticity 140, 157, 184, 185
horiz= in **barplot**() 60, 62
horizontal lines, see h=
hornbills 57–60, 332, 333
hypotenuse 299, 367

identical() 267
if(){} 14, 175
ifelse(){} 24, 26, 88, 216
ignore.case= in **grep**() 271, 273
indenting code 175
independent variable see explanatory variable
indices in lists and nested lists
    [] 14, 15
    [[]] 270
Inf 11
infinity 11, 127, 158, 164, 340, 343, 365
InsectSprays dataset 155–157
inset= in **legend**() 43, 44, 48, 343

install.packages() 8, 51, 67, 82, 85, 110, 116
integer
    testing for 294
interactive file selection 351
intercept 119, 122, 130–134, 136, 155, 156
intercepts, comparing 130–133
inter-quartile distance/range 71, 73
intersect() 87, 92, 94, 234
intToUtf8() 286, 287
inverted commas 6, 8, 16, 20, 35, 51, 276
iris dataset 197–200
is.array() 36
is.data.frame() 362
is.matrix() 36
is.na() 46, 165, 221
is.ordered() 161, 165
is.table() 36
is.whole() 294, 295
italics in **plot**(), **legend**() 43–45
IUPAC DNA codes 287

jitter() 174–176, 186
jpeg("") 247, 332, 334, 354
jpeg() 355

killdeer 218–226
Kolmogorov-Smirnov test 99, 116–118
krill 140–144
kruskal.test() 117
ks.test() 99, 118
kurtosis 82, 103, 104, 116, 117

lac= in **plot**() 358
ladybird beetles 211–214
las= in **par**(), **barplot**(), **plot**() 48, 60
latitude and longitude 244, 247–250, 252, 255
layout() 359
LCM() 362
LD_{50} 176
legend() 43, 44, 51, 55, 56, 59, 63, 84, 166, 308, 343, 346, 357
legend= in **legend**() 56, 58
legend.text= in **barplot** 59
lend= in **lines**() 211, 213, 236, 328, 327, 358
length() 8, 17, 19, 45, 63, 67, 75, 78, 87, 92, 93, 190, 201, 203, 209, 212–214, 216, 221, 230, 236, 262–264, 267, 270–272, 295, 296, 308, 313, 319
length=
    in **arrows**() 24–26, 63, 109
    in **seq**() 135
lengths() 29, 30, 324
LETTERS[] 75, 76, 190
leuk dataset in **MASS** library 226
levels() 78, 165, 212
leverage 125–129
library() 7, 8, 51, 67, 79, 82, 84, 85, 110, 116, 129, 150, 182, 198
limma("") 85, 89
linear models 98–102
linear regression 120, 125, 137
line end shape, *see* lend=
line feed *see* \n
line= in **mtext**(), **title**() 48, 61
lines() 42, 63, 91, 108, 135, 174, 207, 213, 224, 230, 238, 301, 309, 310, 316, 328, 336–339, 343, 358
list() 30, 324
list.files(), list files in a directory 361
list function names in a package, *see* ls("package:")
lists 10, 17, 19, 38, 42, 63, 84, 85, 103, 214, 241, 257, 288, 324
list variables/objects in memory 360
lm() 98, 100, 101, 119, 121, 125, 127–130, 132–134, 137, 138, 141–144, 152, 153, 156, 167
Loblolly dataset 79–81
loess() 239
log() 142, 144, 146, 201, 208-210
log2() 122, 146, 211
log10() 50, 122, 140, 146
logarithm of zero 127, 164
logarithmic axes 49–52
logical(0) 266
logical mathematical operators 362
logistic distribution 341, 342
logistic equations 251, 341
logistic regression 136, 172–176
logit analysis 126, 176
log-log plots 121, 123, 140
log-normal distribution 201, 335, 340, 341
log-series distribution 201
loops *see* **for**(){}; **repeat**(){}; **while**(){}
loops within loops *see* nested loops

Index

lowess() 238, 239
ls() 360
ls("package:") 360
ls.str() 360
lty= in **lines()**, **plot()**, **abline()** 358
lwd= in **lines()**, **plot()**, **abline()** 26, 56

main= in plotting functions 39, 48
mammals 27, 84, 117, 125
Mann-Whitney U test 99, 114–116
mantids 178, 182
map[] 286
mapply() 322
mar= in **par()** 23, 61, 67, 178, 319, 357
masked objects 275
MASS("") 52, 83, 95, 103, 115, 176, 182
matrices 10, 14, 31, 56, 263, 293, 313, 352
matrix() 30, 199, 318
max() 17, 21, 29, 41, 46, 63, 64, 68, 197, 215, 219, 221, 230, 235, 236, 248, 308, 312, 315
maximal model 19
mean() 19, 30, 33, 34, 46, 50, 149, 152, 237, 300, 301
 trimmed mean 18
mean with **apply** 72, 156, 323
median 50, 71, 73, 99, 103, 106, 125
median() 17, 19, 50, 106
menarche dataset in **MASS** library 176
metacercaria 69–73, 324, 325
metacharacters 259, 274, 296
method=
 in **cor.test()** 82
 in **goodfit()** 150
 in **p.adjust()** 78
mfrow= in **par()** 22, 23, 50, 54
min() 19, 41, 46, 48, 63, 197, 215, 230, 235, 236, 248, 262, 299
minimum adequate model 144
mode() 293
model inspection 149–172
model simplification with **anova()** test 144–146
modulus *see* %%
Monte Carlo methods 148, 187–191
months 262
mtext() 48, 89, 252, 256
multiple pairwise comparisons

Bonferroni 77, 78
Tukey's honest significant difference 74–78, 156

NA 20, 29, 31, 32, 219–221
na.omit= 46
na.rm= 30, 46, 219
names() 31
names=
 in **barplot()**, **boxplot()** 63, 64
 in **vennDiagram()** 89
NaN 38
ncol() 30, 203, 205
ncol= in **matrix()** 29, 36, 205, 221, 299
nearest neighbour distances 297, 299, 301
Negate() 226
negative binomial errors in **glm()** 101
negative binomial regression 101, 133, 148
nested factors 27, 28, 133, 260, 270
nested loops 174, 205, 213
new= in **par()** 48, 343
nls() 134–136, 146
nodelabels() in **phytools** library 277
non-parametric tests 98, 104
noquote() 192, 193
normal distribution 171, 322, 325, 335, 337, 338, 340, 341, 365
nortest("") 116, 118
notch= in **boxplot()** 78
nrow() 29, 212, 315
nrow= in **matrix()** 107, 199
NULL 20, 23, 26, 106, 180, 181
null deviance 180, 181
null hypothesis/model 95, 96, 106, 188

objects() 360
odds ratio 108
Ochlerotatus albifasciatus 130–134
Oecophylla smaragdina 53–56, 123
one sample tests 111, 301
one-tailed tests 340
oneway.test() 155
openxlsx("") 352, 355
order() 38, 42, 60, 121, 195, 204
ordering factors 27
oval, function to draw *see* **ellipses()**
over-dispersion 101, 148, 150, 152, 182

p.adjust() 77, 78
paired= see t.test()
pairs() 138
palette() 68, 226, 310, 315
PAOD 105–106
par() 22, 48, 61, 67, 71, 140,
    332–334, 357
    see also layout
parametric statistics 98
Passiflora 151–154
paste() 189, 190, 193, 263, 286, 330
paste multiple items in with lists 287
pat= in ls.str() 360
path 348, 360
path= in list.files() 361
pattern= in gregexpr() 266
pch= in plot(), points() 39, 42, 91
pch, list of symbols 39
pdf() 25
pdf files 44, 354
pdf_text() in pdftools library 354
pdftools("") 354
Pearson's correlation 82, 365
PERL 268, 271, 272
phylogenetic independence 157
phylogenetics 275
picante("") 191
pie() 64, 69
pie charts 26, 37, 64–69
pin= in par() 154, 332, 334, 357
photographs 332–334
phylogram, phylogenetic tree
    276, 280, 281
plot()
    in base R 19, 20
    in ape 279, 283
plot model, selecting which plot 125
plotTree() in phytools library 277, 278
png() 354
pnorm() 99, 116, 337
points() 39, 40
Poisson distribution 342, 343
Poisson errors 148–150, 152, 169, 181
pollination 147, 151, 177
polygon() 248, 249, 256, 334
polynomial regression 120, 140–143
population dynamics 306–317
positions
    of elements in a vector 19, 24
    of patterns in a string 260
power function 120

prcomp() 194, 199
predict() 135, 174, 199, 238
prime number in random generator 187, 188
principal components analysis (PCA)
    194–201
princomp() 194, 199
print() 15, 19, 353
prob= in hist() 345
probit 102, 176
problems with NAs 30, 46
proc.time() 216, 322, 323
proportion data 157–165, 182–186
pseudorandom numbers 188
pt() 306
punctuation 269–271, 272–274
p-value 77, 78, 96, 105, 116, 118, 125, 137,
    157, 163, 165, 182, 226, 335, 346

q() 16
QQ plot 102, 124, 125, 141, 153, 154,
    162, 163, 165, 168, 184
qt() 33, 336
quantile() 17, 362
quantile-quantile plot see QQ plot
quartz() 330, 359
quasi-binomial errors 184
quasi-Poisson errors 148, 152, 154,
    181, 182
quiet= in scan 349
quit(), same as q() 16

radians 90, 362, 367
radians to degrees 367
ragged data 28–30
rainbow() 68, 315
randomizeMatrix() in picante library 191
random numbers
    using sample() 187, 188
    code to generate 187
range() 17, 50, 289, 304
rasterImage() 333
rbind() 34, 45, 53, 58, 63, 107, 113,
    213, 215, 221, 294, 323
r-bloggers xxii
rchisq() 344
read.csv() 32, 149, 349, 351
read.delim() 349, 350
read.dna() in ape library 292
read.fasta() in seqinr library 292

Index 385

read.GenBank() in **ape** library 290
readJPEG() in **jpeg** library 334
readline() 361
readLines() 85, 234, 250, 265, 279, 295, 348, 349
read.table() 32, 158, 172, 177, 202, 246, 348–351
readTIFF() in **tiff** library 334
rect() 298
regexpr() 260, 265, 266
regmatches() 266
regular expressions 260, 265, 268
relative risk 108
remove intercept from model 122
remove.packages() 82
reorder factors 72, 160, 161
rep() 21, 34, 69
repeat{} 14, 262
replace= in **sample**() 189, 204, 205, 212
replace multiple spaces with single space 268
rescale.p= in **chisq**.test() 106
residual deviance 177, 181
response variable 58, 97, 120, 127, 135, 137, 143, 147, 157–165, 172–177
return() 266, 267, 285, 304, 312, 318
rev() 61, 285, 286
reversing a sequence 285
rgb() 26, 241
rgbt() 93, 315
Rgui 4, 6
rm() 84, 360
rnorm() 322, 335, 338
R objects 31, 84
rolling on values when filling an array 45
round() 188, 189, 294, 297, 313, 315, 340
row.names= in **write.csv**() 331, 349
rowMeans() 177, 361
rownames(), also to set row names 58, 195
rowSums() 89, 177, 189, 212, 361
rt() 336
rtree() in **ape** library 280
rug() 213, 236
runif() 304
R.version() 8
rworldmap("") 242, 245

s2c() in **seqinr** library 286
sample() 187, 205, 212
sapply() 193, 221, 285, 322, 324, 325
scan() 234, 349
scatterplot() in **car** library 199
scrolling through the R console 45
sd() 17
search() 360
seasonality 105, 238
seed traits 158–163
select= in **subset**() 255
self-starting functions 136
sep= in **paste**() 32, 190, 263, 264, 266, 267, 271, 286, 295, 339
seq() 90, 91, 108, 135, 174, 178, 201, 238
seqinr("") 285–287, 292
set diagrams, *see* Venn diagrams
setdiff() 87
sets 34, 87, 115, 246, 312
setwd() 1, 360
sex ratio 100
sfsmisc("") 51
Shannon–Weiner index 208
shapiro.test() 99, 116
side= in **mtext**() 48
sigmoidal functions 366 *see also* logistic regression; Weibull; Gompertz
Simpson's index 200, 208, 209
simulate() 122
sin() 14, 354, 367
Sirindhornia 168
slope 98, 119, 125, 132–134
slopes, comparing 130–134
    extracting from model summary 129
smoothers, non-parametric 238
smooth.spline() 204, 207
snails dataset in **MASS** library 182
sort() 72, 212, 213, 215, 217, 264, 301
sound alerts 306
source() 57
space character 10, 13, 14, 65, 172, 237, 246, 250, 252, 269, 270, 293
spar= in **smooth.spline**() 217
spatial polygons and maps 249
Spearman rank correlation 82
species accumulation curve 202–207
species diversity 208
species-richness 140, 200, 208
speed 20, 216, 317, 322, 323
sperm concentration 28

split() 183, 288, 289
split.screen() 359, 361
splitting a string see **strsplit()**
sqrt() 100, 133, 158, 163, 165, 299
srt= in **plot()** 61, 154, 358
Stack Overflow 8, 233, 268
standard deviation (SD) 62, 101, 338, 344
standard error of mean 62, 364
    definition 62
    extracting from model summary 301
    formula 132, 364
str() 276, 290
stratified experimental design 192
stratified random sampling 192
strsplit() 86, 87, 250, 261–263, 269, 270, 284–286, 288, 296
structure() 90
Student's *t* see *t* distribution; t.test()
Styraceaceae 214–216
sub() 233, 264, 290
sub= in **barplot()** 56
subset() 133, 255
substr() 188, 228, 231, 262, 264, 270, 272, 274, 292
subscript in plots 322
sum() 19, 89, 178, 210
summary() 19, 121, 128, 133, 135, 141–143, 223, 225, 276
superscript in plots 108
survival("") 101, 218–220
survminer("") 226
switch() 285
symbols see pch= in **plot()**, **points()**
symbols() 39, 91
syntax 6, 24, 82, 284
Sys.Date() 227, 228
Sys.getlocale() 362
system() 306
Sys.time() 228
Sys.timezone() 228

T, see TRUE
t() 58, 195, 216
t.test() 99, 111, 117, 155, 365, 366
    paired= 99, 117
tab character see \t
table() 27, 28, 149, 212
tables 34, 106, 107, 293
tail() 50, 84
tan() 362, 367

tapply() 68, 72, 156, 324, 325
t distribution 338–340
terrain.colors 68
text= in regular expressions 266, 267
text() 21, 22, 24, 39, 61, 91–93, 108, 109, 114, 174, 190, 197, 230, 231, 237, 308, 316, 319, 327, 337–339
textConnection() 32, 33, 172, 219, 246, 265, 293
text.font= in **legend** 55
text manipulation 257–273
tibbles 352, 353
tick marks, see **rug()**
tiff("") 332
tiff() 332, 334, 354
tilde character see ~
time 8, 11, 23, 67, 108, 130, 131, 135, 182, 188, 220, 228
timing execution of code 6, 323
tiplabels() in **phytools** library 277
title() 61
translate DNA to amino acids see **translate()**
translate DNA to RNA 285
translate() in **Biostrings** library 292
transparency in graphics 298
trigonometric functions 367
trim= in **mean()** 18
trimmed means 18
trimws() 268
TRUE 10, 11, 14, 24, 30, 31, 139
truehist() in **MASS** library 115
try.all.packages= in **help()** 16
tuatara 295
TukeyHSD() 74, 75, 138, 156
two-tailed tests 130, 133, 339, 340
type=
    in **plot()** 67, 337, 341, 344, 358
    in **points()** 43
type I and type II statistical errors 42, 77, 96
typeof() 31

under-dispersion 101, 141, 148, 154
unequal variances 111
union() 87
unique() 8, 87, 202
unlist() 85, 86
unname() 285, 293, 313, 350

usr= in **par**(), **plot**() 154
utf8ToInt() 286

v= in **abline**() 178
value= in **grep**() 272
var() 17, 99, 114
variance 17, 62, 74, 364
var.test() 99
vcd("") 150
vectorized approach 10, 211, 214
vectors 10, 29, 31, 38, 41, 56, 61, 107, 112, 175, 231, 307, 327, 349
vegan("") 201, 208, 210
VennCounts() in **limma** library 89, 90
vennDiagram() in **limma** library 85, 89, 90
Venn diagrams 85–93
vertical lines, *see* v=
vignette() 242, 329, 330, 358

warning messages *see* error and warning messages
Weibull 136, 335, 336, 366
Welch's t-test 99, 111
which() 18, 19, 190, 209, 216, 262, 299, 314, 327, 328

while() 14, 15
wilcox.test() 99, 116, 117
wildcards 257, 259–262, 269, 272–274
with() 33, 34, 131, 164, 166
write() 234, 349
write.dna() in **ape** library 292
write.fasta() in **seqinr** library 292
write.list() in **erer** library 234
write.table() 349
write.tree() in **ape** library 279

xlab= in **plot**() 20, 21, 39, 41, 42, 46, 48, 50, 51, 54–56
xlim= in **plot**() 20, 21, 39, 48, 51, 54, 67, 90, 108, 134
xpd=NA in **plot**() 22, 56, 114

ylab= in **plot**() 20, 21, 38, 39, 41, 42, 46, 48, 50, 51, 54, 55
ylim= in **plot**() 20, 39, 41, 46, 48, 51, 63, 64, 90, 108, 134

z, Z, standard normal variate 132, 133, 301
Zamia, 350

CABI – who we are and what we do

This book is published by **CABI**, an international not-for-profit organisation that improves people's lives worldwide by providing information and applying scientific expertise to solve problems in agriculture and the environment.

CABI is also a global publisher producing key scientific publications, including world renowned databases, as well as compendia, books, ebooks and full text electronic resources. We publish content in a wide range of subject areas including: agriculture and crop science / animal and veterinary sciences / ecology and conservation / environmental science / horticulture and plant sciences / human health, food science and nutrition / international development / leisure and tourism.

The profits from CABI's publishing activities enable us to work with farming communities around the world, supporting them as they battle with poor soil, invasive species and pests and diseases, to improve their livelihoods and help provide food for an ever growing population.

CABI is an international intergovernmental organisation, and we gratefully acknowledge the core financial support from our member countries (and lead agencies) including:

Discover more

To read more about CABI's work, please visit: **www.cabi.org**

Browse our books at: **www.cabi.org/bookshop**,
or explore our online products at: **www.cabi.org/publishing-products**

Interested in writing for CABI? Find our author guidelines here:
www.cabi.org/publishing-products/information-for-authors/